INTRODUCTORY
TECHNICAL
MATHEMATICS

Prentice-Hall Series in Technical Mathematics
Frank L. Juszli, Editor

INTRODUCTORY TECHNICAL MATHEMATICS

John Christopher
Durham Technical Institute
Durham, NC

PRENTICE-HALL, INC., *Englewood Cliffs, NJ* 07632

Library of Congress Cataloging in Publication Data

CHRISTOPHER, JOHN.
   Introductory technical mathematics.

   (Prentice-Hall series in technical mathematics)
   Includes index.
   1. Shop mathematics. I. Title. II. Series.
TJ1165.C54   510   81-1263
ISBN 0-13-501635-5   AACR2

Editorial/production supervision
  and interior design by Karen Skrable
Manufacturing buyer: Gordon Osbourne
Cover design by Jorge Hernandez
Art production by Diane Doran Sturm and Steven Frim

Printed in the United States of America

10 9 8 7 6 5 4 3 2

Prentice-Hall International, Inc., *London*
Prentice-Hall of Australia Pty. Limited, *Sydney*
Prentice-Hall of Canada, Ltd., *Toronto*
Prentice-Hall of India Private Limited, *New Delhi*
Prentice-Hall of Japan, Inc., *Tokyo*
Prentice-Hall of Southeast Asia, Pte. Ltd., *Singapore*
Whitehall Books Limited, *Wellington, New Zealand*

To Philip, Titika, and Maria

Contents

Preface

In preparing INTRODUCTORY TECHNICAL MATHEMATICS my purpose has been to provide a text marked by readability and clarity of treatment. My extensive experience in teaching technical mathematics has impressed upon me the overriding need for a book from which the instructor can teach and the student can learn useful mathematics most effectively.

Mathematics is the foundation of all engineering and technical education. It is, as well, the language of physics, an important part of the training of a modern technician. The mathematics in this book is designed for the student in a technical institute, and it includes a review of some elementary topics. It is also suitable for the older student in a technical program who has forgotten algebra or trigonometry, or never had a clear understanding of these subjects before. The stress is not on derivation and mathematical rigor, but rather on how mathematics is used in solving technical problems. Students understand and remember better and recall more easily concepts, rules, and formulas which are related to or "derived" from previously covered material. Therefore, every effort was made to introduce new concepts, rules, and principles by relating them as simply as possible to mathematics already established.

Not all the material included in this book is needed for all technical programs. The instructor will decide which topics to include in the course beyond those that will give to the student an adequate background in algebra. As is often the case, however, the instructor may elect to start with fractions, decimals, and computations with exact and approximate numbers. Chapters 1 and 2 are included for this reason. Chapter 2 covers the metric system and stresses precision, accuracy, significant figures, exact and approximate numbers, all of which have been introduced in chapt. 1. The geometry in chapts. 15 and 16 is also of great value to the student. The applied and practical geometry in these two chapters, as well as the arithmetic review of the first two chapters, are both of vital importance for the students of technical programs.

There are more than 400 exercises and illustrative problems worked out in detail and supplemented by almost 320 drawings. The almost 3000 appropriate exercises and technical word-problems provide the student an abundance of material for the necessary experience and drill. All chapters end with review questions and review exercises.

Answers to most odd-numbered exercises are provided at the end of the book. In cases of computed results, the answers were arrived at and "rounded off" in accordance with the rules for computations with exact and approximate numbers explained in chapts. 1 and 2. Also, answers to trigonometric problems are in accord with corresponding accuracy of sides and angles explained in chapt. 17 (Table 17-1).

Unlike many other texts, which assume too much and expect too much from the student, this book assumes little, starts with elementary topics, explains in detail and repeats the explanation from a second point of view for certain important subjects. It is hoped that students will enjoy this book, and that it fully and fairly addresses their needs.

JOHN CHRISTOPHER
Chapel Hill, NC

INTRODUCTORY
TECHNICAL
MATHEMATICS

ARITHMETIC

Technical mathematics starts with arithmetic. Neither algebra, geometry, nor trigonometry can be studied successfully without a good background in arithmetic. This is a good reason to start our study with a chapter reviewing this subject. It is assumed that the student knows the basic operations with whole numbers, and so our review will concentrate on a detailed study of fractions, decimals, significant digits, powers, and roots. These subjects are very important in technical problems. This review will also be of great value to the student when algebraic fractions, prime expressions, fractional equations, and related topics are treated in later chapters.

1

An Arithmetic Review

I-I Basic Ideas About Common Fractions

There are two kinds of fractions: **common fractions** (or just fractions), and **decimal fractions** (or just decimals). We start with the common fractions.

If a circle, a rectangular object, a stick, or any article of regular shape is divided into five equal parts (Fig. 1-1), each of these parts is one fifth of the original whole.

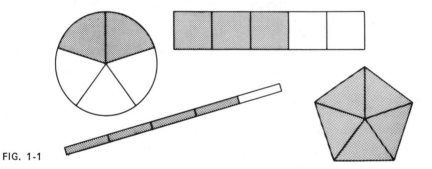

FIG. 1-1

The symbol to denote this quantity is the fraction $\frac{1}{5}$. In Fig. 1-1, $\frac{2}{5}$ of the circle are shaded, and $\frac{3}{5}$ are not. Also shaded are $\frac{3}{5}$ of the rectangle, $\frac{4}{5}$ of the stick, and $\frac{5}{5}$, or the **whole** pentagon, which is a name for a five-sided shape.

In the fraction $\frac{2}{5}$, 2 is the **numerator** and 5 is the **denominator**. In the case of our circle, $\frac{2}{5}$ means that the object was divided into five equal parts and we used, removed, shaded, or are talking about two such parts. In short, the denominator shows into how many equal parts the whole has been divided, and the numerator shows how many such parts we are talking about. Thus, $\frac{2}{3}$ stands for two of three equal parts.

3

In a fraction, the numerator and the denominator together are called its **terms**. Thus, the terms of the fraction $\frac{2}{3}$ are 2 and 3.

A fraction is a way of expressing division. In other words, *a fraction implies the division of the numerator by the denominator.* Thus, $\frac{1}{2}$ means $1 \div 2$. Also, $\frac{3}{5} = 3 \div 5$, and $\frac{8}{9} = 8 \div 9$. Similarly, $4 \div 5$ can be written as the fraction $\frac{4}{5}$. Also, $7 \div 11 = \frac{7}{11}$, and $25 \div 7 = \frac{25}{7}$.

Since a fraction always implies division of the numerator by the denominator, $\frac{5}{5} = 1$, as we have already seen in the case of the pentagon of Fig. 1-1. Similarly, $\frac{15}{15} = \frac{4}{4} = \frac{101}{101} = \frac{17}{17} = 1$.

A fraction in which the numerator is less than the denominator is called a **proper fraction**, since the fraction is really a **fragment** of the whole. A fraction in which the numerator is greater than the denominator is called an **improper fraction**, since it is actually more than just a part of a whole. Thus, $\frac{2}{3}$, $\frac{3}{5}$, and $\frac{11}{17}$ are all proper fractions, while $\frac{16}{5}$, $\frac{7}{3}$, and $\frac{21}{15}$ are all improper fractions.

Improper fractions can be changed to another form of number. For example, the improper fraction $\frac{5}{3}$ implies the division $5 \div 3$, in which the quotient is 1 and the remainder is 2. The remainder 2, however, is still to be divided by 3, and to express this, we write the fraction $\frac{2}{3}$. Thus, $\frac{5}{3} = 1 + \frac{2}{3}$, or just $1\frac{2}{3}$ (read "one and two thirds"). Numbers like $1\frac{2}{3}$ are called **mixed numbers**.

To change an improper fraction to a mixed number, perform the division implied by the fraction and write the quotient followed by a proper fraction formed by placing the remainder over the original denominator. Thus, $\frac{14}{5} = 14 \div 5 = 2\frac{4}{5}$, and $\frac{61}{25} = 2\frac{11}{25}$.

Often we need to change a whole number into a fraction with a desired denominator. Since $5 = 5 \times 1$, and $1 = \frac{3}{3}$, it follows that

$$5 = 5 \times 1$$
$$= 5 \times \frac{3}{3}$$
$$= \frac{15}{3}$$

In the place of 1, we can use $\frac{8}{8}$, $\frac{11}{11}$, $\frac{17}{17}$, or any fraction where both terms are exactly the same. Therefore, *to change a whole number into an improper fraction having a desired denominator, multiply the whole number by a fraction, both terms of which are equal to the desired denominator.* For example, to change the whole number 4 into an improper fraction with 3 as its denominator, we multiply 4 by $\frac{3}{3}$. Thus, $4 = 4 \times \frac{3}{3} = \frac{12}{3}$. Similarly, 5 can be changed to elevenths, and 17 can be changed to eighteenths as follows:

$$5 = 5 \times \frac{11}{11} = \frac{55}{11} \quad \text{and} \quad 17 = \frac{17 \times 18}{18} = \frac{306}{18}$$

Since the whole number part of the mixed number $5\frac{1}{3}$ is equal to $\frac{15}{3}$, we know that the mixed number $5\frac{1}{3}$ is equal to $\frac{15}{3}$ plus $\frac{1}{3}$, or the improper fraction $\frac{16}{3}$. In a similar way, any mixed number can be changed into an improper fraction. *To convert a mixed number to an improper fraction, multiply the whole number by the denominator, and to that product add the numerator of the fractional part of the mixed number. Then place this sum over the denominator.* For example, to change $2\frac{3}{4}$ into an improper fraction, the whole number 2 is multiplied by the denominator 4, and to that product the numerator, 3, is added. Then this sum is placed over the denominator 4. Thus,

$$2\frac{3}{4} = \frac{2 \times 4 + 3}{4} = \frac{8 + 3}{4} = \frac{11}{4}$$

Similarly,

$$5\frac{1}{6} = \frac{5 \times 6 + 1}{6} = \frac{30 + 1}{6} = \frac{31}{6}$$

and

$$3\frac{4}{13} = \frac{3 \times 13 + 4}{13} = \frac{39 + 4}{13} = \frac{43}{13}$$

EXERCISE 1-1

Each square in Fig. 1-2 is divided into a certain number of equal parts. Sketch each square and shade the part represented by the corresponding fraction.

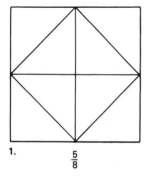

1. $\frac{5}{8}$

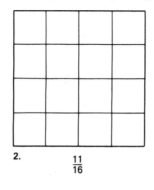

2. $\frac{11}{16}$

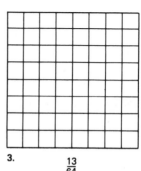

3. $\frac{13}{64}$

FIG. 1-2

Basic Ideas About Common Fractions

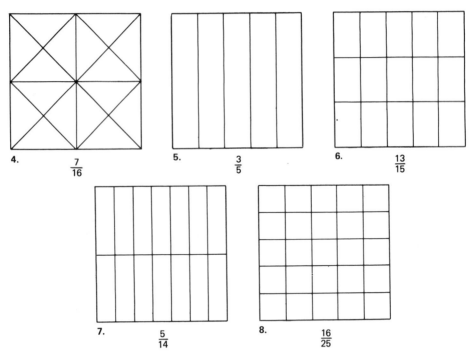

4. $\frac{7}{16}$ **5.** $\frac{3}{5}$ **6.** $\frac{13}{15}$

7. $\frac{5}{14}$ **8.** $\frac{16}{25}$

FIG. 1-2 (continued)

9. Name the terms of each of the following fractions:

$$\frac{5}{6} \qquad \frac{3}{8} \qquad \frac{17}{25} \qquad \frac{11}{16} \qquad \frac{23}{48}$$

10. Name the numerator and the denominator of each of the following fractions:

$$\frac{5}{11} \qquad \frac{12}{23} \qquad \frac{38}{59} \qquad \frac{112}{119} \qquad \frac{7}{8}$$

11. Write each of the following divisions as a fraction:

$$5 \div 7 \qquad 28 \div 15 \qquad 8 \div 15 \qquad 1 \div 25 \qquad 14 \div 3$$

12. Write each of the following fractions as a division:

$$\frac{16}{25} \qquad \frac{187}{94} \qquad \frac{9}{13} \qquad \frac{8}{37} \qquad \frac{27}{8}$$

13. Name the whole number that equals each of the following fractions:

$$\frac{47}{47} \qquad \frac{3}{1} \qquad \frac{23}{1} \qquad \frac{23}{23} \qquad \frac{213}{1}$$

Change each of the following improper fractions to a whole number or a mixed number:

14. $\frac{17}{4}$ **15.** $\frac{18}{5}$ **16.** $\frac{60}{5}$

An Arithmetic Review

17. $\frac{11}{7}$ 18. $\frac{81}{9}$ 19. $\frac{71}{14}$

20. $\frac{97}{16}$ 21. $\frac{83}{13}$ 22. $\frac{125}{15}$

23. $\frac{134}{10}$ 24. $\frac{263}{33}$ 25. $\frac{170}{18}$

26. $\frac{197}{32}$ 27. $\frac{329}{15}$ 28. $\frac{311}{12}$

29. $\frac{862}{29}$ 30. $\frac{748}{47}$ 31. $\frac{1247}{77}$

32. $\frac{3176}{83}$ 33. $\frac{4128}{87}$ 34. $\frac{6308}{102}$

Change each of the following whole numbers to both thirds and sixteenths:

35. 3   36. 9   37. 12

38. 15   39. 23   40. 25

Change each of the following mixed numbers into improper fractions:

41. $3\frac{1}{2}$ 42. $5\frac{2}{3}$ 43. $4\frac{5}{8}$

44. $8\frac{1}{3}$ 45. $6\frac{5}{7}$ 46. $9\frac{3}{8}$

47. $11\frac{7}{3}$ 48. $17\frac{1}{3}$ 49. $22\frac{3}{4}$

50. $31\frac{8}{15}$ 51. $14\frac{1}{24}$ 52. $38\frac{3}{18}$

53. $111\frac{32}{55}$ 54. $88\frac{5}{41}$ 55. $39\frac{23}{27}$

1-2 Equivalent Fractions and Lowest Terms

If a circle is divided into six equal parts, two such parts are $\frac{2}{6}$ of the circle. It is clear, however, from Fig. 1-3, that $\frac{2}{6}$ of the circle are also $\frac{1}{3}$ of it, and we can write $\frac{2}{6} = \frac{1}{3}$. The fractions $\frac{1}{3}$ and $\frac{2}{6}$ are called **equivalent** fractions because they have the same value, in spite of the fact that their terms are different. Notice that the fraction $\frac{2}{6}$ is the result of multiplying the terms of the fraction $\frac{1}{3}$ by 2, that is

$$\frac{2}{6} = \frac{1 \times 2}{3 \times 2}$$

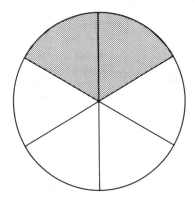

FIG. 1-3

Equivalent Fractions and Lowest Terms

7

Also notice that the fraction $\frac{1}{3}$ is the result of dividing both the terms of the fraction $\frac{2}{6}$ by 2, that is

$$\frac{1}{3} = \frac{2 \div 2}{6 \div 2}$$

In general, equivalent fractions result when the terms of a fraction are either multiplied or divided by the same nonzero number. This means that a given fraction can be expressed in higher or in lower terms. Thus, given the fraction $\frac{21}{28}$, we can say that

$\frac{21}{28} = \frac{42}{56} = \frac{63}{84}$, because $\frac{42}{56} = \frac{21 \times 2}{28 \times 2}$, and $\frac{63}{84} = \frac{21 \times 3}{28 \times 3}$. Also, $\frac{21}{28} = \frac{3}{4}$, because $\frac{3}{4} = \frac{21 \div 7}{28 \div 7}$.

That all fractions maintain their value when their terms are multiplied or divided by the same nonzero number is a very useful and very important property. It is called **the fundamental principle of fractions** and we will first use it in reducing a fraction to its **lowest terms**.

A fraction is said to be in its lowest terms, or its simplest form, when there is no number, other than 1, which divides both its terms exactly. Thus $\frac{3}{4}$ is in its lowest terms, while $\frac{12}{16}$ is not, since the terms 12 and 16 can be divided exactly by 4.

The rule for reducing a fraction to its lowest terms is suggested by the following examples:

$$\frac{8}{16} = \frac{8 \div 8}{16 \div 8} = \frac{1}{2} \qquad \frac{9}{12} = \frac{9 \div 3}{12 \div 3} = \frac{3}{4} \qquad \frac{42}{91} = \frac{42 \div 7}{91 \div 7} = \frac{6}{13}$$

It must be evident from the above that *to reduce a fraction to its lowest terms, divide both numerator and denominator by the largest number which will divide both of them exactly.* If no such number exists, the fraction is in its lowest terms.

To apply this rule in reducing fractions like $\frac{12}{48}, \frac{15}{25}$, and $\frac{14}{49}$ is simple, but it is not as easy to find the largest divisor of the terms of a fraction like $\frac{65}{143}$, and reduce it to $\frac{5}{11}$. A systematic method, using **prime numbers** and **factoring** exists, and makes reducing fractions a routine procedure even for cases that are not simple. First prime numbers and factoring will be explained, and then the method will be illustrated.

A prime number is a whole number that is divisible only by itself and by 1. The numbers 2, 5, and 13 are prime numbers, while neither 4, nor 21 is. A few of the first prime numbers are 2, 3, 5, 7, 11, 13, 17, 19, 23, and 29.

When a number is expressed as the product of two or more whole numbers, each of these numbers is called a **factor**, and we say that the number has been **factored**. For instance, below are two different ways in which 24 has been factored.

$$\text{(a) } 24 = 2 \times 3 \times 4 \qquad \text{(b) } 24 = 2 \times 2 \times 2 \times 3$$

The numbers 2, 3, and 4 are factors of 24 in (a). The expression of a number as the product of prime numbers *only* is called **complete factorization**, or **prime factorization**. Case (b) above is complete factorization since all the factors are prime numbers. Also, each of the following is complete factorization: $28 = 2 \times 2 \times 7$, $42 = 2 \times 3 \times 7$, and $68 = 2 \times 2 \times 17$. Notice that all the factors in the factoring of 28, 42, and 68 are prime numbers.

NOTE: The whole number 1 is a special prime number. It is a factor of every number. Thus, $4 = 4 \times 1$, and $5 = 5 \times 1$.

Any number can be expressed as the product of prime factors, and this can be done in only one way, except for the order of the factors. Thus, in the three possible forms of prime factorization of the number 12, shown below, the prime factors are the same, and only their order differs:

$$12 = 2 \times 2 \times 3$$
$$12 = 2 \times 3 \times 2$$
$$12 = 3 \times 2 \times 2$$

To factor a given number completely, start with the smallest prime number. Use **2** *or* **3** *or* **5** *as a divisor as many times as it will work. Continue dividing in the same way using the other prime numbers in order until division is no longer possible.*

NOTE: The following three guides for the divisibility of a given number by the three first prime numbers 2, 3, and 5, will assist the student greatly in factoring a number completely.

—Divisibility by 2: All even numbers are divisible by two.
   Examples: 14, 32, 528, 96, and 130 are divisible by 2.
—Divisibility by 3: The number which will give the sum of either 3, 6, or 9, when its digits are added until a one-digit number is obtained, is divisible by 3.
   Examples: The numbers 12, 201, 24, 81, and 3771 are divisible by 3, because; in 12, $1 + 2 = 3$; in 201, $2 + 0 + 1 = 3$; in 24, $2 + 4 = 6$; in 81, $8 + 1 = 9$; and in 3771, $3 + 7 + 7 + 1 = 18$, and $1 + 8 = 9$.
—Divisibility by 5: The number, the last digit of which is either 5 or 0, is divisible by 5.
   Examples: 20, 35, 15, 1225, 170, 70, are divisible by 5.

EXAMPLE 1-1 Factor completely each of the following numbers:

$$\text{(a) } 30 \qquad \text{(b) } 84 \qquad \text{(c) } 315$$

SOLUTION

(a) The number 30 is even, therefore it is divisible by 2. Thus, $30 = 2 \times 15$. Since $15 = 3 \times 5$, $30 = 2 \times 3 \times 5$.
(b) Since 84 is even, it is divisible by 2. So, $84 = 2 \times 42$. The number 42 is once more divisible by 2, and we have $84 = 2 \times 2 \times 21$. It is easy now to see that $21 = 3 \times 7$; therefore, $84 = 2 \times 2 \times 3 \times 7$.

(c) $315 = 3 \times 105$
$= 3 \times 3 \times 35$
$= 3 \times 3 \times 5 \times 7$

EXAMPLE 1-2 Factor 270 completely.

SOLUTION

$$270 = 2 \times 135$$
$$= 2 \times 3 \times 45$$
$$= 2 \times 3 \times 3 \times 15$$
$$= 2 \times 3 \times 3 \times 3 \times 5$$

We are ready now to apply prime numbers and factoring in reducing a given fraction to its lowest terms. Here is the method:

To reduce a fraction to its lowest terms, first factor both the terms of the fraction completely. Then, cross out (cancel) all pairs of factors common to both numerator and denominator in the factored terms of the fraction.

EXAMPLE 1-3 Reduce the following fractions to their lowest terms:

(a) $\dfrac{24}{120}$ (b) $\dfrac{18}{210}$ (c) $\dfrac{216}{420}$

SOLUTION

(a) $\dfrac{24}{120} = \dfrac{2 \times 2 \times 2 \times 3}{2 \times 2 \times 2 \times 3 \times 5} = \dfrac{\cancel{2} \times \cancel{2} \times \cancel{2} \times \cancel{3}}{\cancel{2} \times \cancel{2} \times \cancel{2} \times \cancel{3} \times 5} = \dfrac{1}{5}$

NOTE: Since 1 is a factor of all numbers, 1 is a factor of 24. Therefore, when all the factors of 24 in the above example are cancelled, 1 remains in the numerator.

(b) $\dfrac{18}{210} = \dfrac{2 \times 3 \times 3}{2 \times 3 \times 5 \times 7} = \dfrac{\cancel{2} \times \cancel{3} \times 3}{\cancel{2} \times \cancel{3} \times 5 \times 7} = \dfrac{3}{35}$

(c) $\dfrac{216}{420} = \dfrac{2 \times 2 \times 2 \times 3 \times 3 \times 3}{2 \times 2 \times 3 \times 5 \times 7} = \dfrac{\cancel{2} \times \cancel{2} \times 2 \times \cancel{3} \times 3 \times 3}{\cancel{2} \times \cancel{2} \times \cancel{3} \times 5 \times 7} = \dfrac{18}{35}$

EXERCISE 1-2

Each square in Fig. 1-4 is divided into a certain number of equal parts. Sketch each square and shade the part specified by the corresponding fraction.

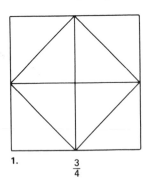

1. $\dfrac{3}{4}$

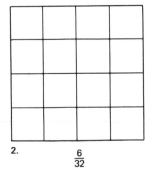

2. $\dfrac{6}{32}$

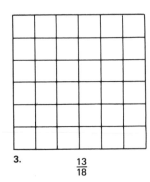

3. $\dfrac{13}{18}$

FIG. 1-4

An Arithmetic Review

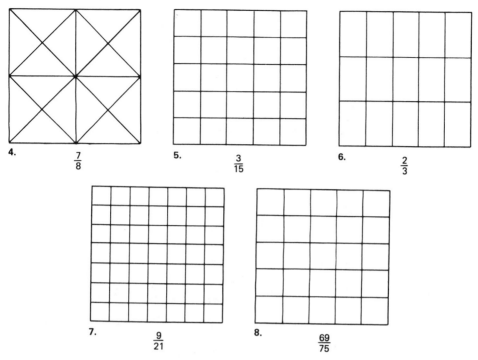

4. $\frac{7}{8}$ 5. $\frac{3}{15}$ 6. $\frac{2}{3}$

7. $\frac{9}{21}$ 8. $\frac{69}{75}$

FIG. 1-4   (continued)

In exercises 9–17, find the fraction which is equivalent to the given fraction by multiplying both its terms by the number in parentheses.

9. $\frac{3}{8}$(5)   10. $\frac{7}{11}$(3)   11. $\frac{16}{7}$(11)

12. $\frac{14}{15}$(4)   13. $\frac{17}{23}$(2)   14. $\frac{5}{6}$(6)

15. $\frac{31}{7}$(9)   16. $\frac{28}{33}$(3)   17. $\frac{7}{12}$(8)

In exercises 18–26, find the fraction which is equivalent to the given fraction by dividing both its terms by the number in parentheses.

18. $\frac{21}{24}$(3)   19. $\frac{14}{35}$(7)   20. $\frac{8}{32}$(2)

21. $\frac{42}{36}$(3)   22. $\frac{7}{21}$(7)   23. $\frac{24}{32}$(4)

24. $\frac{24}{48}$(6)   25. $\frac{18}{45}$(9)   26. $\frac{15}{25}$(5)

Determine the value of the missing term in each of the following fractions, so that the three fractions in each case will be equivalent:

27. $\frac{2}{5} = \frac{}{15} = \frac{12}{}$ 28. $\frac{10}{22} = \frac{}{11} = \frac{}{55}$ 29. $\frac{3}{7} = \frac{9}{} = \frac{18}{}$

30. $\frac{4}{} = \frac{36}{63} = \frac{}{49}$ 31. $\frac{}{108} = \frac{3}{} = \frac{27}{36}$ 32. $\frac{}{14} = \frac{12}{7} = \frac{36}{}$

Tell whether each factored number below is a case of prime factorization or not. Justify your answers.

Equivalent Fractions and Lowest Terms

II

33. $32 = 4 \times 8$

34. $8 = 2 \times 2 \times 2$

35. $77 = 7 \times 11$

36. $90 = 2 \times 9 \times 5$

37. $120 = 2 \times 2 \times 2 \times 3 \times 5$

38. $28 = 2 \times 14$

Factor each of the following numbers completely:

39. 12	**40.** 18	**41.** 26	**42.** 54
43. 78	**44.** 15	**45.** 50	**46.** 22
47. 160	**48.** 112	**49.** 153	**50.** 114
51. 175	**52.** 88	**53.** 121	**54.** 86

Reduce each of the following fractions to lowest terms:

55. $\dfrac{12}{18}$	**56.** $\dfrac{10}{15}$	**57.** $\dfrac{5}{25}$	**58.** $\dfrac{21}{33}$
59. $\dfrac{16}{24}$	**60.** $\dfrac{18}{24}$	**61.** $\dfrac{35}{42}$	**62.** $\dfrac{32}{48}$
63. $\dfrac{52}{65}$	**64.** $\dfrac{36}{81}$	**65.** $\dfrac{35}{105}$	**66.** $\dfrac{35}{57}$
67. $\dfrac{91}{104}$	**68.** $\dfrac{132}{154}$	**69.** $\dfrac{35}{140}$	**70.** $\dfrac{72}{96}$
71. $\dfrac{45}{135}$	**72.** $\dfrac{54}{126}$	**73.** $\dfrac{51}{85}$	**74.** $\dfrac{65}{117}$
75. $\dfrac{24}{144}$	**76.** $\dfrac{198}{204}$	**77.** $\dfrac{52}{117}$	**78.** $\dfrac{66}{528}$

79. In a certain alloy, $\dfrac{4}{7}$ of the weight is copper. In the case of another alloy of the same size, $\dfrac{8}{14}$ of its weight is copper. Do the two alloys contain the same amount of copper?

80. Are $\dfrac{5}{8}$ in. and $\dfrac{24}{64}$ in. the same length?

81. Is $\dfrac{7}{8}$ gal the same as $\dfrac{3}{4}$ gal?

1-3 Addition of Fractions

A basic fact about addition is that only like quantities may be added. The same is true about fractions, and **like fractions** are those which have the same denominator. Thus the fractions $\dfrac{3}{4}, \dfrac{7}{4}$ and $\dfrac{11}{4}$ are like fractions, since all have 4 as their denominator. However, $\dfrac{5}{9}, \dfrac{3}{4}$, and $\dfrac{17}{29}$ are unlike fractions since their denominators are not the same.

Remembering that a fraction is a certain number of equal parts of a given unit, we understand that we can add $\dfrac{2}{7} + \dfrac{3}{7}$ to get $\dfrac{5}{7}$. In like manner $\dfrac{5}{19} + \dfrac{2}{19} + \dfrac{8}{19} = \dfrac{15}{19}$. From these two cases we see that *to add two or more like fractions we add their numerators, and write this sum over their common denominator.*

EXAMPLE 1-4 Add each of the following:

(a) $\dfrac{1}{3} + \dfrac{1}{3}$ (b) $\dfrac{2}{5} + \dfrac{1}{5}$ (c) $\dfrac{1}{8} + \dfrac{3}{8}$ (d) $\dfrac{4}{12} + \dfrac{5}{12}$

An Arithmetic Review

(a) $\dfrac{1}{3}+\dfrac{1}{3}=\dfrac{2}{3}$ (b) $\dfrac{2}{5}+\dfrac{1}{5}=\dfrac{3}{5}$

(c) $\dfrac{1}{8}+\dfrac{3}{8}=\dfrac{4}{8}=\dfrac{1}{2}$ (d) $\dfrac{4}{12}+\dfrac{5}{12}=\dfrac{9}{12}=\dfrac{3}{4}$

Unlike fractions cannot be added directly. In order to add unlike fractions they must first be converted to equivalent fractions having the same denominator.

To understand better the method of changing two or more fractions to equivalent fractions with a common denominator, we should first examine the concepts of **multiple, common multiple,** and **least common multiple.**

The product of two numbers is a multiple of each. Expressed in another way, *a number is a multiple of its factors.* Thus, 16 is a multiple of 2 and 8, because $16 = 2 \times 8$. Similarly 24 is a multiple of 4, since 4 is a factor of 24, and 38 is a multiple of 19, since 19 is a factor of 38.

Since a given number and its multiple have a factor–product relation, it follows that *the multiple of a given number is divisible by that number.* Thus, since 16 is a multiple of 2, 16 is divisible by 2. Similarly, since 18 is a multiple of 2, 3, 6 and 9, 18 is divisible by 2, 3, 6 and 9.

A common multiple of two or more numbers is that number which is a multiple of each one of them. Thus, 24 is a common multiple of each of the numbers 2, 3, 4, 6 and 8. Also, notice that because 24 is a multiple of all these numbers, 24 is divisible by each one of them. Similarly, 66 is a common multiple of 2, 3, 6, 11, 22, and 33, and therefore 66 is divisible by 2, 3, 6, 11, 22, and 33.

The Least Common Multiple (LCM) of two or more numbers is the smallest of their common multiples. Thus, while any of the numbers 12, 24, 36, 48, etc. is a common multiple of 3 and 4, the LCM of 3 and 4 is 12. Similarly, while 24, 48, 72, 96, etc. are common multiples of 3, 6, and 8, the LCM of 3, 6, and 8 is 24.

It follows from the definitions of multiple and common multiple that *the LCM of two or more numbers is divisible by each one of them.* For instance, the LCM of 24 and 28 is 168. Then, $168 \div 24 = 7$, and $168 \div 28 = 6$.

In simple cases the LCM of numbers can be determined at once. For instance, the LCM of 3, 4, and 6 is 12. However, in cases which are not so simple, the LCM can be found by the following two-step method:

1. *Arrange the numbers in a column and factor each number completely.*
2. *Multiply all the different prime factors, using each one the greatest number of times it occurs in any one of the numbers factored. This product is the LCM.*

EXAMPLE 1-5   Find the LCM of 12, 10, and 24.

SOLUTION

$$12 = 2 \times 2 \times 3$$
$$10 = 2 \times 5$$
$$24 = 2 \times 2 \times 2 \times 3$$

The three different prime numbers in the factorization of 12, 10, and 24 are 2, 3, and 5. All three will be factors of the LCM. The number 2 will be used three times, because three is the largest number of times the prime factor 2 appears in the factorization of the numbers 12, 10, and 24. The prime factor 3 will be used once, and the prime factor 5 once also, since both of them appear only once in any of the above factorizations. Thus, the LCM $= 2 \times 2 \times 2 \times 3 \times 5 = 120$. This means that there is no number smaller than 120 which is divisible by 12, 10, and 24.

EXAMPLE 1-6 Find the LCM of 24, 36, and 51.

SOLUTION

$$24 = 2 \times 2 \times 2 \times 3$$
$$36 = 2 \times 2 \times 3 \times 3 \qquad \text{LCM} = 2 \times 2 \times 2 \times 3 \times 3 \times 17 = 1224$$
$$51 = 3 \times 17$$

EXAMPLE 1-7 Find the LCM of 14, 21, and 18.

SOLUTION

$$14 = 2 \times 7$$
$$21 = 3 \times 7 \qquad \text{LCM} = 2 \times 3 \times 3 \times 7 = 126$$
$$18 = 2 \times 3 \times 3$$

To change two or more unlike fractions to equivalent fractions with the smallest common denominator, so that the rule for addition already given can be used, the student should follow this two-step procedure.

1. *Find the LCM of the denominators of the fractions. This LCM is the least common denominator (LCD) of the fractions.*
2. *Change each fraction to an equivalent fraction by multiplying both its terms by the number of times its denominator will go into the LCD.*

Notice that since the LCD is the LCM of the denominators of the fractions, the LCD is divisible by each denominator.

EXAMPLE 1-8 Change the fractions $\frac{5}{6}, \frac{3}{4}$, and $\frac{4}{7}$ to equivalent like fractions with the LCD.

SOLUTION

$$6 = 2 \times 3$$
$$4 = 2 \times 2 \qquad \text{LCM} = \text{LCD} = 2 \times 2 \times 3 \times 7 = 84$$
$$7 = 7$$

Then the terms of the first fraction $\frac{5}{6}$ are multiplied by 14, because 6 will go 14 times into 84. Also, the terms of the second fraction are both multiplied by 21, since 4 will go 21 times into 84. Finally, the terms of the fraction $\frac{4}{7}$ are both multiplied by 12. Thus,

$$\frac{5}{6} = \frac{5 \times 14}{6 \times 14} = \frac{70}{84}$$

$$\frac{3}{4} = \frac{3 \times 21}{4 \times 21} = \frac{63}{84}$$

$$\frac{4}{7} = \frac{4 \times 12}{7 \times 12} = \frac{48}{84}$$

NOTE: It is very simple to find the number of times a given denominator will go into the LCD without actual division. Just cross out from the factored form of the LCD those factors which comprise that particular denominator. The product of the remaining factors is the required quotient. Thus, since for the denominators of example 1-8 above, 6, 4, and 7, the LCD = $2 \times 2 \times 3 \times 7 = 84$

$$84 \div 6 = 2 \times \cancel{2} \times \cancel{3} \times 7 = 2 \times 7 = 14$$

$$84 \div 4 = \cancel{2} \times \cancel{2} \times 3 \times 7 = 3 \times 7 = 21$$

$$84 \div 7 = 2 \times 2 \times 3 \times \cancel{7} = 2 \times 2 \times 3 = 12$$

EXAMPLE 1-9 Change the fractions $\frac{5}{6}, \frac{7}{8}$, and $\frac{4}{9}$ to equivalent like fractions with the LCD.

SOLUTION

$$6 = 2 \times 3$$
$$8 = 2 \times 2 \times 2 \qquad \text{LCM} = \text{LCD} = 2 \times 2 \times 2 \times 3 \times 3 = 72$$
$$9 = 3 \times 3$$

Then

$$\frac{5}{6} = \frac{5 \times 12}{72} = \frac{60}{72} \qquad \frac{7}{8} = \frac{7 \times 9}{72} = \frac{63}{72} \qquad \frac{4}{9} = \frac{4 \times 8}{72} = \frac{32}{72}$$

We are ready now to add unlike fractions.

EXAMPLE 1-10 Add $\frac{3}{4} + \frac{5}{9} + \frac{7}{12}$.

SOLUTION

$$4 = 2 \times 2$$
$$9 = 3 \times 3 \qquad \text{LCD} = 2 \times 2 \times 3 \times 3 = 36$$
$$12 = 2 \times 2 \times 3$$

Then

$$\frac{3}{4} + \frac{5}{9} + \frac{7}{12} = \frac{3 \times 9}{36} + \frac{5 \times 4}{36} + \frac{7 \times 3}{36} = \frac{27}{36} + \frac{20}{36} + \frac{21}{36} = \frac{68}{36} = 1\frac{8}{9}$$

EXAMPLE 1-11 Add $\frac{5}{18} + \frac{7}{27}$.

SOLUTION

$$18 = 2 \times 3 \times 3 \qquad \text{LCD} = 2 \times 3 \times 3 \times 3 = 54$$
$$27 = 3 \times 3 \times 3$$

Then

$$\frac{5}{18} + \frac{7}{27} = \frac{5 \times 3}{54} + \frac{7 \times 2}{54} = \frac{15}{54} + \frac{14}{54} = \frac{29}{54}$$

Addition of Fractions

To add two or more mixed numbers, add the whole numbers together separately, add the fractions together separately, and then combine the partial sums.

EXAMPLE 1-12 Add $9\frac{1}{6} + 7\frac{7}{18} + 5\frac{1}{4}$.

SOLUTION

$$6 = 2 \times 3$$
$$18 = 2 \times 3 \times 3 \qquad \text{LCD} = 2 \times 2 \times 3 \times 3 = 36$$
$$4 = 2 \times 2$$

Then

$$9\frac{1}{6} + 7\frac{7}{18} + 5\frac{1}{4} = 9\frac{6}{36} + 7\frac{14}{36} + 5\frac{9}{36} = 9 + 5 + 7 + \frac{6 + 14 + 9}{36} = 21\frac{29}{36}$$

EXERCISE 1-3

1. Is 17 a multiple of 3? Why?
2. Is 36 a common multiple of 3 and 4? Why?
3. Is 36 the LCM of 3 and 4? Why?
Find the LCM of the following groups of numbers.

4. 8, 6, 12	**5.** 6, 18, 22	**6.** 24, 14, 6, 18
7. 9, 12, 22, 4	**8.** 6, 12, 15, 24	**9.** 5, 7, 8, 10
10. 75, 9, 12, 15	**11.** 32, 42, 60	**12.** 54, 24, 26
13. 20, 15, 84, 90	**14.** 28, 49, 35, 21	**15.** 18, 30, 42, 8
16. 21, 33, 25, 22	**17.** 40, 26, 8, 4	**18.** 160, 330, 315

Change the fractions in each of the following problems to equivalent fractions with the LCD.

19. $\frac{4}{9}, \frac{5}{14}, \frac{2}{21}$	**20.** $\frac{1}{4}, \frac{7}{8}, \frac{3}{10}$	**21.** $\frac{3}{18}, \frac{5}{24}$
22. $\frac{7}{8}, \frac{5}{12}, \frac{3}{18}$	**23.** $\frac{9}{40}, \frac{7}{45}, \frac{8}{9}$	**24.** $\frac{5}{12}, \frac{7}{27}, \frac{1}{3}$
25. $\frac{11}{16}, \frac{7}{18}, \frac{3}{22}$	**26.** $\frac{2}{55}, \frac{4}{77}, \frac{3}{11}$	**27.** $\frac{3}{10}, \frac{3}{16}, \frac{1}{9}, \frac{2}{5}$

Add the fractions in each of the following problems. Express all fractional answers in lowest terms, and all improper fractions as mixed numbers.

28. $\frac{2}{5} + \frac{1}{5}$	**29.** $\frac{3}{7} + \frac{2}{7} + \frac{1}{7}$
30. $\frac{1}{16} + \frac{3}{16} + \frac{5}{16}$	**31.** $\frac{3}{4} + \frac{5}{6} + \frac{5}{12}$
32. $\frac{5}{6} + \frac{3}{8} + \frac{1}{16}$	**33.** $\frac{5}{12} + \frac{7}{22} + \frac{2}{33}$
34. $\frac{3}{10} + \frac{7}{20} + \frac{2}{5}$	**35.** $\frac{4}{15} + \frac{7}{16} + \frac{5}{18}$
36. $\frac{5}{21} + \frac{7}{18} + \frac{7}{12}$	**37.** $\frac{29}{49} + \frac{5}{21} + \frac{3}{14} + \frac{11}{28}$
38. $\frac{15}{16} + \frac{3}{4} + \frac{1}{8} + \frac{1}{2}$	**39.** $\frac{5}{9} + \frac{7}{12} + \frac{1}{8} + \frac{6}{7}$
40. $\frac{5}{12} + \frac{1}{5} + \frac{4}{27} + \frac{2}{3}$	**41.** $\frac{3}{25} + \frac{7}{30} + \frac{2}{3} + \frac{8}{9}$

42. $\dfrac{1}{3} + \dfrac{11}{14} + \dfrac{6}{35} + \dfrac{5}{18}$ **43.** $\dfrac{45}{81} + \dfrac{3}{4} + \dfrac{2}{27} + \dfrac{5}{8}$

44. $6\frac{3}{5} + 5\frac{7}{8}$ **45.** $3\frac{7}{12} + 8\frac{5}{16}$

46. $1\frac{1}{4} + 2\frac{7}{9}$ **47.** $3\frac{9}{24} + 8\frac{11}{18}$

48. $2\frac{3}{4} + 1\frac{1}{7} + 3\frac{5}{8}$ **49.** $7\frac{2}{3} + 5\frac{4}{11} + 3\frac{3}{4}$

50. $16\frac{2}{3} + \frac{6}{17} + \frac{5}{9}$ **51.** $5\frac{6}{7} + \frac{13}{15} + 1\frac{4}{9}$

52. Three pieces of wrought iron pipe weigh $5\frac{1}{8}$ lb, $2\frac{1}{3}$ lb, and $7\frac{5}{6}$ lb, respectively. What is the total weight of the three pieces of pipe?

53. A person traveled by car from New York to Durham, NC in $10\frac{3}{4}$ hr, and from Durham to Asheville, NC in $4\frac{5}{8}$ hr. What was the total travel time from New York to Asheville?

54. A rectangular room is $21\frac{3}{8}$-ft long and $17\frac{5}{12}$-ft wide. What is the total distance around the room?

55. A boy weighing $106\frac{7}{12}$ lb rides a bicycle weighing $49\frac{5}{18}$ lb. What is the total weight of rider and bicycle?

56. A person wants to mail an electric drill weighing $4\frac{5}{8}$ lb, and a set of woodboring bits weighing $1\frac{5}{12}$ lb. If the cardboard box used for the package weighs $\frac{5}{6}$ lb, what is the total weight of the package?

1-4 Subtraction of Fractions

As with addition, only like fractions may be subtracted. Thus, *to subtract two like fractions, the numerators are subtracted, and this difference is placed over their common denominator.*

EXAMPLE 1-13   Subtract $\dfrac{9}{15} - \dfrac{7}{15}$.

SOLUTION

$$\frac{9}{15} - \frac{7}{15} = \frac{9-7}{15} = \frac{2}{15}$$

To subtract two unlike fractions, first their LCD is determined and then the fractions are converted to equivalent fractions with the LCD. Then, the above rule for the subtraction of fraction with the LCD can be applied.

EXAMPLE 1-14   Subtract $\dfrac{4}{5} - \dfrac{3}{10}$.

SOLUTION

$$\text{LCD} = 10$$

$$\frac{4}{5} - \frac{3}{10} = \frac{8}{10} - \frac{3}{10} = \frac{5}{10} = \frac{1}{2}$$

EXAMPLE 1-15   Simplify $\dfrac{5}{12} + \dfrac{7}{30} - \dfrac{9}{20}$.

SOLUTION

$$12 = 2 \times 2 \times 3$$
$$30 = 2 \times 3 \times 5 \qquad \text{LCD} = 2 \times 2 \times 3 \times 5 = 60$$
$$20 = 2 \times 2 \times 5$$

Then,

$$\frac{5}{12} + \frac{7}{30} - \frac{9}{20} = \frac{25}{60} + \frac{14}{60} - \frac{27}{60} = \frac{25 + 14 - 27}{60} = \frac{12}{60} = \frac{1}{5}$$

To subtract mixed numbers, subtract the whole numbers separately, subtract the fractions separately, and then combine the partial differences.

EXAMPLE 1-16   Subtract $5\frac{2}{3} - 2\frac{1}{3}$.

SOLUTION

$$
\begin{array}{c|c}
5 & \dfrac{2}{3} \\[2mm]
2 & \dfrac{1}{3} \\[2mm]
\hline
3 & \dfrac{1}{3} = 3\dfrac{1}{3}
\end{array}
$$

EXAMPLE 1-17   Subtract $6\frac{2}{3} - 2\frac{1}{4}$.

SOLUTION

$$\text{LCD} = 12$$

Then,

$$6\frac{2}{3} - 2\frac{1}{4} = 6\frac{8}{12} - 2\frac{3}{12} = 4\frac{5}{12}$$

EXAMPLE 1-18   Subtract $5\frac{2}{3} - 2\frac{7}{8}$.

SOLUTION

$$\text{LCD} = 24$$

Then,

$$5\frac{2}{3} - 2\frac{7}{8} = 5\frac{16}{24} - 2\frac{21}{24}$$

It is obvious at this point that trying to use the rule for subtraction would require us to subtract 21 from 16. To make the subtraction possible, we will change the first of the mixed numbers as follows: We borrow 1 from 5, which now becomes 4. Then the borrowed 1 is expressed as $\frac{24}{24}$ and is added to the existing $\frac{16}{24}$. Thus, $5\frac{16}{24} = 4 + \frac{24}{24} + \frac{16}{24}$ $= 4\frac{40}{24}$. Now the rule can be used because the original subtraction is of the form $4\frac{40}{24} - 2\frac{21}{24}$, and the final answer is $2\frac{19}{24}$.

EXAMPLE 1-19   Subtract $16 - 6\frac{7}{8}$.

SOLUTION

$$16 - 6\frac{7}{8} = 15\frac{8}{8} - 6\frac{7}{8} = 9\frac{1}{8}$$

EXERCISE 1-4

Subtract the fractions in each of the following problems. Express all fractional answers in lowest terms:

1. $\dfrac{2}{3} - \dfrac{1}{3}$ 　　　　　　　**2.** $\dfrac{7}{8} - \dfrac{3}{8}$ 　　　　　　　**3.** $\dfrac{13}{32} - \dfrac{5}{32}$

4. $\dfrac{7}{10} - \dfrac{2}{5}$ 5. $\dfrac{1}{3} - \dfrac{1}{4}$ 6. $\dfrac{5}{8} - \dfrac{1}{6}$

7. $\dfrac{14}{15} - \dfrac{3}{4}$ 8. $\dfrac{13}{35} - \dfrac{5}{21}$ 9. $\dfrac{11}{56} - \dfrac{3}{16}$

10. $\dfrac{12}{27} - \dfrac{1}{6}$ 11. $12\frac{3}{10} - 3\frac{7}{15}$ 12. $25\frac{3}{16} - 6\frac{7}{20}$

13. $6\frac{1}{2} - 3\frac{7}{10}$ 14. $11\frac{3}{5} - 6\frac{8}{9}$ 15. $18\frac{7}{12} - 9\frac{5}{8}$

16. $6 - 2\frac{7}{8}$ 17. $17 - 11\frac{7}{17}$ 18. $5\frac{3}{4} - 2$

Simplify each of the following:

19. $\dfrac{2}{9} + \dfrac{7}{15} - \dfrac{1}{3}$ 20. $\dfrac{5}{8} + \dfrac{7}{15} - \dfrac{3}{4}$

21. $2\frac{3}{4} + 3\frac{1}{5} - 1\frac{2}{3}$ 22. $18\frac{5}{8} + 5 - 3\frac{1}{7}$

23. $264\frac{3}{16} + 13 - \dfrac{3}{5}$

24. The diameter of a $\frac{7}{8}$-in. steel shaft is reduced by $\frac{5}{64}$-in. What is the new diameter of the shaft?

25. One cubic foot of lead weighs $710\frac{3}{4}$ lb, and one cubic foot of steel weighs $489\frac{7}{18}$ lb. How much heavier is one cubic foot of lead than one cubic foot of steel?

26. A planer takes a $\frac{1}{16}$-in. cut from a brass casting $\frac{3}{4}$-in. thick. What is the final thickness of the casting?

1-5 Multiplication of Fractions

If we buy 3 cartons of 6 bottles of soda, we buy 3 × 6 or 18 bottles. If a car's tank holds 18 gal, and there are 5 identical cars, then there are 5 cars of 18 gal, or 5 × 18 = 90 gal. It is clear that 3 "of" something means to multiply by 3, and 5 "of" something means to multiply by 5.

The same is true of fractions also. Consider once more a circle divided into 6 equal parts. Two such parts are $\dfrac{2}{6}$ of the circle as shown in Fig. 1-5. We know from the figure that half of the shaded part of the circle is $\dfrac{1}{6}$ of it. We also know that $\dfrac{1}{2}$ of $\dfrac{2}{6}$ is $\dfrac{1}{2} \times \dfrac{2}{6}$. Therefore, $\dfrac{1}{2} \times \dfrac{2}{6} = \dfrac{1}{6}$. At this point we notice that the above multipli-

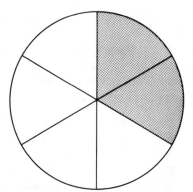

FIG. 1-5

cation of $\frac{1}{2} \times \frac{2}{6}$ will give the same result, $\frac{1}{6}$, if we multiply the numerators and the denominators together and reduce to lowest terms. Thus,

$$\frac{1}{2} \times \frac{2}{6} = \frac{1 \times 2}{2 \times 6} = \frac{2}{12} = \frac{1}{6}$$

We conclude that *to multiply two or more fractions, all the numerators are multiplied together, and this product is placed over the product of all the denominators.*

Since every whole number may be changed into a fraction by placing the whole number over 1, and since every mixed number may be changed into an improper fraction, the above rule is good for all cases of multiplication with fractions.

EXAMPLE 1-20   Multiply $\frac{3}{10} \times \frac{9}{5}$.

SOLUTION

$$\frac{3}{10} \times \frac{9}{5} = \frac{3 \times 9}{10 \times 5} = \frac{27}{50}$$

EXAMPLE 1-21   Multiply $\frac{2}{9} \times \frac{13}{4}$.

SOLUTION

$$\frac{2}{9} \times \frac{13}{4} = \frac{26}{36}$$

Then, $\frac{26}{36}$ reduced to lowest terms is $\frac{13}{18}$.

EXAMPLE 1-22   Multiply $\frac{9}{22} \times \frac{8}{27}$.

SOLUTION

$$\frac{9}{22} \times \frac{8}{27} = \frac{72}{594}$$

Then, $\frac{72}{594}$ reduced to lowest terms is $\frac{4}{33}$.

The reduction, however, of the fraction $\frac{72}{594}$ is not easy, and it is time consuming. The procedure for the multiplication of fractions given below will make the reduction unnecessary, because the result will always be in lowest terms: *Indicate the multiplication of the numerators, and the multiplication of the denominators, and factor completely these indicated products. Then cross out (cancel) any pair of factors common to both numerator and denominator, and multiply the remaining factors.*

This procedure is illustrated by reworking example 1-22.

$$\frac{9}{22} \times \frac{8}{27} = \frac{9 \times 8}{22 \times 27} = \frac{3 \times 3 \times 2 \times 2 \times 2}{2 \times 11 \times 3 \times 3 \times 3} = \frac{\cancel{3} \times \cancel{3} \times \cancel{2} \times 2 \times 2}{\cancel{2} \times 11 \times \cancel{3} \times \cancel{3} \times 3} = \frac{4}{33}$$

NOTE: The student should realize that we can use cancellation because each cancelled pair of factors common to both numerator and denominator is equal

An Arithmetic Review

to 1. As we have seen already, 1 is a factor of every number without affecting its value. Therefore, each such pair of factors can be omitted or cancelled.

EXAMPLE 1-23 Multiply $12 \times \frac{5}{9}$.

SOLUTION

$$12 \times \frac{5}{9} = \frac{12}{1} \times \frac{5}{9} = \frac{12 \times 5}{1 \times 9} = \frac{2 \times 2 \times 3 \times 5}{3 \times 3} = \frac{2 \times 2 \times \cancel{3} \times 5}{\cancel{3} \times 3} = \frac{20}{3} = 6\frac{2}{3}$$

EXAMPLE 1-24 Multiply $6\frac{1}{4} \times 5\frac{3}{5} \times 2\frac{5}{14}$.

SOLUTION

$$6\frac{1}{4} \times 5\frac{3}{5} \times 2\frac{5}{14} = \frac{25}{4} \times \frac{28}{5} \times \frac{33}{14} = \frac{25 \times 28 \times 33}{4 \times 5 \times 14} = \frac{5 \times 5 \times 2 \times 2 \times 7 \times 3 \times 11}{2 \times 2 \times 5 \times 2 \times 7}$$

$$= \frac{\cancel{5} \times 5 \times \cancel{2} \times \cancel{2} \times \cancel{7} \times 3 \times 11}{\cancel{2} \times \cancel{2} \times \cancel{5} \times 2 \times \cancel{7}} = \frac{165}{2} = 82\frac{1}{2}$$

Do each of the following multiplications of fractions. Use cancellation to have all fractional results reduced to their lowest terms. Change all improper fractions to mixed numbers.

1. $\frac{3}{7} \times \frac{2}{5}$ 2. $\frac{3}{5} \times \frac{5}{6}$ 3. $\frac{2}{3} \times \frac{5}{8}$

4. $\frac{9}{15} \times \frac{5}{6}$ 5. $\frac{3}{8} \times \frac{1}{3}$ 6. $\frac{9}{12} \times \frac{22}{12}$

7. $\frac{12}{21} \times \frac{3}{4}$ 8. $\frac{13}{51} \times \frac{34}{39}$ 9. $\frac{1}{7} \times \frac{1}{7}$

10. $\frac{5}{12} \times \frac{6}{15}$ 11. $\frac{2}{3} \times \frac{5}{7}$ 12. $\frac{5}{8} \times \frac{11}{30}$

13. $5 \times \frac{7}{35}$ 14. $8 \times \frac{3}{4}$ 15. $2 \times \frac{5}{4}$

16. $\frac{3}{11} \times 44$ 17. $\frac{1}{4} \times \frac{1}{4} \times \frac{1}{4}$ 18. $\frac{5}{9} \times \frac{18}{25} \times \frac{3}{20}$

19. $\frac{7}{11} \times \frac{33}{49} \times \frac{11}{14}$ 20. $\frac{2}{7} \times \frac{3}{8} \times \frac{21}{18}$ 21. $14 \times \frac{16}{21} \times \frac{9}{24}$

22. $\frac{18}{45} \times \frac{66}{77} \times \frac{35}{48} \times \frac{33}{34}$ 23. $1\frac{2}{3} \times 3\frac{3}{4}$ 24. $\frac{3}{8} \times 2\frac{5}{6}$

25. $9 \times 4\frac{1}{2} \times 6\frac{8}{15}$ 26. $9\frac{3}{4} \times 2\frac{1}{3} \times 2\frac{2}{13}$ 27. $1\frac{3}{7} \times 2\frac{3}{4} \times \frac{1}{3}$

28. $12 \times 5\frac{7}{8} \times 2\frac{2}{3}$ 29. $\frac{7}{15} \times 5 \times 3\frac{1}{3}$ 30. $8 \times 2\frac{3}{11} \times \frac{121}{125}$

31. $6 \times 6\frac{3}{8} \times \frac{5}{34}$ 32. $3 \times \frac{7}{18} \times 3\frac{3}{5}$ 33. $\frac{3}{7} \times 6\frac{1}{8} \times 7$

34. $\frac{8}{11} \times 2\frac{5}{14} \times \frac{21}{24}$ 35. $5\frac{1}{3} \times 1\frac{1}{4} \times 1\frac{4}{5}$ 36. $7\frac{3}{5} \times 2\frac{3}{8} \times 1\frac{38}{51}$

37. How much is $\frac{4}{5}$ of $\frac{7}{16}$?

38. A pie is divided into 6 equal slices. What part of the pie is $\frac{1}{3}$ of a slice?

39. A car gets $17\frac{1}{4}$ miles per gallon. How many miles will it travel on $10\frac{2}{3}$ gal?

Multiplication of Fractions   **21**

40. When a car's tank is full, it contains 22 gal of gas. How many gallons does it contain when it is $\frac{4}{5}$ full?

41. A certain screw has 12 threads per inch. How many threads are there in $2\frac{3}{4}$ in.?

42. The *pitch* of a thread is the distance between corresponding points of two successive threads (see Fig. 1-6). With a complete turn the screw advances a distance equal to the pitch of the screw. If the pitch of a screw is $\frac{2}{7}$ in., and the screw is rotated $8\frac{2}{3}$ turns, how many inches will the screw advance?

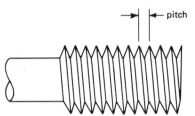

FIG. 1-6

I-6 Division of Fractions

Division of fractions, like multiplication, is simple. In fact, division of fractions is converted to multiplication. Let us see what the reasoning is behind the simple phrase "invert and multiply."

To find a method to perform the division $\frac{2}{7} \div \frac{3}{4}$, using nothing more than what we already know, we proceed as follows: First, since division can be indicated by a fraction, we express the above division as the fraction

$$\frac{\dfrac{2}{7}}{\dfrac{3}{4}}.$$

Now using the fundamental principle of fractions, we multiply both numerator and denominator of this fraction by the fractional number $\frac{4}{3}$ (Notice that $\frac{4}{3}$ is the denominator $\frac{3}{4}$ inverted.) and then simplify. Thus,

$$\frac{\dfrac{2}{7}}{\dfrac{3}{4}} = \frac{\dfrac{2}{7} \times \dfrac{4}{3}}{\dfrac{3}{4} \times \dfrac{4}{3}} = \frac{\dfrac{2}{7} \times \dfrac{4}{3}}{1} = \frac{8}{21}$$

Notice that the denominator of the fraction $\frac{\dfrac{2}{7}}{\dfrac{3}{4}}$ became 1, and the division indicated by the fraction was transformed into the simple multiplication $\frac{2}{7} \times \frac{4}{3}$.

From the above we conclude that *to divide two fractions, invert the second fraction (the divisor) and multiply*, using the multiplication procedure and the cancellation technique. This rule is good for all cases of division with fractions, since whole numbers and mixed numbers can be changed into fractions.

EXAMPLE 1-25 Divide $\frac{3}{5} \div \frac{2}{3}$.

SOLUTION

$$\frac{3}{5} \div \frac{2}{3} = \frac{3}{5} \times \frac{3}{2} = \frac{9}{10}$$

EXAMPLE 1-26 Divide $\frac{5}{8} \div \frac{3}{4}$.

SOLUTION

$$\frac{5}{8} \div \frac{3}{4} = \frac{5}{8} \times \frac{4}{3} = \frac{5 \times \cancel{2} \times \cancel{2}}{\cancel{2} \times \cancel{2} \times 2 \times 3} = \frac{5}{6}$$

EXAMPLE 1-27 Divide $\frac{5}{9} \div 3$.

SOLUTION

$$\frac{5}{9} \div 3 = \frac{5}{9} \div \frac{3}{1} = \frac{5}{9} \times \frac{1}{3} = \frac{5}{27}$$

EXAMPLE 1-28 Divide $2\frac{3}{7} \div 3\frac{5}{21}$.

SOLUTION

$$2\frac{3}{7} \div 3\frac{5}{21} = \frac{17}{7} \div \frac{68}{21} = \frac{17}{7} \times \frac{21}{68} = \frac{\cancel{17} \times 3 \times \cancel{7}}{\cancel{7} \times 2 \times 2 \times \cancel{17}} = \frac{3}{4}$$

EXERCISE 1-6

Perform each of the following divisions:

1. $\frac{4}{5} \div \frac{3}{4}$

2. $\frac{5}{9} \div \frac{2}{3}$

3. $\frac{7}{18} \div \frac{5}{9}$

4. $\frac{5}{12} \div \frac{3}{16}$

5. $\frac{3}{4} \div \frac{7}{8}$

6. $\frac{9}{15} \div \frac{2}{3}$

7. $\frac{7}{8} \div \frac{21}{12}$

8. $\frac{4}{5} \div \frac{1}{2}$

9. $\frac{8}{15} \div \frac{5}{12}$

10. $3 \div \frac{1}{4}$

11. $\frac{4}{5} \div 2$

12. $\frac{5}{7} \div 4$

13. $9 \div \frac{5}{8}$

14. $3\frac{1}{4} \div \frac{2}{3}$

15. $4\frac{1}{3} \div \frac{2}{3}$

16. $\frac{11}{12} \div 2\frac{1}{8}$

17. $\frac{3}{7} \div 3\frac{5}{7}$

18. $2\frac{1}{4} \div 3\frac{1}{8}$

19. $1\frac{1}{4} \div 3$

20. $2\frac{1}{2} \div 3\frac{3}{4}$

21. $3\frac{3}{8} \div 2\frac{1}{16}$

22. $13\frac{1}{4} \div 3\frac{1}{8}$

23. $3\frac{1}{2} \div 1\frac{15}{16}$

24. $3\frac{1}{4} \div 2\frac{3}{8}$

25. A car will travel $18\frac{3}{4}$ miles per gallon of gas. How many gallons will the car use to go $46\frac{7}{8}$ mi?

Division of Fractions

26. A steel bar $5\frac{5}{8}$-in. long is to be divided into 15 equal parts. What is the length of each part?

27. The approximate diameter of a circle is found by dividing its circumference by $\frac{22}{7}$. What is the diameter of a circle whose circumference is $6\frac{7}{8}$ in.?

28. With a complete turn, a screw advances a distance equal to the pitch of the screw (see Fig. 1-6). If the pitch of the screw is $\frac{2}{9}$ in., and the screw was advanced $3\frac{1}{8}$ in., how many times was the screw turned?

29. From the definition of the pitch, it follows that

$$\text{pitch} = \frac{1}{\text{number of threads per inch}}$$

If a certain screw has $5\frac{1}{2}$ threads per inch, find its pitch.

30. Find the pitch of a screw with $2\frac{1}{4}$ threads per inch.

1-7 Complex Fractions and Reciprocals

Fractions like $\dfrac{\frac{3}{4}}{\frac{5}{7}}$, where one or both of the terms of the fraction are fractions themselves are called **complex fractions**. Complex fractions can easily be simplified if we recall that a fraction always implies division of the numerator by the denominator. Thus, to simplify the above complex fraction we proceed as follows:

$$\frac{\frac{3}{4}}{\frac{5}{7}} = \frac{3}{4} \div \frac{5}{7} = \frac{3}{4} \times \frac{7}{5} = \frac{21}{20} = 1\frac{1}{20}$$

EXAMPLE 1-29 Simplify the complex fraction $\dfrac{\frac{5}{18}}{\frac{20}{21}}$.

SOLUTION

$$\frac{\frac{5}{18}}{\frac{20}{21}} = \frac{5}{18} \div \frac{20}{21} = \frac{5}{18} \times \frac{21}{20} = \frac{\cancel{5} \times \cancel{3} \times 7}{2 \times \cancel{3} \times 3 \times 2 \times 2 \times \cancel{5}} = \frac{7}{24}$$

EXAMPLE 1-30 Simplify the complex fraction $\dfrac{\frac{3}{2} + \frac{3}{7}}{\frac{2}{3}}$.

SOLUTION The LCD of the fractions which form the numerator of the complex fraction is 14. Then,

$$\frac{\frac{3}{2} + \frac{3}{7}}{\frac{2}{3}} = \frac{\frac{21}{14} + \frac{6}{14}}{\frac{2}{3}} = \frac{\frac{27}{14}}{\frac{2}{3}} = \frac{27}{14} \times \frac{3}{2} = \frac{81}{28} = 2\frac{25}{28}$$

When the product of two numbers is 1, the two numbers are said to be **recipro-cals** of each other. For instance, since $3 \times \frac{1}{3} = 1$, 3 is the reciprocal of $\frac{1}{3}$, and $\frac{1}{3}$ is the reciprocal of 3. Similarly, the reciprocal of 17 is $\frac{1}{17}$ since $17 \times \frac{1}{17} = 1$. Notice that *the reciprocal of a whole number is 1 over that number.* Thus, the reciprocal of 5 is $\frac{1}{5}$, and the reciprocal of 23 is $\frac{1}{23}$. The same is true of fractional numbers. Therefore, the reciprocal of $\frac{3}{4}$ is $\frac{1}{\frac{3}{4}}$. To simplify this complex fraction we follow our procedure:

$$\frac{1}{\frac{3}{4}} = 1 \div \frac{3}{4} = 1 \times \frac{4}{3} = \frac{4}{3}$$

Similarly, the reciprocal of $\frac{7}{12} = \frac{1}{\frac{7}{12}}$. Then,

$$\frac{1}{\frac{7}{12}} = 1 \div \frac{7}{12} = 1 \times \frac{12}{7} = \frac{12}{7}$$

From these two cases of reciprocals of fractions, we notice that *the reciprocal of a fraction is equal to the same fraction with its terms inverted.* Thus, the reciprocal of $\frac{9}{16}$ is $\frac{16}{9}$.

Reciprocals are frequently used in mathematics. We already used a reciprocal (an inverted fraction) in developing a method of dividing fractions, and we will use more reciprocals in later chapters.

EXERCISE 1-7

1. What is the reciprocal of a whole number?
2. What is the reciprocal of a fraction?
Simplify each of the following complex fractions:

3. $\dfrac{\frac{6}{45}}{\frac{5}{4}}$

4. $\dfrac{\frac{3}{11}}{\frac{33}{44}}$

5. $\dfrac{\frac{33}{24}}{\frac{11}{28}}$

6. $\dfrac{\frac{2}{3} + \frac{4}{5}}{\frac{3}{5}}$

7. $\dfrac{\frac{8}{15} - \frac{1}{6}}{\frac{4}{45}}$

8. $\dfrac{\frac{5}{16} + \frac{2}{40}}{\frac{11}{30} - \frac{2}{21}}$

9. $\dfrac{\frac{5}{18} - \frac{4}{45}}{\frac{22}{63} + \frac{4}{15}}$

Complex Fractions and Reciprocals

Write the reciprocal of each of the following numbers:

10. 8 **11.** 24 **12.** 9 **13.** 11

14. $\dfrac{3}{11}$ **15.** $\dfrac{2}{3}$ **16.** $\dfrac{7}{8}$ **17.** $2\frac{2}{3}$

18. $1\frac{1}{3}$

1-8 Basic Ideas About Decimals

We have already seen in the beginning of this chapter that decimals are fractions. However, they are fractions of a special kind. Let us see why. If asked to write the number five hundredths, you will be equally correct in writing either $\dfrac{5}{100}$ or 0.05. They sound alike when read, and they have the same value. We can say that $\dfrac{5}{100} = 0.05$.

It follows that $\dfrac{5}{100}$ and 0.05 are two different ways of writing the same number. The same is true for $\dfrac{3}{1000}$ and 0.003 or $\dfrac{17}{10,000}$ and 0.0017. Notice that the denominators of the fractional form of these three numbers is 10×10, $10 \times 10 \times 10$, and $10 \times 10 \times 10 \times 10$, respectively. Fractions with denominators of 10, 100, 1000 etc. are special fractions called **decimal fractions**.

The decimal numbers extend the number system to the right of the units' place. The names of the positions of the digits of the number 37,834.19783 are shown in Fig. 1-7. Notice first that a **decimal point** separates the units' place from the fractional part of the number. Then notice that the digit 1 placed in the first position to the right of the units' place takes the value $\dfrac{1}{10}$, the digit 9 placed in the second position to the right of the units' place takes the value $\dfrac{9}{100}$ (or $9/(10 \times 10)$), and so on. This means that for each place a digit is moved to the right, the digit is divided by one more 10 than before.

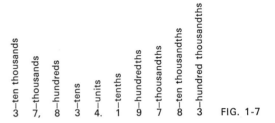

FIG. 1-7

Notice now, that all decimals are not only fractions of a special denominator, but they are all proper fractions. This means that all decimals are less than 1. Numbers, however, like 3.14 or 17.023, which can also be written as $3\frac{14}{100}$ and $17\frac{23}{1000}$, are mixed numbers.

NOTE: Any whole number is understood to have a decimal point to the right of its units' place. Thus, $5 = 5.$, $27 = 27.$, and $238 = 238.$. This means that any

An Arithmetic Review

whole number is a "mixed number" with any number of zeros to the right of the decimal point. Thus, $7 = 7.000\ldots$, and $318 = 318.000\ldots$.

1-9 Addition and Subtraction of Decimals

In arranging decimal numbers for addition or subtraction, care must be taken to form columns of the same kind. This will be done automatically by simply putting the decimal points one under the other. Everything else is like the addition or subtraction of whole numbers.

EXAMPLE 1-31 Add 2.64, 175.9, 0.721, and 9.203.

SOLUTION The numbers are arranged and added as shown below.

$$\begin{array}{r} 2.64 \\ 175.9 \\ 0.721 \\ \underline{9.203} \\ 188.464 \end{array}$$

To keep the columns straight, and thus avoid mistakes of slipping to the wrong column, we can add zeros to the end of the decimals which are to be added or subtracted. Thus, the above addition can be written as shown below.

$$\begin{array}{r} 2.640 \\ 175.900 \\ 0.721 \\ \underline{9.203} \\ 188.464 \end{array}$$

EXAMPLE 1-32 Subtract 27.238 from 51.16.

SOLUTION

$$\begin{array}{r} 51.160 \\ -\ 27.238 \\ \hline 23.922 \end{array}$$

EXAMPLE 1-33 Perform each of the following subtractions:

(a) $6 - 4.27$ (b) $5.71 - 4$

SOLUTION

$$\begin{array}{rr} \text{(a)} & 6.00 \\ & -\ 4.27 \\ \hline & 1.73 \end{array} \qquad \begin{array}{rr} \text{(b)} & 5.71 \\ & -\ 4.00 \\ \hline & 1.71 \end{array}$$

EXERCISE 1-8

Perform each of the following additions of decimals:
1. $126.17 + 93.052 + 7.72 + 801.037$ 2. $23.018 + 3.19 + 134.03 + 75.004$
3. $1389.76 + 2347 + 17.063 + 9.2$ 4. $73.32 + 289.6 + 1.0317 + 8$
5. $28.987 + 2856.0935 + 134 + 0.8456$ 6. $6.017 + 3 + 27.1003 + 25.09$

Addition and Subtraction of Decimals

Perform each of the following subtractions of decimals:

7. $0.107 - 0.034$ 8. $0.2015 - 0.0789$

9. $7 - 0.371$ 10. $19 - 0.37$

11. $264.13 - 65$ 12. $34.461 - 3$

13. The bimetallic element (two strips of different metal riveted together, that expand at different rates) of a thermostat is made of a strip of brass 0.073-in. thick, and a strip of iron 0.125-in. thick. What is the thickness of the bimetallic element?

14. What is the missing dimension in Fig. 1-8?

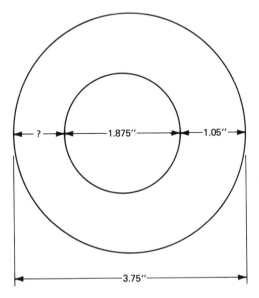

FIG. 1-8

15. What is the thickness of the pipe shown in cross section in Fig. 1-9?

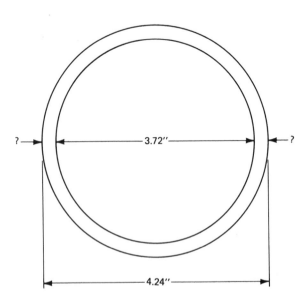

FIG. 1-9

An Arithmetic Review

16. If the major diameter of a screw (shown in Fig. 1-10) is 0.625 in., and the minor diameter is 0.5234 in., find the depth of the screw.

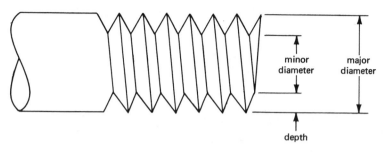

FIG. 1-10

17. If the major diameter of a screw is 0.864 in., and the depth of the screw is 0.1858 in., find the minor diameter (see Fig. 1-10).

18. If the minor diameter of a screw is 0.654 in., and the depth is 0.0913 in., find the major diameter (see Fig. 1-10).

1-10 Multiplication of Decimals

Suppose that we are asked to multiply 0.17×0.2. We certainly can get the correct product using fractions as follows:

$$0.17 \times 0.2 = \frac{17}{100} \times \frac{2}{10} = \frac{34}{1000} = 0.034$$

Also,

$$3.7 \times 5.3 = 3\frac{7}{10} \times 5\frac{3}{10} = \frac{1961}{100} = 19\frac{61}{100} = 19.61$$

The product in each case has as many decimal places as the sum of the decimal places in the two factors that were multiplied. This suggests the following rule for the multiplication of decimals: *To multiply two decimal numbers, ignore, momentarily, the decimal points and multiply the numbers as if they were whole numbers. Then put the decimal point in the product as many places from the right as the sum of all the decimal places in both numbers being multiplied.*

EXAMPLE 1-34 Multiply each of the following:

(a) 5.12×0.17 (b) 704.3×20.4

SOLUTION

	(a)	5.12	(b)	70 4.3
		0.17		2 0.4
		35 84		281 7 2
		51 2		14086 0
		0.87 04		14367.7 2

EXAMPLE 1-35 Multiply each of the following:

$$\text{(a) } 0.18 \times 0.3 \qquad \text{(b) } 3.14 \times 9$$

SOLUTION

$$\begin{array}{cc}
\text{(a)} & \begin{array}{r} 0.18 \\ 0.3 \\ \hline 0.054 \end{array} & \text{(b)} & \begin{array}{r} 3.14 \\ 9 \\ \hline 28.26 \end{array}
\end{array}$$

EXERCISE 1-9

Perform each of the following multiplications:

1. 0.2×0.3	**2.** 0.14×0.3	**3.** 0.8×0.17
4. 0.205×0.6	**5.** 0.118×0.07	**6.** 0.625×0.31
7. 5.34×2.3	**8.** 76.134×0.018	**9.** 0.081×0.043
10. 76.24×12.7	**11.** 0.87×9.4	**12.** 7.243×1.8
13. 0.003×0.026	**14.** 0.767×0.007	**15.** 0.73×5

16. During a certain week a factory worker worked 37.7 hr at the rate of $5.72 per hour. What are her gross earnings?

17. How many miles can a car travel on 31.7 gal at the rate of 17.4 miles per gallon?

18. If concrete weighs 143.7 pounds per cubic foot, how many pounds do 32.4 cubic feet weigh?

19. If gasoline costs $1.29/gal, what is the cost of 18 gal?

20. Find the weekly earnings (gross) of a worker who worked 49 hr at the rate of $5.82 per hour. The rate of pay for hours over 40 is time and a half.

1-11 Division of Decimals

Divisions involving decimals are of two kinds:

1. Divisions with a *whole number divisor*.
   Examples: $14.49 \div 7$; $7 \div 8$
2. Divisions with a *decimal divisor*.
   Examples: $27 \div 0.3$; $7.8 \div 0.6$; $66.456 \div 3.12$

To divide by a whole number divisor, proceed as in the division of whole numbers. Any decimal point in the quotient will be placed directly above the decimal point in the dividend.

EXAMPLE 1-36 Divide $6.4 \div 8$.

SOLUTION The division is arranged as shown in (a), with the decimal point in the quotient as shown in (b), and completed as shown in (c).

$$\text{(a) } 8\overline{)6.4} \qquad \text{(b) } 8\overline{)6.4}^{\,\cdot} \qquad \text{(c) } \begin{array}{r} 0.8 \\ 8\overline{)6.4} \\ 6.4 \\ \hline 0 \end{array}$$

EXAMPLE 1-37 Divide $37.2 \div 3$.

SOLUTION The arrangement of the division with the decimal point in the correct place of the quotient is shown in (a), and the division completed is shown in (b).

$$
\begin{array}{r}
 . \\
\text{(a)}\quad 3\,\overline{)37.2}
\end{array}
\qquad
\begin{array}{r}
12.4 \\
\text{(b)}\quad 3\,\overline{)37.2} \\
3 \\
\overline{07} \\
6 \\
\overline{1\,2} \\
1\,2 \\
\overline{0}
\end{array}
$$

EXAMPLE 1-38 Divide $43.96 \div 14$.

SOLUTION

$$
\begin{array}{r}
3.14 \\
14\,\overline{)43.96} \\
42 \\
\overline{1\,9} \\
1\,4 \\
\overline{56} \\
56 \\
\overline{0}
\end{array}
$$

Consider now the division $3.14\,\overline{)5.652}$. This is a division of the second kind because it has a decimal divisor. However, here also we can use the above rule for division if we change our decimal divisor, 3.14, to a whole number divisor. As we know, this is very easy to do using the fundamental principle of fractions.

$$
5.652 \div 3.14 = \frac{5.652}{3.14} = \frac{5.652 \times 100}{3.14 \times 100} = \frac{565.2}{314}
$$

Then,

$$
\frac{565.2}{314} = 565.2 \div 314
$$

Now we can rewrite our division and locate the decimal point as shown in (a), and we can complete it as shown in (b) using our rule for division by a whole number divisor.

$$
\begin{array}{r}
 . \\
\text{(a)}\quad 314\,\overline{)565.2}
\end{array}
\qquad
\begin{array}{r}
1.8 \\
\text{(b)}\quad 314\,\overline{)565.2} \\
314 \\
\overline{251\,2} \\
251\,2 \\
\overline{0}
\end{array}
$$

From the above illustration, it follows that *to divide by a decimal divisor move the decimal point of both the divisor and the dividend as many places to the right as are necessary to make the divisor a whole number. Locate the decimal point in the quotient directly over the new place of the decimal point in the dividend, and proceed as before.*

EXAMPLE 1-39   Divide 2.331 ÷ 0.37.

SOLUTION   To change the divisor 0.37 into a whole number it is necessary to move its decimal point two places to the right, and so the decimal points of both divisor and dividend are moved two places to the right. Then the decimal point in the quotient is put directly above the new place of the decimal point in the dividend. The original form of the division is shown in (a), the switching of the decimal points of the divisor and the dividend is shown in (b), the place of the decimal point in the quotient is shown in (c), and the division by the whole number 37 is completed as shown in (d).

$$\text{(a) } 0.37 \overline{\smash{)}\,2.331} \qquad \text{(b) } 0.37. \overline{\smash{)}\,2.33.1} \qquad \text{(c) } 37. \overline{\smash{)}\,233.1} \qquad \text{(d) } 37 \overline{\smash{)}\,233.1}$$

$$
\begin{array}{r}
6.3 \\
37 \overline{\smash{)}\,233.1} \\
\underline{222} \\
11\,1 \\
\underline{11\,1} \\
0
\end{array}
$$

Our examples of divisions up to this point have been exact divisions, but most divisions with decimals have a remainder. In such cases, the division is carried to the desired degree of accuracy by supplying the necessary number of zeros.

EXAMPLE 1-40   Divide 2.367 ÷ 0.37. Continue the division until there are two decimal digits in the quotient.

SOLUTION $0.37 \overline{\smash{)}\,2.367}$ becomes $37 \overline{\smash{)}\,236.70}$. The division is completed as shown below.

$$
\begin{array}{r}
6.39 \\
37 \overline{\smash{)}\,236.70} \\
\underline{222} \\
14\,7 \\
\underline{11\,1} \\
3\,60 \\
\underline{3\,33} \\
27
\end{array}
$$

EXAMPLE 1-41   Divide 7.36 ÷ 13.9. Continue the division until there are four decimal digits in the quotient.

SOLUTION   The division is of the form shown in (a), and is completed as in (b).

$$
\text{(a) } 139 \overline{\smash{)}\,73.6000}
\qquad
\text{(b) }
\begin{array}{r}
0.5294 \\
139 \overline{\smash{)}\,73.6000} \\
\underline{69\,5} \\
4\,10 \\
\underline{2\,78} \\
1\,320 \\
\underline{1\,251} \\
690 \\
\underline{556} \\
134
\end{array}
$$

An Arithmetic Review

Perform each of the following divisions. Wherever applicable, continue the division until there are three decimal digits in the quotient:

1. $21.35 \div 7$ **2.** $2.67 \div 3$ **3.** $11 \div 23$

4. $7 \div 9.8$ **5.** $38 \div 24$ **6.** $3.7 \div 6.78$

7. $1 \div 3$ **8.** $2 \div 3$ **9.** $1 \div 16$

10. $0.87 \div 0.63$ **11.** $79 \div 4.05$ **12.** $0.75 \div 3.45$

13. $0.078 \div 0.053$ **14.** $0.014 \div 0.37$ **15.** $0.71 \div 0.029$

16. How many gallons of gas does a car use to go 319.14 mi, if it gets 21.3 miles per gallon?

17. If wrought iron weighs 481.2 pounds per cubic foot, how many cubic feet of wrought iron are there in 1636.18 lb?

18. If a plane consumes 494.08 gal of gas during a 12.8-hr flight, what is the average hourly consumption of gas?

19. Jim Foster worked for 7 hr and earned $29.05. What was his hourly rate?

20. A copper rod 57.66-in. long is cut into 13 pieces of equal length. Allowing 0.06 in. of waste for each cut, how long is each piece?

1-12 Exact and Approximate Numbers

It must be evident at this point that decimal fractions make it possible to express measurements and to carry divisions to any degree of accuracy. In practice, however, there are limits to the accuracy that can be used, as there are limits to the reliability of measurements. This is why measurements and calculated results to technical problems are **rounded** to suit practical limitations.

The student will understand better the purpose and the rules for rounding off, if the related subjects of **exact** and **approximate numbers**, and of **significant digits** are explained first.

Numbers are either exact or approximate. A number is said to be exact when it is exactly equal to the quantity it represents. A number is said to be approximate when it approximately equals the quantity it represents. In general, numbers expressing counting are exact, while numbers representing measurements are approximate numbers.

EXAMPLES

1. The number 23 in the statement, "This class has 23 students." is an exact number.
2. The number 3 in the statement, "The triangle has 3 angles." is an exact number.
3. The number 38 in the statement, "The distance between city A and city B is 38 miles." is an approximate number.
4. The number 17 in the statement, "This typewriter weighs 17 lb." is an approximate number.

1-13 Significant Digits, Accuracy, and Precision

The **accuracy** of a number depends on the number of reliable digits in that number. The digits in a number which are known to be reliable, (certainly correct) are called **significant digits**. They have "significance" because they are used to "signify," or to show, how accurate the number is. In general, the accuracy of the last significant digit of an approximate number is questionable. Suppose, for instance, that we are asked to measure the sides of a rectangular steel plate using a steel rule graduated into hundredths of an inch. If the dimensions obtained are 7.38 and 3.24, the correctness of the last digits of both dimensions is very questionable, since *the exact length cannot be determined.* The "actual" length of the steel plate was either almost 7.38, or just over 7.38, and the last digit 8, was "rounded off" either upward or downward. The same is true for the "actual" width of the steel plate, and for the last digit of 3.24.

The following list of rules is given for recognizing the significant digits in a given number:

1. *All nonzero digits are significant.* For instance, in the following three numbers:

$$283, 0.12, 2865$$

283 has three significant digits, 0.12 has two, and 2865 has four significant digits.

2. *Zeros between nonzero digits are significant.* For instance, in the following three numbers:

$$1003, 1.07, 208$$

1003 has four significant digits, 1.07 has three, and 208 has three significant digits.

3. *Zeros following a decimal point, or at the end of a decimal number are significant.* For instance, in the following five numbers:

$$5.00, 170.00, 923.0, 2.180, 0.31900$$

5.00 has three significant digits, 170.00 has five, 923.0 has four, 2.180 has four, and 0.31900 has five significant digits.

4. *Zeros at the end of a whole number are usually place holders, and are not significant.* For instance, in the following three numbers:

$$20, 34,000, 170$$

20 has one significant digit, 34,000 has two, and 170 has two significant digits.

5. *Zeros at the beginning of a decimal number are place holders and are not significant.* For instance, in the following three numbers:

$$0.002, 0.092, 0.0008$$

0.002 has one significant digit, 0.092 has two and 0.0008 has one significant digit.

EXAMPLES

1. 6.125 is accurate to four significant digits.
2. 612.5 is accurate to four significant digits.
3. 7200 is accurate to two significant digits.

An Arithmetic Review

4. 6120 is accurate to three significant digits.
5. 6.120 is accurate to four significant digits.
6. 0.0070300 is accurate to five significant digits.
7. 123 is accurate to three significant digits.
8. 18.0 is accurate to three significant digits.
9. 50.0 is accurate to three significant digits.

Zeros at the end of a whole number are place holders and are not significant, unless an overbar ($\overline{}$) is written over one of the zeros. The overbar is placed over the last zero which is significant. Thus, in the numbers

$$20, \; 2\overline{0}0, \; 26\overline{0}0, \; 720\overline{0}$$

there are one, two, three, and four significant digits, respectively.

The zero at the end of the whole number 50 ft is simply a place holder and is not significant. All that zero does is locate the decimal point and in this way show that the digit 5 is in the tens' place. The zero of the number 50 ft shows that the ones' place is uncertain, and that 50 ft is between 49.5 and 50.5 ft. The number 50.0 ft, however, is not the same as 50 ft. The length which is expressed by 50.0 ft has been measured by a smaller unit, and we say that 50.0 ft is more **precise** than 50 ft. In the number 50.0 ft, the uncertainty is now 0.05 ft and 50.0 ft is between 49.95 and 50.05 ft.

The **precision** of a number depends on the place value of the righthand-most significant digit. This means that the more digits a number has beyond the decimal point, the more precise it is. For instance, the precision of the number 3.14 is to the nearest hundredth, the precision of the number 340 is to the nearest ten, and the precision of the number 11.8 is to the nearest tenth. In other words, the number 3.14 is more precise than the number 5847.3. Also, 3.14 is more precise then 3.1, and 3.1 is more precise than 3.

EXERCISE 1-11

Identify each number as either exact or approximate in each of the following statements:
 1. This car has four wheels, and its motor has six cylinders.
 2. The acceleration due to gravity is 32 feet per second each second.
 3. A pentagon has five sides.
 4. This man weighs 165 lb.
 5. They met at 5:00 P.M.
 6. They paid $26.03.
 7. He is 7 in. taller than his son.
 8. The left wall of this room contains 923 bricks.
 9. The height of this tree is 43 ft.
10. He covered a distance of 425 mi, and used 21 gal of gas.

Give both the accuracy and the precision of each of the following numbers:

11. 0.3214	**12.** 52	**13.** 200	**14.** 5.38
15. 7.034	**16.** 0.7200	**17.** 0.005	**18.** 0.17008
19. 1200	**20.** 4.51	**21.** 235,720	**22.** 0.34138
23. 500.0	**24.** 0.0070	**25.** 3.10	**26.** 18.20

1-14 Rounding Off Numbers

A common practice is to round off numbers, or computed results to technical problems, either because the accuracy of some digits is doubtful, or because great accuracy cannot be used.

A number is rounded off by dropping one or more digits from its right end according to certain rules, which are listed below. In other words, to round off a number is to write that number with fewer significant digits. For example, to round off 5.823 to two decimal places means to drop the 3, and to write 5.823 as the nearest two-place decimal number. Figure 1-11 shows that the number is 5.82, which is only 0.003 away from 5.823, and therefore closer to it than 5.83, which is 0.007 away from it.

FIG. 1-11

The rules for rounding off numbers are as follows:

1. *Determine the position of the last significant digit to be retained.*
2. *The digit retained is increased by 1 if the digit to its right is 5 or more.*
3. *The digit retained remains unchanged if the digit to its right is less than 5.*

Digits dropped from the end of a whole number, unlike digits dropped from the end of a decimal, are replaced by zeros to keep the decimal point in its proper place.

EXAMPLES

	Number	Rounded off to 4 s.d.	Rounded off to 3 s.d.	Rounded off to 2 s.d.
1.	38,381	38,380	38,400	38,000
2.	285.35	285.4	285	290
3.	33,194	33,190	33,200	33,000
4.	0.68003	0.6800	0.680	0.68
5.	17.0067	17.01	17.0	17
6.	912.00	912.0	912	910

EXERCISE 1-12

Round each of the following numbers to four, three, and two significant digits:

1. 162.067	**2.** 927.85	**3.** 23.935	**4.** 8080.5
5. 21.134	**6.** 8505.03	**7.** 618.54	**8.** 1.2895
9. 9.3239	**10.** 690.04	**11.** 834.35	**12.** 5.173

1-15 Basic Operations with Approximate Numbers

It should be expected that the result of an operation with approximate numbers must be an approximate number also. Since no result can be more accurate than the least accurate number involved in the operation, the result *should be rounded off to the accuracy of the least accurate number involved.*

An Arithmetic Review

The rules for computation with approximate numbers are as follows:

1. *In addition or subtraction, round off each number one more place of precision (decimal place) than the least precise number. Perform the operation, and round off the result to the precision of the least precise number.*
2. *In multiplication or division, round off each factor to one more significant digit than the factor with the least significant digits. Perform the operation, and round off the product or the quotient to the same number of significant digits as the factor with the least number of significant digits.*

EXAMPLE 1-42 Add 14.3697, 7.8, 353.03, and 26.271.

SOLUTION Each of the four addends is rounded off to the hundredths place, one more place than the precision of 7.8, the least precise number. The addition of the rounded-off addends and their sum are shown below.

$$
\begin{array}{r}
14.37 \\
7.8 \\
353.03 \\
26.27 \\
\hline
401.47
\end{array}
$$

Finally, the sum is rounded off to the tenths place, the precision of 7.8, the least precise addend. Thus, the sum is 401.5.

EXAMPLE 1-43 Subtract 27.58 − 12.3081.

SOLUTION Round off 12.3081 to the thousandths position, one more place than the precision of 27.58, the least precise number. The subtraction of the rounded-off numbers and the difference are shown below.

$$
\begin{array}{r}
27.58 \\
-\ 12.308 \\
\hline
15.272
\end{array}
$$

Finally, the difference is rounded off to hundredths, the precision of 27.58, the least precise of the two numbers. Thus, the difference is 15.27.

EXAMPLE 1-44 Multiply 51.163 × 2.16.

SOLUTION The factor 51.163 is rounded off to 51.16, one more significant digit than 2.16, the factor with the least significant digits. The multiplication and the product are shown below.

$$
\begin{array}{r}
51.16 \\
\times\ \ 2.16 \\
\hline
306\ 96 \\
511\ 6\ \ \\
10232\ \ \ \ \\
\hline
110.5056
\end{array}
$$

Finally, the product is rounded to three significant digits, the accuracy of 2.16, the least accurate factor. Thus, the product is 111.

EXAMPLE 1-45 Multiply 2.3168 × 18.6 × 7.113.

SOLUTION All the factors are rounded off to four significant digits, one more than the number of the significant figures of 18.6, the least accurate number. The multiplication and the product are shown below.

$$2.317 \times 18.6 \times 7.113 = 306.54327$$

The final product is 307.

EXAMPLE 1-46 Divide 27.034 ÷ 13.1.

SOLUTION The changed division and the final quotient are shown below.

$$27.03 \div 13.1 = 2.06$$

EXERCISE 1-13

Perform each of the following additions or subtractions. Round the answers to the proper precision (place value) using the rules of this section.

1. 120.05 − 79.103
2. 434.2 + 17.18
3. 524.17 + 17.007 + 70.010 + 3.8256
4. 31.8713 + 3.8 + 126.34 + 58.123 + 6.31
5. 424.208 − 137.09
6. 17.21 − 5.234
7. 78.01 − 56.3
8. 27.68 + 31.314 + 7.1
9. 1275.11 − 1056.72
10. 412.01 − 307.122

Perform each of the following multiplications or divisions. Round the answers to the proper accuracy using the rules of this section.

11. 7.05 × 2.17	12. 3.18 × 0.004	13. 5.2 × 3.14
14. 38.192 × 12.4	15. 11 ÷ 3.3	16. 0.6201 ÷ 0.53
17. $\dfrac{5.4 \times 17.21}{4.08}$	18. $\dfrac{1.18}{5.2 \times 0.17}$	19. 6.3 × 1.12 × 0.7
20. 3.114 × 8.03		

1-16 Changing a Common Fraction to a Decimal

We recall that a fraction always indicates the division of the numerator by the denominator. Therefore, *to convert a fraction into a decimal, we need only carry out the division implied by the fraction.*

We can carry the division to any practical degree of accuracy since the terms of the fraction are exact numbers. We simply supply enough zeros after the decimal point in the numerator.

EXAMPLE 1-47 Convert the fraction $\dfrac{7}{8}$ into a decimal.

SOLUTION By the fraction $\dfrac{7}{8}$, the division 7 ÷ 8 or 7.000 ÷ 8 is implied, and we continue as follows:

An Arithmetic Review

$$
\begin{array}{r}
0.875 \\
8\,\overline{)7.000} \\
6\,4 \\
\hline
60 \\
56 \\
\hline
40 \\
40 \\
\hline
0
\end{array}
$$

Therefore, $\dfrac{7}{8} = 0.875$ evenly

EXAMPLE 1-48 Convert the fraction $\dfrac{3}{7}$ into a decimal. Round your answer to the nearest hundredth.

SOLUTION By the fraction $\dfrac{3}{7}$, the division $3 \div 7$ or $3.000 \div 7$ is implied, and we proceed as follows:

$$
\begin{array}{r}
0.428 \\
7\,\overline{)3.000} \\
2\,8 \\
\hline
20 \\
14 \\
\hline
60 \\
56 \\
\hline
4
\end{array}
$$

Therefore, $\dfrac{3}{7} = 0.43$ to the nearest hundredth

EXAMPLE 1-49 Convert the fractional mixed number $8\frac{5}{9}$ to a decimal mixed number. Round your answer to the nearest hundredth.

SOLUTION The decimal part of the mixed number will be the quotient of the division $5 \div 9$. Since $5 \div 9 = 0.555$ to the third decimal place, $8\frac{5}{9} = 8.56$ to the nearest hundredth.

1-17 Changing a Decimal to a Fraction

To write any decimal in its fractional form, just read the decimal (correctly, of course) and write the fraction as the decimal sounds. Thus, the decimal 0.019 is read "nineteen thousandths" and is written as the fraction $\dfrac{19}{1000}$.

EXAMPLE 1-50 Write the decimal 0.18 as a fraction.

SOLUTION The decimal 0.18 is read "eighteen hundredths," and is written as the fraction $\dfrac{18}{100}$ or $\dfrac{9}{50}$ reduced to lowest terms.

1. How is a decimal changed into a common fraction?

2. How is a common fraction changed into a decimal?

Change each of the following decimals to a common fraction. Reduce to lowest terms.

3. 0.34   **4.** 0.1   **5.** 0.268   **6.** 0.032

7. 0.108   **8.** 0.125   **9.** 0.625   **10.** 0.064

Change each of the following decimal mixed numbers into a fractional mixed number. Reduce to lowest terms.

11. 3.14   **12.** 5.07   **13.** 2.5   **14.** 17.75

15. 81.25   **16.** 2.15   **17.** 31.038   **18.** 6.03

Change each of the following fractions to a decimal.

19. $\dfrac{1}{4}$ **20.** $\dfrac{3}{4}$ **21.** $\dfrac{3}{5}$ **22.** $\dfrac{7}{8}$

23. $\dfrac{1}{20}$ **24.** $\dfrac{3}{25}$ **25.** $\dfrac{3}{16}$ **26.** $\dfrac{5}{32}$

27. $\dfrac{5}{20}$ **28.** $\dfrac{1}{64}$ **29.** $\dfrac{5}{16}$ **30.** $\dfrac{7}{64}$

Convert each of the following fractions or mixed numbers into a decimal or a decimal mixed number. Give answers accurate to hundredths wherever appropriate.

31. $\dfrac{5}{9}$ **32.** $\dfrac{3}{7}$ **33.** $\dfrac{19}{4}$ **34.** $7\frac{2}{5}$

35. $8\frac{3}{7}$ **36.** $5\frac{8}{15}$ **37.** $16\frac{3}{11}$ **38.** $23\frac{5}{23}$

1-18 Powers and Roots

To square a given number is to multiply that number by itself. To find the square root of a given number is to find that number which when multiplied by itself will produce the given number. For example, the square of 3 is $3 \times 3 = 9$. The square root of 49 is 7, since 7×7 is 49. In these examples, 9 is called the **square** or second power of 3, and 7 is called the **square root** of 49.

The square of a number is indicated by a small 2, called an **exponent**, written to the right and above the number. Thus, the square of 3 is indicated by 3^2. In this example 3 is called the **base**. The exponent indicates the number of times the base must be multiplied by itself. Thus, $2^3 = 2 \times 2 \times 2 = 8$, and $2^5 = 2 \times 2 \times 2 \times 2 \times 2 = 32$. In these two illustrations, 8 is called the **cube**, or the third power of 2, and 32 is the fifth power of 2.

EXAMPLE 1-51   What is the meaning and the value of 4^5 and 5^3?

SOLUTION

$$4^5 = 4 \times 4 \times 4 \times 4 \times 4 = 1024$$
$$5^3 = 5 \times 5 \times 5 = 125$$

The square root of a number is indicated by writing the number under the sign $\sqrt{}$, called the **radical sign**. Thus, the square root of 81 is indicated by $\sqrt{81}$ (and read "square root of 81"); we write $\sqrt{81} = 9$. In this illustration, 9 is called the square root of 81. The cube root of a given number is that number, which, when multiplied by itself three times will produce the given number. Thus, the cube root of 8 is 2, since $2 \times 2 \times 2 = 8$.

The cube root of 8 is indicated by writing $\sqrt[3]{8}$ and read "cube root of eight." Thus, $\sqrt[3]{8} = 2$. The small number placed at the notch of the radical sign is called the **index** and indicates which root of the number is to be found. Thus, in $\sqrt[3]{8}$, the index is 3 and in $\sqrt[2]{25}$, the index is 2. The square root of a number can be indicated with or without the index. Thus, $\sqrt[2]{49} = \sqrt{49} = 7$.

EXAMPLE 1-52   What is the meaning and the value of $\sqrt{16}$ and $\sqrt[2]{25}$?

SOLUTION

$$\sqrt{16} \text{ means the square root of 16, and } \sqrt{16} = 4.$$

$$\sqrt[2]{25} \text{ means the square root of 25, and } \sqrt[2]{25} = 5.$$

Square roots, and sometimes cube roots, are used in technical problems. To find the square root or the cube root of a number, tables or calculators are used. Tables are very limited since there are no tables for the square roots of all whole numbers, and certainly tables cannot include all decimal numbers. On the other hand, calculators are not restricted in this sense. Hand calculators are very common today, and most of them have square roots.

The square root of a fraction is found by first finding the square root of the numerator, then the square root of the denominator, and finally dividing the two square roots. Thus,

$$\sqrt{\frac{36}{25}} = \frac{\sqrt{36}}{\sqrt{25}} = \frac{6}{5} = 1.2$$

It is simpler, however, to find the square root of a fraction by changing the fraction into a decimal, and then finding its square root. Thus,

$$\sqrt{\frac{3}{4}} = \sqrt{0.75} = 0.8660254 \quad \text{and} \quad \sqrt{1\frac{1}{4}} = \sqrt{1.25} = 1.1180339$$

EXAMPLE 1-53   Find the square root of $3\frac{7}{8}$.

SOLUTION

$$\sqrt{3\frac{7}{8}} = \sqrt{3.875} = 1.9685019$$

What is the meaning and the value of each of the following?

1. 6^2 2. 3^5 3. 7^4 4. 3^3 5. 8^3

Use a calculator to find, to the nearest thousandth, the value of each of the following:

6. $\sqrt[2]{81}$ 7. $\sqrt[2]{121}$ 8. $\sqrt{138}$ 9. $\sqrt{217}$

10. $\sqrt{31}$ 11. $\sqrt[2]{27}$ 12. $\sqrt{16.4}$ 13. $\sqrt{3.27}$

14. $\sqrt{\frac{5}{8}}$ 15. $\sqrt{\frac{2}{3}}$ 16. $\sqrt{5\frac{3}{8}}$ 17. $\sqrt{2\frac{1}{4}}$

18. $\sqrt{6\frac{11}{17}}$

Powers and Roots

REVIEW QUESTIONS

1. Which basic operation is indicated by a fraction?
2. State the fundamental principle of fractions.
3. What are equivalent fractions?
4. When is a fraction in its lowest terms?
5. What is a prime number?
6. What is prime factorization?
7. How is the LCD of two or more fractions determined?
8. What is the reciprocal of a number?
9. Which numbers are exact numbers?
10. Which numbers are approximate numbers?
11. What are significant digits?
12. What does the accuracy of a number depend on?
13. What does the precision of a number depend on?
14. State the rule for rounding sums or differences.
15. State the rule for rounding products or quotients.

REVIEW EXERCISES

Factor each of the following numbers completely:

1. 735   **2.** 924   **3.** 1800

Reduce each of the following fractions to lowest terms:

4. $\dfrac{18}{81}$ **5.** $\dfrac{35}{75}$ **6.** $\dfrac{171}{522}$

In each of the following problems, perform the indicated operation. Reduce all answers to lowest terms.

7. $\dfrac{7}{18} + \dfrac{8}{21} + \dfrac{6}{15}$ **8.** $\dfrac{5}{3} + \dfrac{3}{8} + \dfrac{17}{16} + \dfrac{1}{9}$

9. $1\dfrac{4}{49} + 3\dfrac{7}{9} + \dfrac{13}{21} + \dfrac{16}{81}$ **10.** $\dfrac{18}{33} - \dfrac{8}{21}$

11. $35\dfrac{7}{36} - 22\dfrac{15}{16}$ **12.** $\dfrac{8}{15} \times \dfrac{3}{16} \times \dfrac{5}{9}$

13. $2\dfrac{2}{7} \times 1\dfrac{3}{8} \times 3\dfrac{1}{11}$ **14.** $\dfrac{7}{8} \div \dfrac{3}{5}$

15. $1\dfrac{4}{7} \div 1\dfrac{11}{14}$ **16.** $117\dfrac{1}{3} + 2\dfrac{1}{8} - 53\dfrac{7}{8}$

Find the reciprocal of each of the following:

17. $\dfrac{7}{8}$ **18.** 23   **19.** $\dfrac{1}{16}$

Simplify each of the following complex fractions:

20. $\dfrac{\dfrac{5}{11}}{\dfrac{10}{33}}$ **21.** $\dfrac{\dfrac{3}{4} - \dfrac{5}{7}}{\dfrac{3}{7} + \dfrac{1}{2}}$

Perform each of the indicated operations. Do not round results.

22. $0.17 + 0.264 + 1.01 + 19.313$ **23.** $3.84 - 1.9$

24. $17 - 5.034$ **25.** 3.14×7.3

26. 5×2.08 **27.** 0.03×0.01

28. $92.4 \div 7$ **29.** $37.912 \div 13.54$

Identify the number in each of the following statements as either exact or approximate:

30. This is a five-bedroom house.

31. The height of this desk is 32 in.

32. In most colleges a full load is five classes.

Give the accuracy and the precision of each of the following numbers:

33. 0.08   **34.** 130.0   **35.** 11.3

Perform each of the following operations. Round the answers to the proper precision.

36. $3.17 + 0.181 + 18.9$ **37.** $0.03 + 0.127 + 1.3 + 17$

38. $17.3 - 2.58$ **39.** $12.3 - 5.8$

Perform each of the following operations. Round the answers to the proper accuracy.

40. 8.5×3.14 **41.** 18.71×6.8

42. $16.28 \div 81.7$ **43.** $92.8 \div 15$

Change each of the following fractions to a decimal or a decimal mixed number:

44. $\dfrac{1}{8}$ **45.** $\dfrac{6}{5}$ **46.** $\dfrac{9}{25}$ **47.** $\dfrac{17}{8}$

Use a calculator to find each square root in exercises 62–67 to the nearest hundredth.

48. $\sqrt{134}$ **49.** $\sqrt{169}$ **50.** $\sqrt{12.5}$

51. $\sqrt{0.17}$ **52.** $\sqrt{\dfrac{11}{16}}$ **53.** $\sqrt{3\tfrac{7}{15}}$

54. The area of Sweden is about 170,000 square miles, and the Scandinavian peninsula is about 300,000 square miles in area. What fractional part of the Scandinavian peninsula is Sweden?

55. A motor turns at 1850 revolutions per minute. How many minutes will it take the motor to make $14,337\tfrac{1}{2}$ revolutions?

56. A piece of leather $23\tfrac{1}{2}$-in. wide is cut into strips $3\tfrac{1}{4}$-in. wide. How many strips can be cut, and what is the width of the remaining strip?

57. A full revolution of the earth around its axis takes 24 hr. How many hours does the earth take to make $3\tfrac{5}{8}$ revolutions?

58. In a certain alloy, $\dfrac{2}{5}$ of its weight is tin, $\dfrac{7}{16}$ copper, and $\dfrac{13}{80}$ zinc. How many pounds of each metal are there is 350 lb of the alloy?

59. Eight sheets of steel, 0.036-in. thick, are piled together. What is the total thickness of the pile?

60. How far can a motorcycle travel on 6.7 gal, if it averages 58.3 miles per gallon?

61. The area of a square is given by the formula $A = s^2$, where A is the area and s is the length of one side of the square. This means that one side of the square is equal to the square root of its area. What is the length of the side of a square room whose area is 248.1 sq ft?

2

The Metric System
of Measurement.
More About Exact
and Approximate Numbers

2-1 The English and the Metric Systems

The first units of length people used were the sizes of members of their bodies, like feet and arms. Legend has it that the yard is the length of the arm of the twelfth century English king Henry I. It is very probable that after the yard was defined in this way, the foot was arbitrarily made one third of the yard. Such seems to be the beginning of the development of the **English system**, and thus, it is of medieval and arbitrary origin.

Just as there is no logical relation between the lengths of the human foot and the human arm, there is no logical reason that there should be 5280 feet in a mile, 231 cubic inches in a gallon, or $272\frac{1}{4}$ square feet in a square rod. A table of measures of the English system will show that the numbers of multiples and submultiples of the units of this system are awkward and arbitrary. It is true that many of the numbers of this system are highly divisible. Divisibility, however, does not make numbers like 144, 1728, 1760, and 5280 less confusing, or easier to remember. Also, divisibility is limited, and in a number system based on 10, like ours, it is difficult to work with the fractions and the decimal equivalents of the English system.

The **metric system** is a logical system. It was developed by a commission of French scientists at the time of the French Revolution, when reason and logic were almost deified. Being a decimal system, like our number system, the metric system has units, the multiples and submultiples of which are round, logical, and easy to remember numbers. All the numbers are powers of 10, and no fractions and awkward decimal equivalents are needed in the system. To change from a given unit of the system to a larger or smaller unit of the same kind, all one has to do is to move the decimal point to the left or right the necessary number of places. The metric system is much simpler and much easier to use.

After World War II, almost all countries adopted the metric system. England did so in the 1960's, and Canada changed to the metric system recently. The United States is now the only major country still using the English system. In science in this country, the metric system is used exclusively, and there are signs all over that the use of the metric system in engineering, industry, and trade is only a matter of time.

The meter, abbreviated m, is the basic unit of length in the metric system. It is about 39.37 inches, or slightly longer than a yard. Below is a list of the multiples and the submultiples of the meter with the name, the symbol, and the relation of each to the meter.

	Name	Symbol	Relation to Meter	Prefix and Meaning
multiples	kilometer	km	1000 m	kilo: one thousand times
	hectometer	hm	100 m	hecto: one hundred times
	dekameter	dam	10 m	deka: ten times
	meter	m		BASIC UNIT
submultiples	decimeter	dm	0.1 m	deci: one tenth of
	centimeter	cm	0.01 m	centi: one hundredth of
	millimeter	mm	0.001 m	milli: one thousandth of

The student should memorize the names and the symbols of these multiples and submultiples, and should know the relation of each to the meter.

The convenience, simplicity, and interrelationship of the metric system must be evident from the above table. As has already been pointed out, all one has to do to change from a given unit of the chart to the next higher or the next lower unit, is to move the decimal point one place to the right or to the left. Of course, one should not forget to indicate the new name of the unit after the numerical change. For instance, 123 m = 12.3 dam, and 123 m = 1230 dm. It should be said at this point that the decimeter, the dekameter, and the hectometer are rarely used, and that the units which are used much more often are the millimeter, the centimeter, the meter, and the kilometer. A briefer and more direct form of the relations of the units which are used most often is given below:

$$1 \text{ km} = 1000 \text{ m}$$

$$1 \text{ m} = 100 \text{ cm}$$

$$1 \text{ m} = 1000 \text{ mm}$$

$$1 \text{ cm} = 10 \text{ mm}$$

EXAMPLE 2-1 Change 2365 m to kilometers.

SOLUTION Using the relation 1 km = 1000 m, we divide 2365 by 1000. Thus,

$$2365 \div 1000 = 2.365 \text{ km}$$

NOTE: In Chap. 9, we will use a more technical method of performing such changes.

EXAMPLE 2-2 How many centimeters are there in 3.12 m?

SOLUTION Using the relation 1 m = 100 cm, we multiply 3.12 by 100. Thus,

$$3.12 \times 100 = 312 \text{ cm}$$

2-3 Applications of the Metric Units of Length.
Computations with Approximate Numbers

As the metric system becomes more and more a part of everyday life in the United States, it will become necessary during this transition period to convert from the English system to the metric system, and from the metric system to the English system. There is a table of units of the English and the metric systems of measurement inside the front cover. A convenient table of **conversion factors** for linear measurements between the two systems is given below:*

$$1 \text{ in.} = 2.54 \text{ cm}$$
$$1 \text{ cm} = 0.3937 \text{ in.}$$
$$1 \text{ ft} = 0.3048 \text{ m}$$
$$1 \text{ m} = 3.281 \text{ ft}$$
$$1 \text{ m} = 39.37 \text{ in.}$$
$$1 \text{ mi} = 1.609 \text{ km}$$
$$1 \text{ km} = 0.621 \text{ mi}$$

In working with measurement problems in this chapter, we will use the two rules for computations with approximate numbers of the previous chapter, since measurements in any system are approximate numbers. When, however, an exact number is involved in such a computation, the situation changes, because these rules do not apply to exact numbers. Remembering that the exact number 7 is 7.00000 . . . , and the exact number 13 is 13.00000000 . . . , *we realize that in a computation with exact and approximate numbers, the approximate number is always the least accurate.* We can state then that *the product (or quotient) of an approximate number with an exact number is an approximate number, and that this product (or quotient) should be rounded off to the accuracy of the approximate factor.* For instance, suppose we are told to multiply the exact number 15 and the approximate number 2.17 in. Because the number 15 has as many significant digits as we want (remember, $15 = 15.00000$ . . .), the approximate number 2.17 is the less accurate of the two, and the final product will be rounded to the accuracy of 2.17. Thus,

$$15 \times 2.17 = 32.55$$
$$= 32.6 \text{ in.}$$

Similarly, if a bolt is 2.25-cm long, the total length of five such bolts will be

$$5 \times 2.25 = 11.25$$
$$= 11.3 \text{ cm}$$

*There are conversion factors inside the back cover also.

Notice that the final result is rounded off to the accuracy of 2.25, not to the accuracy of five.

In the same way, 3.78 m ÷ 3 = 1.26 m.

Another very important point for the student to remember is that *conversion factors are considered exact numbers*. Thus, all the conversion factors which have already been introduced and those which will follow, will be treated as exact numbers. For instance, suppose we are given 1114.7 in. to convert to centimeters. Using the relation 1 in. = 2.54 cm, we have the multiplication 1114.7 × 2.54. Because the conversion factor 2.54 is considered an exact number, the approximate number 1114.7 in. is less accurate, and the final product is rounded off to its accuracy. Thus,

$$1114.7 \text{ in.} = 1114.7 \times 2.54$$
$$= 2831.338$$
$$= 2831.3 \text{ cm}$$

EXAMPLE 2-3 The distance from Atlanta, GA to Washington DC is 625 mi. How many kilometers is this?

SOLUTION We use the relation 1 mile = 1.609 km. Then,

$$625 \times 1.609 = 1005.625$$
$$= 1010 \text{ km}$$

Since the conversion factor 1.609 is considered an exact number, the final result is rounded to three significant digits.

EXAMPLE 2-4 Seven sheets of a certain grade of paperboard have a combined thickness of 1.0 cm. What is the thickness of each paperboard, and the combined thickness of 216 sheets of that paperboard?

SOLUTION

The thickness of a single sheet is 1.0 ÷ 7 = 0.14285
$$= 0.14 \text{ cm}$$
Rounded off to two significant digits.

The combined thickness of 216 sheets is 216 × 0.14285 = 30.8556
$$= 31 \text{ cm}$$
Rounded off to two significant digits.

Notice that the **unrounded** result of the first part, the number 0.14285 was used in the second part of the problem, and not the rounded result 0.14. This is so because in reality we have the product 216 × (1.0/7), or 216 × 0.14285. At the end, this product was rounded off to two significant digits. In general, *when a result is to be used in further computations, the unrounded form of that result should be used. All results should be rounded off to the accuracy of the least accurate factor in the problem.*

Applications of the Metric Units of Length. Computations with Approximate Numbers

EXAMPLE 2-5 Add 3.257 m and 2.8 m.

SOLUTION The two addends representing measurements are approximate numbers. Therefore, 3.257 is rounded off to one more digit than the precision of 2.8. Thus,

$$3.26$$
$$\underline{2.8}$$
$$6.06 \text{ m}$$

The final sum, rounded off to the precision of 2.8 is 6.1 m.

EXERCISE 2-1

1. Which of the two is shorter, the yard or the meter?
2. State and support some advantages of the metric system over the English system.
3. You are told that there are 5280 ft in a mile, and 1000 m in a kilometer. You are then asked to find the number of inches in 7 mi, and the number of centimeters in 7 km. Which of the two is easier to do?
4. Change 24 cm to millimeters.
5. Change 38 cm to meters.
6. Change 1.352 km to meters.
7. Change 1.17 m to millimeters.
8. Change 263 m to kilometers.
9. Change 471 m to centimeters.
10. Change 7233 m to kilometers.
11. If a certain pipe post is 8.4-ft long, how long are 9 such posts?
12. A 3.4-m long board is cut into 4 equal pieces. How long is each piece?
13. Jim wants to make a square frame of 3.82 ft each side. What is the length of the metal needed for the frame?
14. Add 7.3 m, 1.231 m, and 3.18 m.
15. What is the cost of 275 m of fencing wire at 48 cents per meter?
16. A man and his son are 6 ft 1 in. and 5 ft 2 in. tall, respectively. What is the difference in their heights, in centimeters?
17. Subtract 2.3 km from 7.018 km.
18. Which of the two lengths is larger, 20.0 m or 69.0 ft? Give the difference in inches.
19. A drill is $\frac{1}{8}$ in. in diameter. Find the diameter of the drill in centimeters.
20. Two rectangular solid beams are 8.7-m and 5.35-m long, respectively. (a) What is the combined length of the two beams? (b) How much shorter is the one beam than the other?
21. If 12 sheets of a certain grade of paper have a thickness of 0.024 in., what is the combined thickness of 300 such sheets of paper in (a) centimeters? (b) millimeters?
22. Mount Logan is the highest peak in Canada, and the second highest peak in North America. Find its altitude in meters if it has an elevation of 19,850 ft.
23. A jet airplane is flying at the speed of 610 miles per hour. Find the speed of the jet in (a) kilometers per hour, (b) meters per hour, and (c) meters per minute?
24. A 600-page book (excluding the covers) has a thickness of 1.8 in. Find the average thickness of a single sheet of that paper in millimeters.
25. If the crown of a certain type of staple (see Fig. 2-1) is 1.0-in. long, and the length of each leg is $1\frac{1}{4}$-in. long, how many such staples can be made from 1000 m of wire?

The Metric System of Measurement. More About Exact and Approximate Numbers

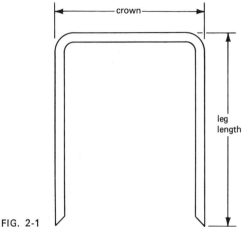

crown

leg
length

FIG. 2-1

2-4 Metric Units of Area and Volume

As in the English system, the units of area and volume in the metric system are obtained from the units of length. The student should recall that the square inch (a unit of area) is a square each side of which is one inch long, the square foot is a square each side of which is one foot long, and so on for the square yard and the square mile. Also, the cubic inch (a unit of volume) is a cube each edge of which is one inch long, the cubic foot is a cube each edge of which is one foot long, and so on for the cubic yard and the cubic mile.

It is common practice to indicate square and cubic units in their power forms. Thus, 1 sq in. is often written as 1 in.^{2}, 5 cu ft is written as 5 ft^3, and so on.

It clearly follows from the above that *the area units are the squares of linear units, and the volume units are the cubes of linear units.* Thus, for the area units we have,

$$1 \text{ ft} = 12 \text{ in., and } 1 \text{ ft}^2 = 12^2 \text{ in.}^2, \text{ or } 1 \text{ ft}^2 = 144 \text{ in.}^2$$
$$1 \text{ yd} = 3 \text{ ft, and } 1 \text{ yd}^2 = 3^2 \text{ ft}^2, \text{ or } 1 \text{ yd}^2 = 9 \text{ ft}^2$$

And for the cubic units we have,

$$1 \text{ ft} = 12 \text{ in., and } 1 \text{ ft}^3 = 12^3 \text{ in.}^3, \text{ or } 1 \text{ ft}^3 = 1728 \text{ in.}^3$$
$$1 \text{ yd} = 3 \text{ ft, and } 1 \text{ yd}^3 = 3^3 \text{ ft}^3, \text{ or } 1 \text{ yd}^3 = 27 \text{ ft}^3$$

In the metric system, square units and cubic units are defined in exactly the same manner as they are in the English system. Thus, for the area units we have,

$$1 \text{ m} = 100 \text{ cm, and } 1 \text{ m}^2 = 100^2 \text{ cm}^2, \text{ or } 1 \text{ m}^2 = 10,000 \text{ cm}^2$$
$$1 \text{ dm} = 10 \text{ cm, and } 1 \text{ dm}^2 = 10^2 \text{ cm}^2, \text{ or } 1 \text{ dm}^2 = 100 \text{ cm}^2$$
$$1 \text{ cm} = 10 \text{ mm, and } 1 \text{ cm}^2 = 10^2 \text{ mm}^2, \text{ or } 1 \text{ cm}^2 = 100 \text{ mm}^2$$

And for the cubic units we have,

$$1 \text{ m} = 100 \text{ cm}, \text{ and } 1 \text{ m}^3 = 100^3 \text{ cm}^3, \text{ or } 1 \text{ m}^3 = 1,000,000 \text{ cm}^3$$
$$1 \text{ dm} = 10 \text{ cm}, \text{ and } 1 \text{ dm}^3 = 10^3 \text{ cm}^3, \text{ or } 1 \text{ dm}^3 = 1000 \text{ cm}^3$$
$$1 \text{ cm} = 10 \text{ mm}, \text{ and } 1 \text{ cm}^3 = 10^3 \text{ mm}^3, \text{ or } 1 \text{ cm}^3 = 1000 \text{ mm}^3$$

For convenient reference, the above area and volume relations of the metric system are listed below in a more compact form.

Area Units of the Metric System	Volume Units of the Metric System.
$1 \text{ m}^2 = 10,000 \text{ cm}^2$	$1 \text{ m}^3 = 1,000,000 \text{ cm}^3$
$1 \text{ dm}^2 = 100 \text{ cm}^2$	$1 \text{ dm}^3 = 1000 \text{ cm}^3$
$1 \text{ cm}^2 = 100 \text{ mm}^2$	$1 \text{ cm}^3 = 1000 \text{ mm}^3$

EXAMPLE 2-6 A rectangular bench is 2.36-m long and 0.90-m wide. Find the area of the bench.

SOLUTION Recalling that the area of a rectangle is length × width, and remembering that 2.36 m and 0.90 m are approximate numbers, we have

$$\text{area} = 2.36 \times 0.90$$
$$= 2.124$$
$$= 2.1 \text{ m}^2$$

EXAMPLE 2-7 Find the volume of a rectangular box 0.35 m by 0.27 m by 0.26 m.

SOLUTION Recalling that the volume of a rectangular solid is length × width × height, we have

$$\text{volume} = 0.35 \times 0.27 \times 0.26$$
$$= 0.02457$$
$$= 0.025 \text{ m}^3$$

Notice that because all three factors are approximate numbers of the same accuracy, none is rounded off before multiplication. The final product, however, is rounded off to the common accuracy of two significant digits.

There should be no difficulty in using the conversion factors of the linear units for finding the conversion factors of the area and the volume units. For instance,

$$1 \text{ in.} = 2.54 \text{ cm}, \text{ and } 1 \text{ in.}^2 = (2.54)^2 \text{ cm}^2, \text{ or } 1 \text{ in.}^2 = 6.45 \text{ cm}^2$$
$$1 \text{ cm} = 0.3937 \text{ in.}, \text{ and } 1 \text{ cm}^2 = (0.3937)^2 \text{ in.}^2, \text{ or } 1 \text{ cm}^2 = 0.1550 \text{ in.}^2$$
$$1 \text{ ft} = 0.3048 \text{ m}, \text{ and } 1 \text{ ft}^3 = (0.3048)^3 \text{ m}^3, \text{ or } 1 \text{ ft}^3 = 0.02832 \text{ m}^3$$

The student should notice that in all of the above powers of approximate numbers, *the power of a number has the same number of significant digits as the number itself*. Thus,

The Metric System of Measurement. More About Exact and Approximate Numbers

$$(2.54)^2 = 2.54 \times 2.54 = 6.4516 = 6.45$$

$$(0.3937)^2 = 0.3937 \times 0.3937 = 0.15499969 = 0.1550$$

$$(0.3048)^3 = 0.3048 \times 0.3048 \times 0.3048 = 0.02831685 = 0.02832$$

For convenient reference, conversion factors derived with the above method are given below in a compact form.

Area Conversion Factors	Volume Conversion Factors
$1 \text{ in.}^2 = 6.45 \text{ cm}^2$	$1 \text{ in.}^3 = 16.4 \text{ cm}^3$
$1 \text{ cm}^2 = 0.1550 \text{ in.}^2$	$1 \text{ cm}^3 = 0.0610 \text{ in.}^3$
$1 \text{ ft}^2 = 0.09290 \text{ m}^2$	$1 \text{ ft}^3 = 0.02832 \text{ m}^3$
$1 \text{ m}^2 = 1550 \text{ in.}^2$	$1 \text{ m}^3 = 61{,}020 \text{ in.}^3$

The student should not be overwhelmed with these many conversion factors, thinking that they should be memorized. Good engineers and good technicians memorize basic principles, basic facts, and basic relations only.

EXAMPLE 2-8   How many square feet are there in a square meter?

SOLUTION   From the relation $1 \text{ m}^2 = 1550 \text{ in.}^2$, and from the fact that $1 \text{ ft}^2 = 144 \text{ in}^2$., we have

$$1550 \text{ in.}^2 = \frac{1550}{144} = 10.76 \text{ ft}^2$$

This is a useful relation to keep in mind.

EXAMPLE 2-9   How many cubic feet are there in a cubic meter?

SOLUTION   From the relation $1 \text{ m}^3 = 61{,}020 \text{ in.}^3$, and from the fact that $1 \text{ ft}^3 = 1728 \text{ in.}^2$, we have

$$61{,}020 \text{ in.}^3 = \frac{61{,}020}{1728} = 35.31 \text{ ft}^3$$

This, too, is a useful relation to keep in mind.

EXAMPLE 2-10   A water tank is 3.3-m long, 2.5-m wide, and 1.6-m high. (a) How many cubic meters of water can the tank hold when filled to the top? (b) How many cubic feet is that quantity of water?

SOLUTION

$$\text{(a) volume} = \text{length} \times \text{width} \times \text{height}$$
$$= 3.3 \times 2.5 \times 1.6$$
$$= 13.2$$
$$= 13 \text{ m}^3$$

$$\text{(b)} \quad 1 \text{ m}^3 = 35.31 \text{ ft}^3 \text{ (from example 2-9)}$$
$$13.2 \text{ m}^3 = 13.2 \times 35.31$$
$$= 466.092$$
$$= 470 \text{ ft}^3$$

EXAMPLE 2-11 An athletic club's outdoor swimming pool measures 50.0-m long and 22.0-m wide. In three consecutive hot summer days, a layer of water 8-mm thick evaporated from the swimming pool. Find the volume of the water that evaporated during these three days (a) in cubic meters, (b) in cubic feet.

SOLUTION

(a) The water that evaporated is a rectangular volume 50.0 m by 22.0 m by 0.008 m. Therefore,

$$\text{volume} = 50.0 \times 22.0 \times 0.008$$

Since all three factors are approximate numbers, they are rounded off to one more digit than the accuracy of 0.008, the least accurate number, before the multiplication, and to one significant digit, the accuracy of 0.008, after the multiplication. Thus,

$$\text{volume} = 50 \times 22 \times 0.008$$
$$= 8.8$$
$$= 9 \text{ m}^3$$

(b) Using the relation 1 m³ = 35.31 ft³, from example 2-9, and the unrounded product 8.8 m³, we have

$$8.8 \text{ m}^3 = 8.8 \times 35.31$$
$$= 310.728$$
$$= 300 \text{ ft}^3$$

EXAMPLE 2-12 A cylindrical storage tank is 2.0-m high, and the radius of its base is 1.20-m long. How much can this tank hold (a) in cubic meters? (b) in cubic feet?

SOLUTION The formula for the volume of a cylinder is

$$\text{volume} = \pi r^2 h$$

where $\pi = 3.1416$, r is the radius of the base, and h is the height. Thus,

(a) $$\text{volume} = 3.14 \times (1.20)^2 \times 2.0$$
$$= 3.14 \times 1.44 \times 2.0$$
$$= 9.0432$$
$$= 9.0 \text{ m}^3$$

Notice that the value of π was rounded to one more significant digit than the least accurate number.

(b) Using the relation 1 m³ = 35.31 ft³ (from example 2-9) and the unrounded number 9.0432 m³, we have

$$9.0432 \text{ m}^3 = 9.0432 \times 35.31$$
$$= 319.31539$$
$$= 320 \text{ ft}^3$$

EXERCISE 2-2

1. If 836 square asbestos acoustical tiles $\frac{3}{4}$-ft long on a side are needed to cover the ceiling of a room, what is the area of the ceiling of that room?
2. Two rooms are 9.6 m by 5.8 m and 28 ft by 12 ft, respectively. Which of the two rooms has a larger area, and by how much? Give the difference in sq ft.

The Metric System of Measurement. More About Exact and Approximate Numbers

3. A sheet of typing paper is 11-in. long and 8.5-in. wide. What is the area of such a sheet of paper in square centimeters?
4. A driveway measures 11.0-m long and 3.0-m wide. Find the area of the driveway in sq ft.
5. How much concrete was used for the driveway of problem 4 (a) in m³, and (b) in cu ft, if the thickness of the driveway is 12 cm?
6. If one square kilometer is divided into 5 equal tracts of land, what will be the area of each tract in square meters?
7. A rectangular lawn is 58.0-ft long and 32.0-ft wide. Find the area of the lawn in m².
8. The walls and the bottom of an olympic size swimming pool are to be painted with a protective paint. The swimming pool measures 50.0-m long and 22.0-m wide and its depth all around is a diving depth of 3.0 m. How many gallons of paint are needed for the four walls and the bottom of the pool, if 1 gallon of paint covers 400 sq ft?
9. The gasoline tank of a foreign car holds 18.0 gal. If one gallon takes up 231 cu in., what is the volume of the tank (a) in cu ft, and (b) in m³?
10. An outdoor swimming pool is 15.0-m long and 10.0-m wide. During a summer week, a layer 7-mm thick evaporated from the surface of the swimming pool. How much water evaporated (a) in m³, and (b) in cu ft?
11. A roadside table is 5.5-ft long and 2.5-ft wide. Find its area in m².
12. Find the area of an 18-cm square window pane.
13. A garden is 32.0-m long and 18.0-m wide. (a) How much fencing in feet is needed to surround this garden? (b) What is the area of the garden in sq ft?
14. A cylindrical storage tank (shown in Fig. 2-2) is 6.0-m high, and the radius of its circular base is 2.50 m. How much can this tank hold (a) in m³, and (b) in cu ft?

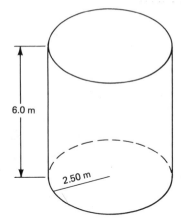

6.0 m

2.50 m

FIG. 2-2

2-5 Metric Units of Weight and Liquid Measure

The unit of weight in the metric system is the **gram**, abbreviated g. It is defined as the weight of one cubic centimeter of water at the temperature in which a given quantity of water is at its smallest volume (39.32°F). A submultiple of the gram is the **milligram**, abbreviated mg. It is 0.001 g. The gram is too small for most common measurements. A multiple of the gram is the **kilogram**, abbreviated kg. It is defined as the weight of 1000 cubic centimeters (one cubic decimeter) of water. From the definition of the kilogram, it follows that the weight of

$$1 \text{ m}^3 \text{ of water} = 1000 \text{ kg}$$

The weight of 1000 kilograms is the **metric ton**, abbreviated t.

The unit for the liquid measure in the metric system is the **liter**, abbreviated ℓ. It is defined as the volume of 1000 cm³ (one cu dm) of water. From the definition of the liter, it follows that the volume of

$$1 \text{ cm}^3 \text{ of water} = 0.001 \ \ell$$

The volume of 0.001 ℓ is called the **milliliter**, abbreviated mℓ.

NOTE: The relation 1 cm³ = 0.001 ℓ is extensively used in chemistry and in pharmacology. Another very common form of this relation is

$$1 \text{ cc} = 1 \text{ m}\ell$$

It is read "one cubic centimeter equals one milliliter."

A list of common weight and liquid measurements is given below:

Units of Weight	Units of Liquid Measure
ton (t) = 1000 kg	1 liter (ℓ) = 1000 milliliters (mℓ)
1 kilogram (kg) = 1000 grams (g)	
1 gram = 1000 milligrams (mg)	

Six useful conversion factors are given below:

$$1 \text{ kg} = 2.205 \text{ lb}$$
$$1 \text{ lb} = 0.4535 \text{ kg}$$
$$1 \text{ g} = 0.03527 \text{ oz}$$
$$1 \text{ oz} = 28.35 \text{ g}$$
$$1 \ \ell = 0.264 \text{ gal}$$
$$1 \text{ gal} = 3.79 \ \ell$$

EXAMPLE 2-13 A man on a diet lost 4.3 kg in a week. If his weight last week was 168.0 lb, what is his present weight in kg?

SOLUTION Using the relation 1 lb = 0.4535 kg, and remembering that conversion factors are considered exact numbers, we have

$$168.0 \text{ lb} = 168.0 \times 0.4535$$
$$= 76.188$$
$$= 76.19 \text{ kg}$$

Then, subtracting the approximate number 4.3 kg from the approximate number 76.19 kg, we have

The Metric System of Measurement. More About Exact and Approximate Numbers

$$76.19$$
$$\frac{4.3}{71.89 \text{ kg}}$$

Finally, 71.89 is rounded off to the precision of 4.3, to give 71.9 kg.

EXAMPLE 2-14 A swimming pool measures 20.0-m long, and 8.0-m wide. At first it was filled to 2.60 m, and then water was added to bring the level to 2.80 m. (a) How many cubic meters of water were added? (b) What is the volume of that water in liters? (c) In gallons?

SOLUTION

(a) The volume of the additional water in cubic meters is

$$\text{volume} = 20.0 \times 8.0 \times 0.20 = 32 \text{ m}^3$$

(b) Since 1 m³ = 1000 ℓ,

$$32 \text{ m}^3 = 32 \times 1000$$
$$= 32,000 \ \ell$$

(c) Since 1 ℓ = 0.264 gal,

$$32,000 \ \ell = 32,000 \times 0.264$$
$$= 8448$$
$$= 8400 \text{ gal}$$

For engineers and technicians, volume and weight are very important, and very closely related. In physics the student will use the concept of **density** which is defined in terms of weight and volume. When we say that aluminum weighs 168 pounds per cubic foot, or that brass weighs 8.5 grams per cubic centimeter, we are talking about the density of aluminum and brass. The formula for density is

$$\text{density} = \frac{\text{weight}}{\text{volume}}$$

NOTE: A more correct formula for density is density = mass/volume, but for industrial purposes and for the problems in this book, the first formula is equally valid. The student will be told, in physics, about the difference between weight and mass.

Density is expressed in units of weight per unit of volume. For instance, the density of water is 62.4 lb/ft³ (read "62.4 pounds per cubic foot"), and the density of copper is 8.9 g/cm³ (read "8.9 grams per cubic centimeter").

EXAMPLE 2-15 The density of lead is 687 pounds per cubic foot. What is the weight of a block of lead 0.50-ft thick, 3.0-ft long, and 2.0-ft wide?

SOLUTION First we find the volume of the piece of lead in cubic feet. Thus,

$$0.50 \times 3.0 \times 2.0 = 3.0 \text{ ft}^3$$

Since 1 ft³ of lead weighs 687 lb, the total weight of this piece of lead is

$$3.0 \times 687 = 2061$$
$$= 2100 \text{ lb}$$

Metric Units of Weight and Liquid Measure

EXAMPLE 2-16 A truck is loaded with 200 pine rafters. Each rafter is 8.0-m long, 0.20-m wide, and 0.10-m thick. What is the load of the truck if the density of pine is 540 kg/m³?

SOLUTION The volume of each rafter is

$$\text{volume} = 8.0 \times 0.20 \times 0.10$$
$$= 0.16 \text{ m}^3$$

The weight of each rafter is

$$\text{weight} = 0.16 \times 540$$
$$= 86.4$$
$$= 86 \text{ kg}$$

The load of the truck is

$$\text{load} = 200 \times 86.4$$
$$= 17{,}280$$
$$= 17{,}000 \text{ kg}$$

EXAMPLE 2-17 Hardwood is heavier than pine. If a 3.2 m by 0.15 m by 0.10 m oak rafter weighs 39 kg, what is the density of oak?

SOLUTION The volume of the rafter is

$$\text{volume} = 3.2 \times 0.15 \times 0.10 = 0.048 \text{ m}^3$$

Since density = weight/volume, we have

$$\text{density} = \frac{39}{0.048}$$
$$= 812.5$$
$$= 810 \text{ kg/m}^3$$

EXERCISE 2-3

1. Change 3.3 kg to pounds.
2. Change 3.5 gal to liters.
3. Change 4.75 ℓ to gallons.
4. Change 23.5 qt to liters.
5. Change 2 qt (a) to milliliters, (b) to cubic centimeters.
6. Change 2.73 ℓ (a) to milliliters, (b) to cubic centimeters.
7. Change 825 mℓ to liters.
8. If gold is 19.30 times heavier than water, what is the density of gold in pounds per cubic foot? (Density of water = 62.4 lb/ft³)
9. The average weight of a man 6.00-ft tall is 160.0 lb. Give this information in meters and kilograms, respectively.
10. What is the weight of the air of a room 9.00-m long, 6.00-m wide, and 3.00-m high, if the density of air is 1.3 kg/m³?
11. If the gas tank of your car holds 20.0 gal, how many liters of gasoline will you need to fill your tank when the metric system is adopted in this country?
12. A building contractor has to move 200 slabs of marble. Each slab is 2.0-m long, 0.080-m

wide, and 0.025-m thick, how many tons of marble need to be moved, if the density of marble is 2.90 tons per cubic meter?

13. An olympic size swimming pool is 50.0-m long, 22.0-m wide, and 3.0-m deep. What is the weight of the water in the pool, in metric tons, when the pool is filled to the top? (Hint: 1 m³ of water = 1000 kg)

14. The radius of the circular base of a cylindrical container is 0.500 ft, and the height of the container is 2.30 ft. (a) What is the weight of the contained water, if water weighs 62.4 lb/ft³? (b) How many kilograms is that? (Remember to use the correct number of digits for the value of π).

15. A bridge over a creek is resting on 3 horizontal hardwood rafters. The length, width, and thickness of each rafter is 7.00 m, 0.30 m, and 0.25 m, respectively. If the density of the wood of the rafters is 820.0 kg/m³, what is the weight of the 3 rafters?

REVIEW QUESTIONS

1. What is the basic unit of length in the metric system?
2. What does each of the prefixes kilo, deci, centi, and milli mean?
3. In computations with exact and approximate numbers, which of the two is always the least accurate?
4. The product of an exact and an approximate number should be rounded to the accuracy of which of the two factors?
5. What are conversion factors considered, exact or approximate numbers?
6. When an approximate number, with a given number of significant digits, is raised to the second or to the third power, to how many significant digits will that power be rounded off?

REVIEW EXERCISES

1. Change 18 cm to millimeters.
2. Change 42 cm to meters.
3. Change 2.318 km to meters.
4. Change 0.23 m to millimeters.
5. Change 398 m to kilometers.
6. Change 76 m to centimeters.
7. Subtract 5.8 m from 8.178 m.
8. If 23 sheets of a certain grade of paper have a combined thickness of 0.058 in., what is the combined thickness of 483 sheets of such paper in (a) centimeters and (b) millimeters?
9. A jet is flying at the speed of 595 mph. Find the speed of the jet in kilometers per hour.
10. How many 7.3-cm long nails can be made from a wire 1288-m long?
11. A piece of plywood is 1.23 m by 0.87 m. What is the area of that plywood in square meters?
12. A driveway measures 12.3-m long, 3.2-m wide, and 0.14-m thick. How many cubic meters of concrete were used in making that driveway?
13. If a piece of land of 630,000 square meters is divided into 3 equal tracts, what will be the area of each tract in square meters?

14. A rectangular lawn is 72.3-ft long, and 28.7-ft wide. Find the area of the lawn in square meters.

15. An oil tank is full with 288.0 gal of oil. If one gallon is 231 cu in., what is the volume of that tank in (a) cu ft, (b) m³?

16. Change 7.8 kg to pounds.

17. Change 5.6 gal to liters.

18. Change 8.23 ℓ to gallons.

19. Change 3 qt to (a) milliliters, (b) cubic centimeters.

20. Air weighs 1.3 kg/m³. Find the weight of the air contained in a room 7.6-m long, 5.8-m wide, and 3.1-m high.

BASIC ALGEBRA

Algebra is arithmetic generalized. It is a method of doing arithmetic by means of letters which represent numbers, and signs which represent their relations. This generalization through symbols and signs makes algebra a very powerful tool in the hands of scientists, engineers, and technicians. Algebra becomes a mathematical language to express science, engineering, and technical relations, and facilitates the solution of problems. A technical problem, which otherwise is complicated, becomes simple and easy when reduced to an algebra problem. This part of this book will lead the student from arithmetic to algebra in clear and gradual steps.

3

Signed Numbers

The numbers of arithmetic are the **natural numbers**, 1, 2, 3, and so on; the **whole numbers**, 0, 1, 2, 3, and so on; the **rational numbers**, $\frac{3}{4}$, 2.5, 8, and so on (numbers which can be expressed as a fraction); and the **irrational numbers**, π, $\sqrt{2}$, and so on (numbers which can not be expressed as a fraction). Of importance to us here are the whole numbers.

3-2 Properties of the Whole Numbers

The whole numbers have some outstanding properties, which might look trivial now, but will prove very useful in algebra, and will be used in many cases. These properties are as follows:

1. *The Commutative Law of Addition.* The sum of two whole numbers is not affected by the order in which they are added. Thus, $3 + 5 = 5 + 3 = 8$.
2. *The Commutative Law of Multiplication.* The product of two whole numbers will be the same no matter which of the numbers is used as the multiplier or the multiplicand. Thus, $3 \times 5 = 5 \times 3 = 15$.
3. *The Associative Law of Addition.* The manner in which whole numbers are grouped will not affect their sum. Thus, $5 + 3 + 4 = (5 + 3) + 4 = 8 + 4 = 12$, and $5 + 3 + 4 = 5 + (3 + 4) = 5 + 7 = 12$.
4. *The Associative Law of Multiplication.* The product of whole numbers is not affected by the manner in which the numbers are grouped. Thus, $3 \times 2 \times 4 = (3 \times 2) \times 4 = 6 \times 4 = 24$, and $3 \times 2 \times 4 = 3 \times (2 \times 4) = 3 \times 8 = 24$.
5. *The Distributive Law.* When the sum of two or more numbers is to be multiplied by a given number, each of the numbers is multiplied by the given number. The sum of these products is the final result. Thus, $2 \times (4 + 3) = 2 \times 4 + 2 \times 3 = 8 + 6 = 14$, and $3 \times (2 + 5 + 4) = 3 \times 2 + 3 \times 5 + 3 \times 4 = 6 + 15 + 12 = 33$.

NOTE: In three of the above illustrations, parentheses were used. It will suffice for the moment if the student keeps in mind that the numbers enclosed by parentheses are to be taken together. In a later chapter we will say more about such grouping symbols.

3-3 Negative Numbers

With only the arithmetic numbers we would be severely restricted in algebra, where we look at various topics from a more general standpoint. For instance, arithmetic has no answer for the subtraction $5 - 8$. That is, there is no arithmetic number which when added to 8 will give 5. This is an algebra problem, and the answer to it is a **negative number**. The negative numbers give algebra great flexibility.

We will now extend the whole numbers of arithmetic to include the negative numbers. Think of a line extending indefinitely in both directions, as shown in Fig. 3-1. Such a line is called the **number line**. Points on this line are marked and numbered

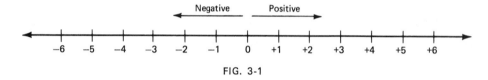

FIG. 3-1

1, 2, 3 . . . at equal intervals on both sides of the number zero. The numbers on the right of zero are the familiar whole numbers of arithmetic, but the numbers on the left of zero are the negative numbers. The student should notice that the numbers to the left of zero are preceded by the $(-)$ sign and are called negative numbers, to be distinguished from the numbers to the right of zero which are called **positive numbers**. Thus, from now on, *the plus sign in front of a number indicates that the number is positive, and the minus sign in front of a number indicates that the number is negative. A number written with no sign is understood to be positive.* Thus, $+5$, 16, 8, and $+12$ are all positive numbers, and -3, -17, -121, and -1 are all negative numbers. The positive and negative numbers together are called **signed numbers**. Because the signed numbers are ordered around zero, which is called the **origin**, and because they extend in opposite directions, the signed numbers are also known as **directed numbers**. From the arrangement of the number line, it must be apparent that *a given number is less than any number to its right.* Thus, any negative number is less than zero, and -7 is less than -5. In mathematics, order relations are expressed briefly with the two **symbols of inequality** $(>)$ (read "greater than"), and $(<)$ (read "less than"). Thus, to express that $+5$ is greater than -11, we write $+5 > -11$ (read "positive five is greater than negative eleven") and to express that -4 is less than -1, we write $-4 < -1$ (read "negative four is less than negative one").

A close examination of Fig. 3-1 will show that to every positive number on the number line, at equal distance from zero, there corresponds a negative number, and to every negative number there corresponds a positive number. Thus, $+5$ corresponds to -5, and -17 corresponds to $+17$. In this sense, -8 and $+8$ are **opposite numbers**.

The signed numbers of the number line of Fig. 3-1 are called the **integers**. The integers, therefore, are the natural numbers, their opposites, and zero. But, of course, these are not the only signed numbers. The student should expect what is shown in Fig. 3-2: there are positive and negative rational numbers, and there are positive and negative irrational numbers. Thus, halfway between 0 and 1 is the positive rational number $+\frac{1}{2}$, and halfway between 0 and -1 is the negative rational number $-\frac{1}{2}$.

Also, the signed rational numbers $+4\frac{3}{4}$ and $-4\frac{3}{4}$ are shown, as well as the irrational signed numbers $+\sqrt{2}$ at the point $+1.4142$, and $-\sqrt{2}$ at the point -1.4142.

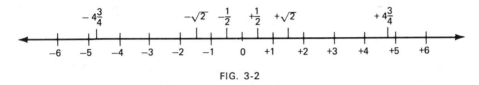

FIG. 3-2

All the different kinds of numbers we have seen in the number line, that is, the integers, the rational signed numbers, and the irrational signed numbers, form the number system of algebra. The algebra numbers are known as **real** numbers.

3-4 The Absolute Value of Numbers

Often we need the value of a signed number independent of its sign; that is, we want the distance of the number from zero on the number line. In such a case, we are asking for the **absolute value** of the signed number. The absolute value of a number is indicated by writing the number between two vertical bars. Thus, the symbol $|-5|$ is read "the absolute value of negative five," and the symbol $|3|$ is read "the absolute value of positive three."

The absolute value of a signed number is the number itself without its sign. Thus, $|-5| = 5$, and $|+5| = 5$. From these two examples, the student can see that the absolute value of a positive number is positive, and that the absolute value of a negative number is positive also. In general, *the absolute value of a number is never negative.*

EXAMPLE 3-1 Find the absolute value of the following signed numbers:

(a) $|-7|$ (b) $|7|$ (c) $|17|$ (d) $|-11|$

SOLUTION

(a) $|-7| = 7$ (b) $|7| = 7$ (c) $|17| = 17$ (d) $|-11| = 11$

EXERCISE 3-1

1. Which property of the whole numbers is illustrated by each of the following examples?
   a. $(3 + 5) + 2 = 3 + (5 + 2)$ b. $7 \times 2 = 2 \times 7$
   c. $2 \times (1 + 3) = 2 \times 1 + 2 \times 3$ d. $8 + 7 = 7 + 8$
   e. $(2 \times 3) \times 5 = 2 \times (3 \times 5)$
2. Do the integers include negative numbers?
3. In how many different ways can you write a negative number?

The Absolute Value of Numbers

4. In how many different ways can you write a positive number?

5. Is this ($<$) the "greater than" symbol?

6. Is this ($>$) the "less than" symbol?

7. What numbers are included in the real numbers?

8. Are there numbers with a positive absolute value?

9. Are there numbers with a negative absolute value?

10. Write down the signed numbers corresponding to points *a, b, c, d,* and *e* on the number line of Fig. 3-3.

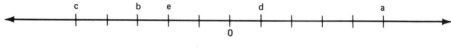

FIG. 3-3

11. Make your own number line on a piece of paper, and locate each of the following signed numbers: $+5, -2, -11, 3, 6$ and -15.

Which of the following are true statements, and which are false?

12. $-14 < 0$	**13.** $-1 > 0$	**14.** $7 < -7$	**15.** $-8 > -28$
16. $-8 < 0$	**17.** $-111 < 0$	**18.** $13 > -264$	**19.** $0 > -79$
20. $-5 > 5$	**21.** $|3| = -3$	**22.** $|-3| = 3$	**23.** $|-18| = 18$
24. $|+23| = +23$	**25.** $|+29| = 29$	**26.** $|-29| = +29$	**27.** $|+1| = -1$
28. $|+74| = 74$	**29.** $|-10| = 10$	**30.** $|-31| = 31$	

3-5 Addition of Signed Numbers

We have seen that the signed numbers are directed numbers in the sense that they are ordered from zero and are along the positive and the negative directions of the number line. If we now move in the positive direction 5 units from zero, and then 3 units in the same direction, we will find ourselves on the point $+8$ (Fig. 3-4). Therefore, $5 + (+3) = +8$. In the same way, we find from the number line that $+5 + (-3) = +2$ (Fig. 3-5), and $-5 + (-3) = -8$ (Fig. 3-6).

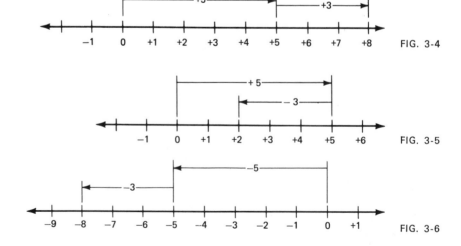

FIG. 3-4

FIG. 3-5

FIG. 3-6

Signed Numbers

From the above examples, we can see that to add two signed numbers:

1. *If their signs are the same, add their absolute values, and put in front of this sum their common sign.*
2. *If their signs are different, subtract their absolute values, and in front of this difference, put the sign of the number with the larger absolute value.*

Notice that according to this rule, *the sum of any pair of opposite numbers is zero.*

EXAMPLE 3-2   Add the following signed numbers:

(a) $3 + (+4)$ (b) $-3 + (-4)$ (c) $-3 + (+4)$

(d) $3 + (-4)$ (e) $-3 + (+3)$

SOLUTION   Problems (a) and (b) are cases of the first part of the rule, and (c) and (d) are cases of the second part. Thus,

(a) The absolute values of $+3$ and $+4$ are added to give 7, and the common sign, $(+)$, is put in front of this sum. Therefore, $3 + (+4) = 7$.
(b) The absolute values of -3 and -4 are added to give 7, and the common sign, $(-)$, is put in front of this sum. Therefore, $-3 + (-4) = -7$.
(c) The absolute values of -3 and $+4$ are subtracted to give 1, and in front of this difference, we put the sign $(+)$, which is the sign of $+4$, the number with the largest absolute value. Therefore, $-3 + (+4) = +1$.
(d) The absolute values of 3 and -4 are subtracted to give 1, and in front of this difference, we put the sign $(-)$, which is the sign of -4, the number with the largest absolute value. Therefore, $3 + (-4) = -1$.
(e) The sum of any two opposite numbers is zero. Therefore, $-3 + (+3) = 0$.

To add together more than two signed numbers, you may add first the two first numbers, and then add to that sum the third number, and so on, or, using the commutative law of addition, you may add all the positive numbers together, and then all the negative numbers together, according to the first part of the rule. Then you will add together the positive and negative partial sums using the second part of the rule.

EXAMPLE 3-3   Add $(+3) + (-6) + (+7) + (-4) + (-8)$.

SOLUTION

$$(+3) + (-6) + (+7) + (-4) + (-8) = (-3) + (+7) + (-4) + (-8)$$
$$= (+4) + (-4) + (-8)$$
$$= (0) + (-8)$$
$$= -8$$

or, by using the Commutative Law of Addition,

$$(+3) + (-6) + (+7) + (-4) + (-8) = (+3) + (+7) + (-6) + (-4) + (-8)$$
$$= (+10) + (-18)$$
$$= -8$$

The student must have noticed that the signs (+) and (−) are used both to indicate addition and subtraction as well as to distinguish between positive and negative numbers. This dual use of the plus and the minus signs causes confusion. To avoid this confusion, parentheses have been used up to this point, but in a subsequent section, we will agree on a notation which will make addition and subtraction easier, and the use of parentheses unnecessary.

EXERCISE 3-2

Add the signed numbers in each of the following problems:

1. $+7 + (-4)$	**2.** $3 + (+8)$
3. $-5 + (-8)$	**4.** $-6 + (+3)$
5. $(-17) + (-11)$	**6.** $-5 + (-2)$
7. $-19 + (-3)$	**8.** $12 + (-13)$
9. $25 + (+32)$	**10.** $-1 + (113)$
11. $34 + (-24)$	**12.** $-86 + (-93)$
13. $-214 + (+79)$	**14.** $+8 + (-9) + (-18)$
15. $-3 + (+17) + (-21)$	**16.** $8 + (+81) + (-19) + (+38)$
17. $-49 + (-63) + (+8) + (-3)$	**18.** $+53 + (-86) + (-44) + (+37)$
19. $-211 + (+127) + (-87) + (-11)$	**20.** $538 + (-432) + (-103) + (+28)$

3-6 Subtraction of Signed Numbers

Remember that subtraction is the inverse operation of addition. Therefore, to subtract a number with a given sign, it will be the same as adding the same number with opposite sign. Consider for example, $5 - 4$. The only sign here is (−). This obviously is a subtraction sign, and since no other sign is left for 5 and 4, both of them are positive numbers. Then, $5 - 4$ means: from the positive number 5, subtract the positive number 4. This is a simple arithmetic subtraction, and $5 - 4 = 1$. Notice now that the result of this subtraction will be the same if we give it the following addition form: $5 + (-4)$. This now means: To the positive number 5, add the negative number -4. But according to the rule for the addition of signed numbers, $5 + (-4) = 1$. Therefore, $5 - 4 = 5 + (-4) = 1$. The above suggests, then, that *to subtract a signed number from another, change the sign of the number to be subtracted* (*subtrahend*), *and then add the numbers following the rule for the addition of signed numbers.*

EXAMPLE 3-4 Subtract $+6 - (-5)$.

SOLUTION The number to be subtracted is -5. Change it to $+5$, and then add $+6$ and $+5$. Thus, $+6 - (-5) = +6 + (+5) = 11$.

EXAMPLE 3-5 Subtract $7 - (+4)$.

SOLUTION The number to be subtracted is $+4$. Change it to -4, and then add 7 and -4. Thus, $7 - (+4) = 7 + (-4) = 3$.

EXAMPLE 3-6 Subtract $-8 - (-5)$.

SOLUTION The number to be subtracted is -5. Change it to $+5$, and then add -8 and $+5$. Thus, $-8 - (-5) = -8 + (+5) = -3$.

Perform the subtraction of the signed numbers in each of the following exercises:

1. $7 - (-8)$
2. $-11 - (+3)$
3. $-8 - (-3)$
4. $5 - (-17)$
5. $-1 - (+21)$
6. $-24 - (35)$
7. $33 - (+78)$
8. $-11 - (+8)$
9. $6 - (+7)$
10. $-98 - (-82)$
11. $193 - (+72)$
12. $137 - (+36)$
13. $-234 - (439)$
14. $+63 - (-357)$
15. $-117 - (314)$
16. $59 - (-813)$
17. $139 - (-63)$
18. $-248 - (-11)$
19. $-61 - (-235)$
20. $+633 - (+147)$

3-7 Addition and Subtraction without Parentheses

Suppose we want to add the following three signed numbers, $(+9) + (-4) + (+3)$. This addition can be written in a much simpler form. Notice first that without any risk of confusion we can drop both the plus signs and the parentheses of all positive signed numbers. Thus, $(+9) + (-4) + (+3)$ can be written as $9 + (-4) + 3$. Notice now that both the $(+)$ sign and the parentheses of $+ (-4)$ can be omitted if from now on we take -4 to mean "add minus four". Thus, $(+9) + (-4) + (+3)$ becomes $9 - 4 + 3$. This last form of our addition is interpreted to mean: To the positive 9 add the negative -4, and to that sum add positive 3. Thus, $9 - 4 + 3 = 8$.

We can simplify in a similar way the indicated addition $(+4) + (-5) + (-9) + (+2)$. Thus,

$(+4) + (-5) + (-9) + (+2) = 4 + (-5) + (-9) + 2$ *[dropping (+) signs and parentheses of all positive numbers.]*

$= 4 - 5 - 9 + 2$ *[dropping parentheses of negative numbers and the (+) signs which are in front of such parentheses.]*

Now, $4 - 5 - 9 + 2$ is interpreted to mean: Add together positive 4 and negative -5, and negative -9, and positive 2.

The student must have noticed that we have simplified only indicated additions of signed numbers. Since, however, subtraction can be changed to addition, the same can be done for indicated subtraction of signed numbers. For instance, if we are given $(+3) + (-8) - (-4) + (-7) - (-9)$, in which both addition and subtraction of signed numbers are involved, we can simplify as follows:

$(+3) + (-8) - (-4) + (-7) - (-9)$

$= (+3) + (-8) + (+4) + (-7) + (+9)$ *(by changing all indicated subtractions to additions)*

$= 3 + (-8) + 4 + (-7) + 9$ *[by dropping the (+) signs and the parentheses of all positive numbers]*

$$= 3 - 8 + 4 - 7 + 9$$

[*by dropping the parentheses of negative numbers, and dropping the* $(+)$ *signs which are in front of each of such parentheses*]

$$= 1$$

(*by addition of signed numbers*)

Of great practical importance in this discussion is the fact that in a "chain" of signed numbers, like $11 - 3 + 5 - 16 - 2 + 5 - 8$, *the* $(+)$ *and the* $(-)$ *signs are not operation signs. They are only signs of signed numbers. The operation, as we have seen, is always understood to be addition.* From now on we will call the result of such an addition the **algebraic sum**.

EXAMPLE 3-7 Find the algebraic sum of each of the following:
(a) $3 + 8 - 2 - 11 + 7 - 18$
(b) $-12 + 18 - 5 - 3 + 6$
(c) $-1 + 17 + 4 - 2 - 26$

SOLUTION In all three cases, addition of signed numbers is implied. Therefore,
(a) $3 + 8 - 2 - 11 + 7 - 18 = -13$
(b) $-12 + 18 - 5 - 3 + 6 = 4$
(c) $-1 + 17 + 4 - 2 - 26 = -8$

EXERCISE 3-4

Find the algebraic sum of each of the following:
 1. $-5 + 2 + 7 - 13 - 8 + 6$
 2. $7 + 8 - 17 - 3 + 14 - 9 - 1$
 3. $-19 - 3 + 6 - 4 + 1 - 18 + 7$
 4. $7 - 11 - 17 + 23 + 2 - 7 + 18 - 5$
 5. $-13 - 6 + 5 - 2 + 19 - 23 + 52 - 3$
 6. $28 - 33 + 4 + 13 - 37$
 7. $-89 + 27 - 3 - 86 + 39$
 8. $17 + 77 - 14 - 11 - 4 + 101$
 9. $33 - 48 + 13 + 27 - 8$
10. $-125 - 31 + 66 - 88$
11. $234 - 118 + 35 + 12 - 36$
12. $51 - 139 - 47 + 58 - 69$
13. $-28 - 117 - 73 - 13 + 88$
14. $32 - 68 - 73 + 124$
15. $43 + 68 + 178 - 11 - 5$

3-8 Multiplication of Signed Numbers

Let us see first the multiplication of two signed numbers of unlike signs, $3 \times (-2)$ and $-3 \times (4)$, for example.

Recalling from arithmetic that multiplication is actually addition of equal

Signed Numbers

numbers, we understand that $3 \times (-2)$ means $(-2) + (-2) + (-2)$, and we know that $(-2) + (-2) + (-2) = -2 - 2 - 2 = -6$.

Also, to find the meaning and the product of $-3 \times (4)$, we apply first the Commutative Law of Multiplication. Thus, $-3 \times (4) = 4 \times (-3)$, the meaning of which has already been established. Therefore, $-3 \times (4) = 4 \times (-3) = (-3) + (-3) + (-3) + (-3) = -3 - 3 - 3 - 3 = -12$. The above examples suggest the following rule: *To multiply two signed numbers with unlike signs, multiply the absolute values of the numbers, and prefix a negative sign to that product.*

EXAMPLE 3-8 Multiply:

(a) $(-7) \times 5$ (b) $5 \times (-8)$ (c) $3 \times (-12)$ (d) $(-5) \times 11$

SOLUTION In all four cases we have the product of signed numbers with unlike signs. Therefore, in all cases, the product will be negative. Thus,

(a) $(-7) \times 5 = -35$ (b) $5 \times (-8) = -40$

(c) $3 \times (-12) = -36$ (d) $(-5) \times 11 = -55$

Let us see next the multiplication of two signed numbers of like signs, $(+3) \times (+4)$ and $(-3) \times (-4)$.

The first multiplication is a familiar arithmetic case: $3 \times 4 = 12$.

To find the product of $(-3) \times (-4)$, we start with what we already know. We know that

$$-3 \times [0] = 0$$

$$-3 \times [(+4) + (-4)] = 0 \quad \textit{(by replacing 0 with the sum of two opposites)}$$

$$(-3) \times (+4) + (-3) \times (-4) = 0 \quad \textit{(by multiplying the above using the Distributive Law)}$$

$$-12 + (-3) \times (-4) = 0 \quad \textit{[by multiplying } (-3) \times (+4)]$$

Notice at this point that $(-3) \times (-4)$ must be equal to a number which when added to -12 will give zero. This number is the opposite of -12, which is 12. Therefore, $(-3) \times (-4) = 12$.

The above examples suggest the following rule: *To multiply two signed numbers with like signs, multiply the absolute values of the numbers, and prefix a positive sign (or no sign) to that product.*

EXAMPLE 3-9 Multiply:

(a) $(+5) \times (+7)$ (b) $(-3) \times (-6)$ (c) $(-5) \times (-7)$ (d) $(+3) \times (+6)$

SOLUTION Since in all cases we have the multiplication of two signed numbers with like signs, all products will be positive. Thus,

(a) $(+5) \times (+7) = 35$ (b) $(-3) \times (-6) = 18$

(c) $(-5) \times (-7) = 35$ (d) $(+3) \times (+6) = 18$

The rule for the multiplication of two signed numbers can be stated very briefly as follows: *To multiply two signed numbers, multiply their absolute values. This product will be positive if the two numbers are of like signs; it will be negative if the two numbers are of unlike signs.* The shortness of the following will make it very easy to remember what the signs of the products of signed numbers should be:

$$(+) \times (+) = + \atop (-) \times (-) = +\Big\} \text{numbers of like signs give positive products}$$

$$(+) \times (-) = - \atop (-) \times (+) = -\Big\} \text{numbers of unlike signs give negative products}$$

EXAMPLE 3-10 Multiply:

(a) $(+3) \times (+8)$ (b) $(-3) \times (-8)$ (c) $(+3) \times (-8)$ (d) $(-3) \times (+8)$

SOLUTION

(a) $(+3) \times (+8) = 24$ (b) $(-3) \times (-8) = 24$

(c) $(+3) \times (-8) = -24$ (d) $(-3) \times (+8) = -24$

When more than two signed numbers are to be multiplied together, multiply first the two first numbers. Then multiply that product with the third number, and so on.

EXAMPLE 3-11 Multiply:

(a) $(-2) \times (+3) \times (-7)$ (b) $(+1) \times (+2) \times (-3) \times (+5)$

SOLUTION

(a) $(-2) \times (+3) \times (-7) = (-6) \times (-7) = 42$

(b) $(+1) \times (+2) \times (-3) \times (+5) = (+2) \times (-3) \times (+5) = (-6) \times (+5) = -30$

EXAMPLE 3-12 Evaluate each of the following:

(a) $(-2)^3$ (b) $(-3)^2$ (c) $(-1)^2 \times (2)^3$

SOLUTION

(a) $(-2)^3 = (-2) \times (-2) \times (-2) = -8$

(b) $(-3)^2 = (-3) \times (-3) = 9$

(c) $(-1)^2 \times (2)^3 = (-1) \times (-1) \times (2) \times (2) \times (2) = 8$

3-9 Division of Signed Numbers

Division is the inverse of multiplication. We use multiplication to check division. Since division and multiplication are so intimately related, we can establish a rule for the division of signed numbers from what has already been established for the multiplication of signed numbers. We know, for instance, that

$$(-3) \times (2) = -6$$

But this multiplication is the check for the following two divisions,

$$\begin{array}{r} -3 \\ 2\overline{\smash{\big)}-6} \\ \underline{-6} \\ 0 \end{array}$$
$$\begin{array}{r} 2 \\ -3\overline{\smash{\big)}-6} \\ \underline{-6} \\ 0 \end{array}$$

because $(-3) \times (2) = -6$ because $(2) \times (-3) = -6$

These two divisions suggest the following rule: *To divide two signed numbers, divide their absolute values, and prefix a positive sign to the quotient if the two numbers are of like signs, or a negative sign to the quotient if the two numbers are of unlike signs.* The shortness of the following will make it very easy to remember what the signs of quotients of signed numbers should be:

$$\left.\begin{array}{l} (+) \div (+) = + \\ (-) \div (-) = + \end{array}\right\} \text{the quotient of like signs is positive}$$

$$\left.\begin{array}{l} (+) \div (-) = - \\ (-) \div (+) = - \end{array}\right\} \text{the quotient of unlike signs is negative}$$

EXAMPLE 3-13 Divide:

(a) $(+12) \div (+4)$ (b) $(-12) \div (-4)$ (c) $(+12) \div (-4)$ (d) $(-12) \div (+4)$

SOLUTION

(a) $(+12) \div (+4) = 3$ (b) $(-12) \div (-4) = 3$

(c) $(+12) \div (-4) = -3$ (d) $(-12) \div (+4) = -3$

EXAMPLE 3-14 Perform each of the following divisions:

(a) $-2\overline{\smash{\big)}+8}$ (b) $-5\overline{\smash{\big)}-30}$ (c) $+4\overline{\smash{\big)}-20}$ (d) $-18\overline{\smash{\big)}+72}$

SOLUTION

(a)
$$\begin{array}{r} -4 \\ -2\overline{\smash{\big)}+8} \\ \underline{+8} \\ 0 \end{array}$$
(b)
$$\begin{array}{r} 6 \\ -5\overline{\smash{\big)}-30} \\ \underline{-30} \\ 0 \end{array}$$
(c)
$$\begin{array}{r} -5 \\ +4\overline{\smash{\big)}-20} \\ \underline{-20} \\ 0 \end{array}$$
(d)
$$\begin{array}{r} -4 \\ -18\overline{\smash{\big)}+72} \\ \underline{+72} \\ 0 \end{array}$$

EXAMPLE 3-15 Perform the division indicated by each of the following fractions:

(a) $\dfrac{+8}{+2}$ (b) $\dfrac{-18}{-3}$ (c) $\dfrac{+21}{-7}$ (d) $\dfrac{-24}{+3}$

SOLUTION

(a) $\dfrac{+8}{+2} = +4$ (b) $\dfrac{-18}{-3} = +6$ (c) $\dfrac{+21}{-7} = -3$ (d) $\dfrac{-24}{+3} = -8$

EXERCISE 3-5

1. State the rule for the multiplication of signed numbers.
2. State the rule for the division of signed numbers.
Multiply:
3. $(-2) \times (+6)$ **4.** $(+3) \times (+5)$

5. $(-3) \times (+7)$ 6. $4 \times (-1)$
7. $(-2) \times (-8)$ 8. $8 \times (+7)$
9. $7 \times (-11)$ 10. $(-5) \times (0)$
11. $(-1) \times (-19)$ 12. $(+1) \times (-3)$
13. $15 \times (-1)$ 14. $(+2) \times (-2)$
15. $(-3) \times (+3)$ 16. $(-2) \times (9)$
17. $(1) \times (23)$ 18. $(0) \times (-7)$
19. $(-4) \times (-3)$ 20. $(-5) \times (2)$
21. $(+3) \times (-2) \times (3)$ 22. $(-5) \times (4) \times (+5)$
23. $(-1) \times (2) \times (-3) \times (-2)$ 24. $(+1)^2 \times (+2)^3$
25. $(+2)^2 \times (-2)^3$

Divide:

26. $(+21) \div (-3)$ 27. $(-12) \div (-2)$ 28. $(+12) \div (+3)$
29. $(-28) \div (+7)$ 30. $(+22) \div (+11)$ 31. $(+69) \div (-23)$
32. $(-14) \div (-7)$ 33. $(-17) \div (-1)$ 34. $(+18) \div (-18)$
35. $(-11) \div (+1)$ 36. $(+12) \div (-1)$ 37. $(-32) \div (-4)$
38. $(-7) \div (-7)$ 39. $(+8) \div (-8)$ 40. $(-21) \div (+21)$
41. $(-342) \div (-38)$ 42. $(-686) \div (+14)$ 43. $-7\,\overline{\smash{\big)}\,+21}$
44. $+3\,\overline{\smash{\big)}\,-27}$ 45. $-8\,\overline{\smash{\big)}\,-48}$ 46. $+18\,\overline{\smash{\big)}\,+72}$
47. $\dfrac{-33}{3}$ 48. $\dfrac{+84}{+21}$ 49. $\dfrac{96}{-8}$
50. $\dfrac{-128}{-32}$

REVIEW QUESTIONS

1. Which numbers are the natural numbers?
2. Which numbers are the whole numbers?
3. State the Commutative Law of Addition.
4. State the Commutative Law of Multiplication.
5. State the Associative Law of Addition.
6. State the Associative Law of Multiplication.
7. State the Distributive Law.
8. Sketch and explain the number line.
9. In how many ways can a positive number be designated? Describe the way(s).
10. In how many ways can a negative number be designated? Describe the way(s).
11. What is the absolute value of a number?
12. State the rule for the addition of signed numbers.
13. State the rule for the subtraction of signed numbers.
14. State the rule for the multiplication of signed numbers.
15. State the rule for the division of signed numbers.

REVIEW EXERCISES

1. Which property of the whole numbers is illustrated by each of the following?
   a. $(3 \times 4) \times 2 = 3 \times (4 \times 2)$
   b. $(1 + 2) + 7 = 1 + (2 + 7)$

c. $17 + 8 = 8 + 17$
d. $6 \times (2 + 7) = 6 \times 2 + 6 \times 7$
e. $21 \times 3 = 3 \times 21$

2. Write down the signed numbers corresponding to points a, b, c, d, e, and f on the number line shown in Fig. 3-7.

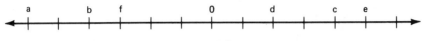

FIG. 3-7

3. Make your own number line on a piece of paper and locate each of the following signed numbers: -7, $+9$, $+2\frac{3}{4}$, $-1\frac{1}{2}$, $+8\frac{1}{3}$, and $-5\frac{1}{4}$.

Which of the following are true statements and which are false?

4. $0 > -11$	5. $-5 < 0$	6. $-11 > -12$
7. $-6 < -7$	8. $8 > 18$	9. $-1 > 1$
10. $|8| = -8$	11. $|-3| = 3$	12. $|-28| = 28$

Perform each of the following operations:

13. $(+7) + (-8)$	14. $27 - 32$
15. $-17 - (-18)$	16. $-1 - (+12)$
17. $-4 + 17 - 28$	18. $-7 + 25 - 37 - 5 + 3$
19. $(+16) \times (-3)$	20. $(-14) \times (+7) \times (-2)$
21. $(-24) \div (-1)$	22. $(-36) \div (1)$

4

Introduction to Algebra. Basic Ideas

4-1 Literal Numbers

In addition to the explicit numbers 0, 1, 2, 3, . . . used in arithmetic, in algebra we also use letters to represent numbers. When a letter is so used in algebra, it is called a **literal number**.

In algebra, like in arithmetic, we use the sign (=) between two numbers to indicate that the numbers are equal. We also use the sign (≠) (read "not equal to") to indicate that the two numbers between which the sign is placed are not equal. The signs of inequality (> and <) have already been introduced in Chap. 3 to be used when we wish to say more than just $x \neq 8$.

EXAMPLE 4-1 Which of the following are true and which are false?

(a) $10 = 2 \times 5$ (b) $4 \neq 2 + 3$ (c) $2 + 5 \neq 7$

SOLUTION

(a) True   (b) True   (c) False

The signs $+$, $-$, $\times$, and $\div$ for the basic operations are also used in algebra, and they have exactly the same meaning as in arithmetic. Thus, if a and b represent any two numbers, $a + b$ represents their sum, $a - b$ their difference, $a \times b$ their product, and $a \div b$ represents their quotient.

To avoid confusion by using x both as a multiplication sign, and as a literal number, x is not really used to show multiplication in algebra. Instead, to indicate the product of a and b, one of the following three notations is used:

1. $(a)(b)$ 2. $a \cdot b$ 3. ab

The elevated dot should be clearly elevated so as not to be mistaken for a decimal point, and it is better not to use it at all when decimals are involved.

Literal numbers are not like arithmetic numbers which can be added, subtracted, multiplied, and divided. The result of a given operation can only be indicated with literal numbers. Thus, with the two explicit arithmetic numbers, 6 and 2, we can say that $6 + 2 = 8$, $6 - 2 = 4$, $6 \times 2 = 12$, and $6 \div 2 = 3$. With the two literal numbers, x and y, however, we can only indicate their sum as $x + y$, their difference as $x - y$, their product as $(x)(y)$, $x \cdot y$, or xy, and their quotient as $x \div y$ or x/y. Explicit and literal numbers are used together in algebra as in the symbols $a/2$, $3x$, $b - 5$, and $3k + 7$. These four symbols indicate the division of a by 2, the product of 3 and x, the difference of b and 5, and the product of 3 and k increased by 7, respectively.

EXAMPLE 4-2 Indicate the following statements algebraically using algebraic numbers and symbols:

(a) Two times a number less a second number.
(b) The sum of a number and three times a second number.
(c) The product of two numbers plus seven times a third number.
(d) The quotient of two numbers decreased by their product.

SOLUTION

(a) $2x - y$ (b) $x + 3y$ (c) $xy + 7z$ (d) $\dfrac{x}{y} - xy$

4-2 Algebraic Terms and Definitions

The formulas the student has used in earlier mathematics courses have both explicit and literal numbers. For instance, in the formula for the area of the triangle, $A = \frac{1}{2}bh$, $\frac{1}{2}$ is an explicit number, while A, b, and h are literal numbers. As we know the value of b (base), h (height), and A (area) varies with different problems, while the value of $\frac{1}{2}$ in all problems remains the same. For this reason, explicit numbers in algebra are called **constants**, and literal numbers are called **variables**. Thus, in the formula for the perimeter of a rectangle, $P = 2L + 2W$, P, L, and W are variables, while 2 is a constant.

NOTE: In the formula for the circumference of a circle, $C = \pi D$, C and D are variables, while π, the approximate value of which is always taken to be 3.14, is a constant, in spite of the fact that π is a literal number.

EXAMPLE 4-3 The formula for the volume of a right cylinder is $V = \pi r^2 h$. Which of the numbers in the formula are constants, and which are variables?

SOLUTION The constants are π and 2, while V, r, and h are variables.

When algebraic numbers, constants and variables are "combined" with indicated algebraic operations such as addition, subtraction, multiplication, and division, they form an **algebraic expression**. The following, for instance, are five algebraic expressions:

(a) $3 + x$ (b) $3 - 5x$ (c) $ad - 8$

(d) $2mk + c - 2$ (e) $7x - 3m + 2n - \dfrac{m}{3}$

As we have already seen with signed numbers, the $(+)$ and the $(-)$ signs which appear in expressions are number signs. They are not signs of operation. The operation is always understood to be addition. Thus,

(a) $3 + x$ means: Add together the positive number 3 and the positive number $+x$.
(b) $3 - 5x$ means: Add together the positive number 3 and the negative product $-5x$.
(c) $ad - 8$ means: Add together the positive product $a \cdot d$, and the negative number -8.
(d) $2mk + c - 2$ means: Add together the positive product $2 \cdot m \cdot k$, the positive number $+c$, and the negative number -2.
(e) $7x - 3m + 2n - m/3$ means: Add together the positive product $7 \cdot x$, the negative product $-3 \cdot m$, the positive product $+2 \cdot n$, and the negative quotient $-m/3$.

NOTE: When grouping symbols and powers are introduced shortly, we will see more kinds of algebraic expressions.

EXAMPLE 4-4 What does each of the following two expressions mean?

$$\text{(a) } 6x - 2y + 3 \qquad \text{(b) } 8 + 2d - e$$

SOLUTION

(a) $6x - 2y + 3$ means: To the positive product $6 \cdot x$, add the negative product $-2 \cdot y$ and then add the positive number 3.
(b) $8 + 2d - e$ means: Add together the positive number 8, the positive product $2 \cdot d$, and the negative number $-e$.

In the indicated multiplication of two or more algebraic numbers, each number is called a **factor**. Thus, in the indicated multiplication $3xy$, 3, x, and y are factors.

An indicated multiplication in an expression, which is to be added to another indicated multiplication in the same expression, is called a **term**. Thus, *a term of an algebraic expression is an addend (a number or quantity to be added to another) consisting of factors only*. For instance, in the expression $7x - 3m + 2n/k - 5$, there are four terms, $7x$, $-3m$, $+2n/k$, and -5. The student should notice that all four terms in this expression are signed addends and that they are indicated multiplications. Notice that $2n/k$ is also an indicated multiplication since $2n/k = 2n \cdot 1/k$.

EXAMPLE 4-5 Give the number of terms in each of the following expressions, name them, and state why they are terms.

$$\text{(a) } b + \frac{3}{c} + 2k \qquad \text{(b) } 5x - \frac{2y}{3} - 3z + 8$$

SOLUTION

(a) There are three terms, b, $\dfrac{3}{c}$, and $2k$. All three are algebraic addends and are indicated multiplications because $b = b \cdot 1$, $\dfrac{3}{c} = 3 \cdot \dfrac{1}{c}$, and $2k = 2 \cdot k$.

(b) There are four terms, $5x$, $-\dfrac{2y}{3}$, $-3z$, and $+8$. All four are algebraic addends and are indicated multiplications because $5x = 5 \cdot x$, $-\dfrac{2y}{3} = -2y \cdot \dfrac{1}{3}$, $-3z = -3 \cdot z$, and $8 = 8 \cdot 1$.

When an expression contains only one term, it is called a **monomial**. When an expression contains two terms, it is called a **binomial**, and when it contains three terms, it is called a **trinomial**. An expression containing two or more terms can also be called a **polynomial**. Thus, 8, $2x$, $\dfrac{3}{a}$, are all monomials, $2x + y$, $7k - n$, $\dfrac{a}{b} - c$, are all binomials, and $x + y - 3z$ and $3x - \dfrac{1}{3}d + R$ are both trinomials.

When the same factor appears more than once in a term, the indicated multiplication of such factors is written in a special way. For instance, $a \cdot a \cdot a$ is written a^3 (read "a to the third power" or "a cube"), and x^4 (read "x to the fourth power") means $x \cdot x \cdot x \cdot x$. In the symbol a^3, 3 is called the **exponent**, and a is called the **base**. The exponent shows how many times the base has to be used as a factor.

EXAMPLE 4-6   What does each of the following mean?

$$\text{(a) } x^2 \qquad \text{(b) } a^5 \qquad \text{(c) } a^4$$

SOLUTION

(a) $x^2 = x \cdot x$
(b) $a^5 = a \cdot a \cdot a \cdot a \cdot a$
(c) $a^4 = a \cdot a \cdot a \cdot a$

EXAMPLE 4-7   Write each of the following products as a power:

$$\text{(a) } m \cdot m \cdot m \qquad \text{(b) } d \cdot d \qquad \text{(c) } r \cdot r$$

SOLUTION

(a) $m \cdot m \cdot m = m^3$
(b) $d \cdot d = d^2$
(c) $r \cdot r = r^2$

A constant present as a factor in a term is always written as the first factor, and is called the **numerical coefficient** of the term. Thus, in $3ab$, 3 is the numerical coefficient. In x^2y, the numerical coefficient is 1, in $-5abc$ the numerical coefficient is -5, and in $-ax^3y^2$, the numerical coefficient is -1. The student should notice that if no constant appears in a term, then the numerical coefficient is understood to be either 1 or -1 depending on the sign of the term.

There are also literal coefficients and in axy, a is the literal coefficient of xy, x is the literal coefficient of ay, and y is the literal coefficient of ax.

EXAMPLE 4-8   Name the numerical coefficient in each of the following:

$$\text{(a) } -3d^2e \qquad \text{(b) } abx^2 \qquad \text{(c) } -t^2s$$

SOLUTION

(a) The numerical coefficient in $-3d^2e$ is -3.
(b) The numerical coefficient in abx^2 is 1.
(c) The numerical coefficient in $-t^2s$ is -1.

Algebraic Terms and Definitions

EXAMPLE 4-9 What is the coefficient of each of the factors in the monomial ABk?

SOLUTION

The coefficient of A is Bk.
The coefficient of B is Ak.
The coefficient of k is AB.

Often we talk about **the degree of a polynomial**. We say, for instance, that the polynomial $3x^4 + 2x^3 + x - 1$ is of degree 4, because 4 is the highest power of any variable in the polynomial. In general, *the degree of a polynomial is the degree of the variable in the polynomial with the highest power.* Thus, the polynomial $2x^2 + 3x + 5$ is of degree 2, and the polynomial $5x^3 + 7x^2 - 2x + 8$ is of degree 3.

If the literal factors of two or more terms of an expression are exactly the same (base and exponent), then the terms are called **similar terms**. Thus, in the polynomial $2x^3 + 5xy - 3x^3 - 7x^2y$, the first and the third terms are similar, but the second and the fourth terms are not. Notice that the numerical coefficients of the first and the third are different, but this makes no difference, and the two terms are similar because the literal factors in both terms is x^3. This is not the case with the literal factors of the second and the fourth terms. Therefore, these two terms are not similar. Here are three more examples:

In $2x + x$	*(both terms are similar)*
In $5m^2k - 3mk^2$	*(there are no similar terms)*
In $7av^2w - 3avw^2 + av^2w$	*(only the first and the third terms are similar)*

EXERCISE 4-1

What does each of the following mean?

1. $\dfrac{2a}{3m}$ **2.** $3d + k$ **3.** $5x - y$ **4.** $5aty$

Indicate each of exercises 5–10 algebraically using algebraic symbols and numbers.

5. Three times a number plus twice a second number.
6. The product of two numbers decreased by a third number.
7. The quotient of two numbers increased by three times their product.
8. The sum of five times a number and two times a second number.
9. One-fourth the difference of k and 3.
10. Five times m divided by 2.
11. Distinguish between a variable and a constant.
12. What is a term?
13. What is a factor?
14. What is a base? An exponent?
15. What is a numerical coefficient?
16. What is a polynomial?

State the number of terms in each of the following expressions:
17. $ab + b$ **18.** $3x + 2y - 7z$

Introduction to Algebra. Basic Ideas

19. $\dfrac{x}{y} - \dfrac{4}{xy}$

20. $5k + 7m - \dfrac{21k}{a} + 2$

Change each of the following fractions (indicated divisions) to indicated multiplications:

21. $\dfrac{5}{a}$

22. $\dfrac{7m}{4}$

23. $\dfrac{15xy}{7d}$

24. $\dfrac{1}{3x}$

25. Given the expression $3k + xy^2 + 8abc - w$, state the numerical coefficient of (a) the first term, (b) the second term, (c) the fourth term.

26. The formula for the volume of a cone is $V = \frac{1}{3}\pi r^2 h$. State which numbers in the formula are constants and which factors are variables.

Which of the following statements are true and which are false?

27. $a^5 = 5a$

28. $x^2 = x \cdot x$

29. $t^2 \neq 2t$

30. $R \cdot R \cdot R \neq 3R$

31. $\dfrac{k^2}{5} = \dfrac{1}{5} \cdot k^2$

32. Write the product $k \cdot k \cdot k \cdot k \cdot k \cdot k$ as a power of k.

33. Which of the following three groups of terms contains all like terms?
  a. $3xy^2 - 3x^2y + 3x^2y^2$
  b. $x^3y^2 + x^2y^3 - xy$
  c. $-4xy + xy + 7xy$

34. Which of the following three groups of terms contains only unlike terms?
  a. $-3a^2c^2 - 3a^2c^2$
  b. $7a^2c^2 + 8a^2 + c^2$
  c. $-5ac + ac$

4-3 Properties of the Algebraic Numbers

We have seen that the numbers of algebra are the numbers of the number line, and that they are called real numbers. We will state now without formal justification that the properties of the real numbers are the same as those of the whole numbers of Sect. 3-2.

Below are the properties of the whole numbers in their more general form. They are now called the properties of the real numbers.

1. *Commutative Law of Addition*

$$a + b = b + a$$

2. *Commutative Law of Multiplication*

$$ab = ba$$

3. *Associative Law of Addition*

$$a + (b + c) = (a + b) + c$$

4. *Associative Law of Multiplication*

$$a(bc) = (ab)c$$

5. *The Distributive Law*

$$a(b + c) = ab + ac$$

4-4 Grouping and Grouping Symbols

Grouping is very frequently used in mathematics. The most commonly used grouping symbols are parentheses ( ), brackets [ ], and braces { }. The fraction bar, (——), also is a grouping symbol, and is equivalent to parentheses.

When more than one of the above grouping symbols are used in an expression, the common practice is to have the parentheses as the innermost grouping symbol. The parentheses are followed by the brackets, which in turn are followed by the braces, as the outer grouping symbol. This is illustrated in the following expressions:

$$\text{(a)} \quad 2 - [x + (3 - 2d)] \qquad \text{(b)} \quad 3\{w - a[a - b(x + k)]\}$$

The grouping symbols are used to indicate that the numbers, or terms which they enclose are to be treated as a whole unit. Let us illustrate this. In the expression $12 - [3 + (2 - 1)]$, the number enclosed by the parentheses is equivalent to 1, and the number enclosed by the brackets is equivalent to 4.

Parentheses are used to separate the signs of operation from the signs of numbers, as we have already done. Thus, we have seen that $2x + (-3) - (-7d)$ means: Add together the positive product $2x$ and the negative number -3, and from that sum, subtract the negative product $-7d$.

Parentheses are used also to enclose numbers which are to be multiplied. Thus, $6(-5)$ means 6 times -5. Also, $3(4 + a)$ (read "three times the quantity 4 plus a") means 3 times the sum of 4 and a. We have seen that this is the case of the Distributive Law, and

$$3(4 + a) = 3 \cdot 4 + 3 \cdot a$$
$$= 12 + 3a$$

EXAMPLE 4-10   Perform the indicated operations for the expression $a(t + a)$.

SOLUTION   This is a case of the Distributive Law. Therefore,

$$a(t + a) = a \cdot t + a \cdot a$$
$$= at + a^2$$

EXAMPLE 4-11   Evaluate each of the following:

$$\text{(a)} \quad (-3)^2 \qquad \text{(b)} \quad -3^2 \qquad \text{(c)} \quad (-2)^2$$

SOLUTION

(a) The base is -3. Therefore, $(-3)^2 = (-3)(-3) = 9$.
(b) Notice that the base in this case is 3, not -3. Therefore, $-3^2 = -(3)(3) = -9$.
(c) The base is -2. Therefore, $(-2)^2 = (-2)(-2) = 4$.

Finally, parentheses are used to make clear an intended meaning in an expression. For instance, look at the two expressions

$$25 - 2 \cdot 3 \qquad \text{and} \qquad (25 - 2) \cdot 3$$

The numbers in both expressions are the same, but the second expression, because of the parentheses, has a different meaning from the first. Thus,

$$25 - 2 \cdot 3 = 25 - 6$$
$$= 19$$

and

$$(25 - 2) \cdot 3 = 3 \cdot (25 - 2) \quad \textit{(by the Commutative Law of Multiplication)}$$
$$= 3 \cdot 25 - 3 \cdot 2 \quad \textit{(by the Distributive Law)}$$
$$= 75 - 6$$
$$= 69$$

It is important to remember that *when parentheses are not used with signed numbers, the indicated multiplications in a term (division also is an indicated multiplication) must be performed first before the term can be added (terms are addends).*

EXAMPLE 4-12   Evaluate each of the following expressions:

(a) $23 - 3 \cdot 5$ (b) $32 + 16 \div 2$

SOLUTION

(a) $23 - 3 \cdot 5 = 23 - 15$ (b) $32 + 16 \div 2 = 32 + 8$
              $= 8$ $= 40$

EXERCISE 4-2

Name the property of the real numbers illustrated by each of the following:
1. $a + (x + 2) = (a + x) + 2$ 2. $x + k = k + x$
3. $3(x + m) = 3x + 3m$ 4. $st = ts$
5. $g(hk) = (gh)k$
Perform the indicated operations in each of the following:
6. $4(2 + t)$ 7. $3(4 + v + x)$ 8. $5(a + 2)$
9. $s(r + t)$ 10. $t(t + u)$
Evaluate each of the following:
11. $(-2)^4$ 12. -2^4 13. -2^6
14. $(-8)^2$ 15. -5^2 16. $(-5)^2$
17. $(-1)^8$ 18. -1^4 19. $41 - 6 \cdot 3$
20. $11 - 2 \cdot 4$ 21. $28 + 8 \div 4$ 22. $5 \cdot 8 - 31$
23. $54 + 4 \cdot 3$ 24. $54 - 18 \div 3 + 7 \cdot 2$ 25. $68 + 3 \cdot 13 - 4 \div 2$
26. $72 - 6 \cdot 11$ 27. $102 + 96 \div 12$ 28. $41 - 8 \div 4$

4-5 Laws of Exponents

We have seen in Sect. 4-2 that when a factor is to be multiplied by itself several times, the product is written in the form of a power. Thus, we have defined a^3 to mean $a \cdot a \cdot a$. In general terms, a^m means: Multiply a times itself m times. We have called a the base and m the exponent. The five laws of exponents which are introduced below follow from this basic definition.

Suppose we want to multiply x^2 and x^3. We know that

$$
\begin{aligned}
x^2 \cdot x^3 &= (x \cdot x)(x \cdot x \cdot x) \quad \textit{(by definition)} \\
&= x \cdot x \cdot x \cdot x \cdot x \quad \textit{(by multiplication)} \\
&= x^5 \quad\quad\quad\quad\quad\quad \textit{(by definition)} \\
&= x^{2+3}
\end{aligned}
$$

Therefore, *to multiply two or more powers of the same base, raise the common base to the sum of the exponents.* The algebraic form of *Law 1* is

$$
a^m \cdot a^n = a^{m+n}
$$

Suppose we want to raise x^3 to the second power. We know that

$$
\begin{aligned}
(x^3)^2 &= (x^3)(x^3) \quad \textit{(by definition)} \\
&= x^{3+3} \quad\quad\; \textit{(by Law 1)} \\
&= x^6 \\
&= x^{3 \cdot 2}
\end{aligned}
$$

Therefore, *to raise a power to another power, raise the base to the product of the exponents.* The algebraic form of *Law 2* is

$$
(a^m)^n = a^{mn}
$$

Suppose we want to raise the product xy to the third power. We know that

$$
\begin{aligned}
(xy)^3 &= (xy)(xy)(xy) \quad\quad\quad\; \textit{(by definition)} \\
&= x \cdot y \cdot x \cdot y \cdot x \cdot y \quad \textit{(by multiplication)} \\
&= x \cdot x \cdot x \cdot y \cdot y \cdot y \quad \textit{(by the Commutative Law} \\
&\quad\quad\quad\quad\quad\quad\quad\quad\quad\; \textit{of Multiplication)} \\
&= x^3 y^3 \quad\quad\quad\quad\quad\quad\; \textit{(by definition)}
\end{aligned}
$$

Therefore, *to raise the product of two or more factors to a power, the exponent of each factor is multiplied by the exponent of the power to which the product is raised.* The algebraic form of *Law 3* is

$$
(ab)^m = a^m b^m
$$

NOTE: When a factor has no exponent, the exponent of that factor is understood to be 1.

Suppose we want to raise the fraction x/y to the second power. We know that

$$\left(\frac{x}{y}\right)^2 = \left(\frac{x}{y}\right)\left(\frac{x}{y}\right) \quad \text{(by definition)}$$

$$= \frac{x \cdot x}{y \cdot y} \quad \text{(by multiplication)}$$

$$= \frac{x^2}{y^2} \quad \text{(by definition)}$$

Therefore, *to raise a fraction to a power, raise both terms of the fraction to that power.* The algebraic form of *Law 4* is

$$\left(\frac{a}{b}\right)^m = \frac{a^m}{b^m}$$

In the division of powers of the same base, there are three cases.
(a) Suppose we want to divide x^5 by x^3. We know that

$$\frac{x^5}{x^3} = \frac{x \cdot x \cdot x \cdot x \cdot x}{x \cdot x \cdot x} \quad \text{(by definition)}$$

$$= \frac{\cancel{x} \cdot \cancel{x} \cdot \cancel{x} \cdot x \cdot x}{\cancel{x} \cdot \cancel{x} \cdot \cancel{x}} \quad \begin{array}{l}\text{(by the fundamental principle}\\ \text{of fractions)}\end{array}$$

$$= x^2 \quad \text{(by definition)}$$

$$= x^{5-3}$$

The algebraic form of *Law 5* is

$$\frac{a^m}{a^n} = a^{m-n}$$

(b) Suppose that we want to divide x^5 by x^5. We know that

$$\frac{x^5}{x^5} = x^{5-5} \quad \text{(by Law 5)}$$

$$= x^0$$

We also know that $\frac{x^5}{x^5} = 1$.

Therefore, since $\frac{x^5}{x^5} = x^0 = 1$, $x^0 = 1$. This is the case of Law 5 in which the exponents m and n are equal, and the formula of Law 5 takes the form

$$\frac{a^m}{a^m} = a^{m-m}$$

But the left side is equal to 1, and the right side is a^0, therefore, $1 = a^0$ and $a^0 = 1$.

(c) Suppose we want to divide x^3 by x^5. We know that

$$\frac{x^3}{x^5} = x^{3-5} \quad (by \ Law \ 5)$$

$$= x^{-2}$$

Also

$$\frac{x^3}{x^5} = \frac{x \cdot x \cdot x}{x \cdot x \cdot x \cdot x \cdot x} \quad (by \ definition)$$

$$= \frac{\not{x} \cdot \not{x} \cdot \not{x}}{\not{x} \cdot \not{x} \cdot \not{x} \cdot x \cdot x} \quad \begin{array}{l}(by \ the \ fundamental \ principle \\ of \ fractions)\end{array}$$

$$= \frac{1}{x^2} \quad (by \ definition)$$

Therefore, since $\dfrac{x^3}{x^5} = x^{-2} = \dfrac{1}{x^2}$,

$$x^{-2} = \frac{1}{x^2}$$

This is the case of Law 5 in which $m < n$. Notice that fraction $\dfrac{a^m}{a^n}$ of the formula of Law 5 can be written as the product $a^m \cdot \dfrac{1}{a^n}$, and the formula can take successively the following three forms:

$$a^m \cdot \frac{1}{a^n} = a^{m-n}$$

$$a^m \cdot \frac{1}{a^n} = a^{m+(-n)}$$

$$a^m \cdot \frac{1}{a^n} = a^m \cdot a^{-n}$$

It should be evident at this point that $\dfrac{1}{a^n}$ and a^{-n} denote the same thing since $a^m \cdot \dfrac{1}{a^n}$ and $a^m \cdot a^{-n}$ are equal. Therefore, we can say that $a^{-n} = \dfrac{1}{a^n}$.

Law 5 can be summarized as follows. *The quotient of two powers of the same base is equal to the common base raised to the difference of the exponents. If this difference is zero or negative, the resulting power should be interpreted according to $a^0 = 1$, or $a^{-x} = \dfrac{1}{a^x}$, respectively.*

For easy reference, the Laws of Exponents are given below in compact form:

LAWS OF EXPONENTS

1. $a^m a^n = a^{m+n}$
2. $(a^m)^n = a^{mn}$
3. $(ab)^m = a^m b^m$

4. $\left(\dfrac{a}{b}\right)^m = \dfrac{a^m}{b^m}$ $(b \neq 0)$

5. $\dfrac{a^m}{a^n} = \begin{cases} a^{m-n} & \text{when } m > n \\ 1 & \text{when } m = n \\ \dfrac{1}{a^{n-m}} & \text{when } m < n \end{cases}$ $(a \neq 0)$

The student should realize that the laws of exponents are only for powers with the same base, and that they are used only for products and quotients of powers; not for sums or differences.

EXAMPLE 4-13 Simplify each of the following:

(a) $x^5 x^4$ (b) $t^2 t^3 t$ (c) $k^2 k^4$

SOLUTION

(a) $x^5 x^4 = x^{5+4} = x^9$
(b) $t^2 t^3 t = t^{2+3+1} = t^6$
(c) $k^2 k^4 = k^{2+4} = k^6$

EXAMPLE 4-14 Simplify each of the following:

(a) $(x^5)^2$ (b) $(d^3)^4$

SOLUTION

(a) $(x^5)^2 = x^{5 \cdot 2} = x^{10}$ (b) $(d^3)^4 = d^{3 \cdot 4} = d^{12}$

EXAMPLE 4-15 Simplify each of the following:

(a) $(m^3 n k^2)^5$ (b) $(v^4 w^2)^3$

SOLUTION

(a) $(m^3 n k^2)^5 = m^{3 \cdot 5} n^5 k^{2 \cdot 5} = m^{15} n^5 k^{10}$
(b) $(v^4 w^2)^3 = v^{4 \cdot 3} w^{2 \cdot 3} = v^{12} w^6$

EXAMPLE 4-16 Simplify each of the following:

(a) $\left(\dfrac{u}{v}\right)^5$ (b) $\left(\dfrac{d^2}{g^3}\right)^5$ (c) $\left(\dfrac{os^2}{t^3 u}\right)^2$

SOLUTION

(a) $\left(\dfrac{u}{v}\right)^5 = \dfrac{u^5}{v^5}$

(b) $\left(\dfrac{d^2}{g^3}\right)^5 = \dfrac{d^{2 \cdot 5}}{g^{3 \cdot 5}} = \dfrac{d^{10}}{g^{15}}$

(c) $\left(\dfrac{os^2}{t^3 u}\right)^2 = \dfrac{o^2 s^4}{t^6 u^2}$

EXAMPLE 4-17 Simplify each of the following:

(a) $\dfrac{g^6}{g^2}$ (b) $\dfrac{k^3}{k^3}$ (c) $\dfrac{z^2}{z^6}$

(d) $\dfrac{a^2 x^3 z}{a^3 x z^5}$ (e) $(4k)^0 + (5m)^0 + x^0$ (f) $4k^0 + 5m^0 + x^0$

SOLUTION

(a) $\dfrac{g^6}{g^2} = g^{6-2} = g^4$

(b) $\dfrac{k^3}{k^3} = k^{3-3} = k^0 = 1$

(c) $\dfrac{z^2}{z^6} = z^{2-6} = z^{-4} = \dfrac{1}{z^4}$

(d) $\dfrac{a^2x^3z}{a^3xz^5} = \dfrac{a^2}{a^3} \cdot \dfrac{x^3}{x} \cdot \dfrac{z}{z^5} = \dfrac{1}{a} \cdot x^2 \cdot \dfrac{1}{z^4} = \dfrac{x^2}{az^4}$

(e) $(4k)^0 + (5m)^0 + x^0 = 1 + 1 + 1 = 3$

(f) $4k^0 + 5m^0 + x^0 = 4 + 5 + 1 = 10$

EXERCISE 4-3

What does each of the following mean?

1. a^3 **2.** 4^5 **3.** y^2

Write each of the following multiplications using exponents:

4. $t \cdot t \cdot t$ **5.** $w \cdot w \cdot w \cdot w \cdot w$

6. $x \cdot x \cdot x \cdot y \cdot y$ **7.** $d \cdot k \cdot k \cdot k \cdot d$

Multiply each of the following:

8. $3^3 \cdot 3^5$ **9.** $2^2 \cdot 2^4$ **10.** $a^7 \cdot a^2$

11. $m^2 \cdot h^3 \cdot m^3 \cdot h^5$ **12.** $x^4 \cdot x^3 \cdot y^2 \cdot x^5$

Divide each of the following:

13. $\dfrac{3^5}{3^2}$ **14.** $\dfrac{2^4}{2^3}$ **15.** $\dfrac{k^2}{k^3}$ **16.** $\dfrac{t^7}{t^7}$

17. $\dfrac{v^3}{v^8}$ **18.** $\dfrac{a^5b^2}{a^3b^4}$ **19.** $\dfrac{k^3m^4}{km^6}$ **20.** $\dfrac{v^2w^3x^2}{v^3wx^5}$

21. $\dfrac{t^7v^3w^3}{t^2v^4w^2}$ **22.** $\dfrac{x^2y^7z}{x^3y^3z^2}$ **23.** $\dfrac{u^2w^3x}{u^5v^2w^2}$

Simplify each of the following:

24. $(3^2)^3$ **25.** $(2^2)^2$ **26.** $(v^3)^7$

27. $(m^2n^3k)^5$ **28.** $(x^2y^3)^0$ **29.** $(2v)^0 + (7w)^0 + x^0$

30. $7x^0 + 8y^0 + 2z^0$

Simplify each of the following:

31. $\left(\dfrac{3^2}{5^3}\right)^2$ **32.** $\left(\dfrac{w^3}{y^5}\right)^3$ **33.** $\left(\dfrac{x^5}{z^3}\right)^4$

4-6 Combining Similar Terms

We have seen that terms with identical literal parts are similar terms. An expression in which such terms occur is simplified when the similar terms are combined. *Similar terms are combined when their numerical coefficients are added and their algebraic sum is placed as the numerical coefficient of their common literal factors.* Remember that the numerical coefficients are signed numbers, and that the signs which precede them indicate only whether they are positive or negative numbers. The operation is understood to be addition in all cases.

EXAMPLE 4-18 Simplify the following expressions by adding similar terms:

(a) $5ab + 2ab + 4ab$
(b) $3xy + 5xy - 4xy$
(c) $2kw + 2mv - 5kw - 4mv$
(d) $3a^2b + 2ab^2 - 5a^2b + 7ab^2$

SOLUTION

(a) $5ab + 2ab + 4ab = 11ab$
(b) $3xy + 5xy - 4xy = 4xy$
(c) $2kw + 2mv - 5kw - 4mv = -3kw - 2mv$
(d) $3a^2b + 2ab^2 - 5a^2b + 7ab^2 = -2a^2b + 9ab^2$

EXERCISE 4-4

Simplify each of the following by combining similar terms:

1. $3t + 2t$ **2.** $2g - 5g$
3. $8x + 2x - 7x$ **4.** $5u - 8u + 2u$
5. $-y + 2y - 7y + 13y$ **6.** $11m - 5m + m - 4m$
7. $-3w + 8w + 5w - w$ **8.** $2x^2 + 5x^2 - x^2 - 3x^2$
9. $k^3 + 3k^3 - 2k^3$ **10.** $16z^5 + 2z^5 - 8z^5 - z$
11. $x^2y - xy^2 + 5x^2y$ **12.** $a^2b^3 - a^3b^2 + 7a^3b^2 + 8a^2b^3$
13. $3d + 6e - 2d + 3f - 5e + 7f$ **14.** $d^2 + de + 5d^2 - 3de - 8$
15. $vw^2z^3 - v^2wz^3 + 3vw^2z^3 + 8v^2wz^3$ **16.** $2km^2 + 3k^2m - 5km^2 + 7k^2m - 9$
17. $6gt^3 - 8k^2m + 4 - 2gt^3 + 3k^2m - 5$
18. $16x^2 - 5x^2y^3 - 11 + 6x^2y^3 + 8$
19. $4x^2 - 3xy + 3y^2 - 5x - 3x^2 + xy - 5y^2 - 2$
20. $5a + 6b - 4c - 3b - a - 3c - 2ab - 5$

4-7 Removing Grouping Symbols

We have seen that in certain cases the use of grouping symbols is necessary. To reduce an expression to its simplest form however, the grouping symbols must be removed and the indicated operations performed.

Suppose we are asked to remove the parentheses from the expression $3(5x + 2y) - 2(3x - y)$. This is not a totally unfamiliar problem because we recognize in the expression two cases of the distributive law. Applying the distributive law, we have $3(5x + 2y) - 2(3x - y) = 15x + 6y - 6x + 2y$. Notice now that the parentheses are gone and that the only thing that is left to do in order to get the original expression in its simplest form is to collect similar terms. Thus, $15x + 6y - 6x + 2y = 9x + 8y$.

It is suggested by the above illustration that *to remove a grouping symbol which is grouping together certain terms, multiply each term of the group by the coefficient of that group.* When there is no coefficient before the grouping symbol, it is understood to be 1 or -1, according to whether the sign in front of the grouping symbol is plus or minus. Remember that this procedure is nothing more than the application of the distributive law, and that we are multiplying signed numbers (explicit or literal).

EXAMPLE 4-19 Remove the parentheses and simplify each of the following:

$$\text{(a) } 4(a + b - c) - 3(a + c) \qquad \text{(b) } 3x + 2(x + 3y - 7z)$$

SOLUTION

(a) $4(a + b - c) - 3(a + c) = 4a + 4b - 4c - 3a - 3c$ (*by the Distributive Law*)

$$= a + 4b - 7c \qquad \text{(*by combining similar terms*)}$$

(b) $3x + 2(x + 3y - 7z) = 3x + 2x + 6y - 14z$ (*by the Distributive Law*)

$$= 5x + 6y - 14z \qquad \text{(*by combining similar terms*)}$$

EXAMPLE 4-20 Remove the parentheses and simplify each of the following:

$$\text{(a) } 3P + 2Q + (P - Q + 6R) \qquad \text{(b) } 5 + m - (k - m + 3n)$$

SOLUTION

(a) $3P + 2Q + (P - Q + 6R) = 3P + 2Q + P - Q + 6R$ (*by multiplying each term enclosed by the parentheses by* $+1$, *the coefficient of the group of terms in the parentheses*)

$$= 4P + Q + 6R \qquad \text{(*by combining similar terms*)}$$

(b) $5 + m - (k - m + 3n) = 5 + m - k + m - 3n$ (*by multiplying each term enclosed by the parentheses by* -1, *the coefficient of the group of terms in the parentheses*)

$$= 5 + 2m - k - 3n \qquad \text{(*by combining similar terms*)}$$

From example 4-20, it is easy to see that the following shortcut is possible:

1. *A grouping symbol preceded by a plus sign may be removed as long as the sign of each term within the grouping symbol remains the same.*
2. *A grouping symbol preceded by a minus sign may be removed as long as the sign of each term within the grouping symbol is changed.*

EXAMPLE 4-21 Remove the parentheses and simplify each of the following:

$$\text{(a) } 3d + (d - 2k + 5) \qquad \text{(b) } 6d - (3d + 4k - 7m)$$

SOLUTION

(a) $3d + (d - 2k + 5) = 3d + d - 2k + 5$ (*by removing the parentheses without changing the signs of the terms enclosed by the parentheses*)

$$= 4d - 2k + 5 \qquad \text{(*by combining similar terms*)}$$

(b) $6d - (3d + 4k - 7m) = 6d - 3d - 4k + 7m$ (*by removing the parentheses and changing the signs of the terms enclosed by the parentheses*)

$$= 3d - 4k + 7m \qquad \text{(*by combining similar terms*)}$$

Introduction to Algebra. Basic Ideas

EXAMPLE 4-22 Remove the parentheses and simplify each of the following:

(a) $3a^2 - 5ab - 2a(a - 7b - 8)$
(b) $8x^2 + 3xy + 5x(2x - 3xy - z)$

SOLUTION

(a) $3a^2 - 5ab - 2a(a - 7b - 8) = 3a^2 - 5ab - 2a^2 + 14ab + 16a$

$$= a^2 + 9ab + 16a$$

(b) $8x^2 + 3xy + 5x(2x - 3xy - z) = 8x^2 + 3xy + 10x^2 - 15x^2y - 5xz$

$$= 18x^2 + 3xy - 15x^2y - 5xz$$

Suppose we are asked to remove the grouping symbols and to simplify the expression $3x - [5x - (6 - 2x)]$. Here we know at least how to remove the parentheses and when we proceed to do that, we have

$$3x - [5x - (6 - 2x)] = 3x - [5x - 6 + 2x]$$
$$= 3x - [7x - 6]$$

At this point, however, $3x - [7x - 6]$ is not totally unfamiliar because the brackets are playing the role of parentheses. If we treat the brackets as parentheses, we have

$$3x - [7x - 6] = 3x - 7x + 6$$
$$= -4x + 6$$

We conclude, then, from the above illustration, that, *in expressions where two or more different kinds of grouping symbols are used, first, the innermost grouping symbol is removed, using the known rule. Then, the innermost symbol of what remains is removed, and so on. All along, combine similar terms as soon as a grouping symbol is removed.*

EXAMPLE 4-23 Remove the grouping symbols from each of the following:

(a) $-13x - [8x - (5 - 2x)]$
(b) $8k - \{4 - [-3k(2 + m)]\}$
(c) $5 - \{3 - [2 - (2 - x) - (5 - x)]\}$

SOLUTION

(a) $-13x - [8x - (5 - 2x)] = -13x - [8x - 5 + 2x]$

$$= -13x - [10x - 5]$$
$$= -13x - 10x + 5$$
$$= -23x + 5$$

(b) $8k - \{4 - [-3k(2 + m)]\} = 8k - \{4 - [-6k - 3km]\}$

$$= 8k - \{4 + 6k + 3km\}$$
$$= 8k - 4 - 6k - 3km$$
$$= 2k - 4 - 3km$$

(c) $5 - \{3 - [2 - (2 - x) - (5 - x)]\} = 5 - \{3 - [2 - 2 + x - 5 + x]\}$
$$= 5 - \{3 - [2x - 5]\}$$
$$= 5 - \{3 - 2x + 5\}$$
$$= 5 - \{8 - 2x\}$$
$$= 5 - 8 + 2x$$
$$= -3 + 2x$$

EXERCISE 4-5

Remove the grouping symbols and combine similar terms in each of the following:

1. $(17 - 6) - 8$ 2. $(8 - 3) - (2 - 5)$
3. $17 - (2 + 9)$ 4. $23 - (18 - 2)$
5. $74 - 3(8 - 5)$ 6. $-27 + 2(5 - 3)$
7. $3(2h - m) - 2(3h + m)$ 8. $2(3x + 2y) - 4(5x - 3y)$
9. $-3c(c^2 - 2cd - d^2) + 2c(c^2 - 5cd - d^2)$
10. $b[b - 3(c - 2d) + 4] + 2b[2x - 2(b - d)]$
11. $m[-2m + 3(m + t) + m] - 2m[m - 2(m - t) - 3m]$
12. $2x - \{4x + 3[2x - 2(3x + 2u) - (x - y) - 7(x + y)] - (7x + 3y)\}$
13. $3d - \{7d + [(5d - 4) - 2(4d - 7) - (d + 1)] - 3\}$
14. $7k^2 - \{3km - [2k(k - m) - 5m(k + 2m)] + 5\}$

REVIEW QUESTIONS

1. State the three ways of indicating multiplication in algebra.
2. Distinguish between constants and variables.
3. What is an algebraic expression?
4. What is a term?
5. What is a polynomial?
6. What is a base? An exponent?
7. What is a numerical coefficient?
8. What is a literal coefficient?
9. What terms are called similar terms?
10. What is A^0 equal to?
11. What is A^{-x} equal to?
12. State the rule for combining similar terms.
13. State the rule for removing a grouping symbol.
14. State the rule for removing more than one kind of grouping symbol from expressions.

REVIEW EXERCISES

Indicate each of exercises 1–4 algebraically using algebraic symbols and numbers.

1. A number increased by five times a second number.
2. The second power of a number decreased by twice a second number.
3. The sum of a number and five times the difference of two other numbers.

Introduction to Algebra. Basic Ideas

4. The difference between a number and the quotient of two other numbers.

Name the property of the real numbers illustrated in each of the following:

5. $2a(3b + c - 5d) = 6ab + 2ac - 10ad$

6. $8a + (m + k) = (8a + m) + k$

7. $t + s = s + t$

8. $vw = wv$

9. $h(jk) = (hj)k$

Evaluate each of the following:

10. $(-5)^2$ **11.** -3^3 **12.** -7^2

13. -1^2 **14.** $2 + 3 \cdot 4$ **15.** $8 - 6 \div 2$

16. $26 - 28 \div 7$

Use the Laws of Exponents to simplify each of the following:

17. $2^3 \cdot 2^4$ **18.** $3^2 \cdot 3^4$ **19.** $a^5 \cdot a^2$

20. $g^2 \cdot h^3 \cdot g \cdot h^2$ **21.** $(3^2)^5$ **22.** $(km^3n)^2$

23. $\dfrac{t^2 s^3 v}{t^3 s v^2}$ **24.** $\left(\dfrac{c^2}{d^3}\right)^3$ **25.** $\left(\dfrac{g^3}{h^4}\right)^2$

Remove grouping symbols and simplify.

26. $(2a - 3d) - d$

27. $x - (2x + y)$

28. $38 - 3(a - 7)$

29. $k + 5(k - 4)$

30. $3a - [5a + (a - 2)]$

31. $2d - \{d + [3d - (-d - 8)]\}$

32. $59 - \{[x + 3y - 5(x - 2y)] - 2x + 5y\}$

5

Basic Operations
with Algebraic Expressions

5-1 Addition of Algebraic Expressions

Suppose we want to add the expressions $8w + 3x - 5y - 3$ and $7w - 4x + 3y - 2$. We understand that the four terms of the first expression have to be added to the four terms of the second expression. Expressing this with parentheses, we have

$$(8w + 3x - 5y - 3) + (7w - 4x + 3y - 2)$$

It should be evident now that to add these two expressions we must remove the grouping symbols and combine like terms. Thus,

$$(8w + 3x - 5y - 3) + (7w - 4x + 3y - 2) = 8w + 3x - 5y - 3$$
$$+ 7w - 4x + 3y - 2$$
$$= 15w - x - 2y - 5$$

It follows then, that if we want to add two or more expressions, we form a single expression from all the expressions to be added and combine similar terms.

EXAMPLE 5-1 Add $2x^3 - 3y^2 + 5x$, $5x^3 + y^2 - 7x$, and $x^3 - 6y^2 + 3x$.

SOLUTION Forming one expression, we have

$$2x^3 - 3y^2 + 5x + 5x^3 + y^2 - 7x + x^3 - 6y^2 + 3x$$

Combining similar terms, the sum is

$$8x^3 - 8y^2 + x$$

Instead of forming one long expression with a great number of terms, it is often much more convenient to use the vertical form of addition. Thus, *to add two or more expressions place them one under the other lining up similar terms in columns. Then, similar terms are combined to give the final sum.* Using this method, the expressions of example 5-1 can be written and added as follows:

$$2x^3 - 3y^2 + 5x$$
$$5x^3 + y^2 - 7x$$
$$\underline{x^3 - 6y^2 + 3x}$$
$$8x^3 - 8y^2 + x$$

EXAMPLE 5-2   Add $ab^2 + 2c^3d + 5$, $-3ab^2 - 2$, and $-c^3d + 4$.

SOLUTION

$$ab^2 + 2c^3d + 5$$
$$-3ab^2 - 2$$
$$\underline{ -c^3d + 4}$$
$$-2ab^2 + c^3d + 7$$

Add the expressions in each of the following problems:

1. $3x + 2y$, $5x - 7y$
2. $5a + 7b$, $4a - 3b$
3. $8t^2 - 9s$, $3t^2 - s$
4. $5w + 4x$, $9v - x$
5. $2g + 3h - k$, $4g - h + 8k$
6. $11g^2 + 8h^3 + k^2$, $5g + 3h^3 + k^2$
7. $2m + r + s + t$, $3m - r + s - 5t$
8. $2ax^2 + 3x + y - z$, $5ax^2 - 7x + 8y + 6z$
9. $9a^2 + 5ab - b$, $7ab + b$
10. $2c^2x - 3xy + z$, $11c^2x + 7xy - 3z + 5$
11. $3k^2 + 4m^2 - 3km + 5n^2$, $6k^2 - 3m^2 + 2km + n^2$, $-3k^2 + 5m^2 - 3km - 2n^2$
12. $2x - 3y + 4z$, $-3x - y - 2z$, $4x - 2y + 3z$
13. $3a^2b - 2ab - 5ab^2$, $4a^2b - 5ab + 7ab^2$, $4a^2b + 5ab - 3ab^2$
14. $5km - 3kn - 7mn$, $-2km - 8kn + 6mn$, $3km + 2kn + 4mn$
15. $ax^3 + 2by + 3$, $-3by - 4$, $-2ax^3 - by$
16. $5r + s$, $-r + 2t$, $4s - 7t$

5-2   Subtraction of Algebraic Expressions

Suppose we are asked to subtract the expression $2x + 4xy - d$ from the expression $12x - 2xy - 3d$. We understand that the three terms of the **subtrahend** (the expression to be subtracted) must be subtracted from the three terms of the **minuend** (the expression from which the subtrahend is subtracted). Expressing this with parentheses, we have

$$(12x - 2xy - 3d) - (2x + 4xy - d)$$

Again, we have to remove the parentheses and form one long expression. Since, however, the second parentheses are preceded by a minus sign, the signs of all the terms of the second expression must be changed. Thus,

Subtraction of Algebraic Expressions

$$(12x - 2xy - 3d) - (2x + 4xy - d) = 12x - 2xy - 3d - 2x - 4xy + d$$
$$= 10x - 6xy - 2d$$

It follows then, that if we want to subtract one expression from another, we form a single expression out of the two, with the sign of each term of the expression to be subtracted changed. The expression resulting after combining similar terms is the difference.

EXAMPLE 5-3 Subtract $7m + 2km - 3n^2$ from $3m - km - 5n^2$.

SOLUTION A single expression is formed with the sign of each term of the expression to be subtracted changed. Thus,

$$3m - km - 5n^2 - 7m - 2km + 3n^2 = -4m - 3km - 2n^2$$

Instead of forming one long expression with many terms, it is more convenient to use the vertical form of subtraction. Thus, *to subtract one expression from another, place the minuend over the subtrahend, lining up similar terms in columns. After this arrangement, change (mentally) the sign of every term of the subtrahend as the term is added to the corresponding term of the minuend.* Using this method, the subtraction of example 5-3 can now be written and performed as follows:

$$
\begin{array}{r}
3m - km - 5n^2 \\
7m + 2km - 3n^2 \\
\hline
-4m - 3km - 2n^2
\end{array}
$$

Notice that the first term of the result, $-4m$, is the algebraic sum of $3m$ and $-7m$. The second term of the result, $-3km$, is the algebraic sum of $-km$ and $-2km$. The third term of the result, $-2n^2$, is the algebraic sum of $-5n^2$ and $+3n^2$.

EXAMPLE 5-4 Subtract $a - 4b + 7c - 3$ from $9a - 2b + 3c$.

SOLUTION

$$
\begin{array}{r}
9a - 2b + 3c \phantom{{}- 3} \\
a - 4b + 7c - 3 \\
\hline
8a + 2b - 4c + 3
\end{array}
$$

EXAMPLE 5-5 Subtract $5g^2 + 7gt$ from $3g^2 - 3gt + v$.

SOLUTION

$$
\begin{array}{r}
3g^2 - 3gt + v \\
5g^2 + 7gt \phantom{{}+ v} \\
\hline
-2g^2 - 10gt + v
\end{array}
$$

EXERCISE 5-2

In each of the following problems subtract the second expression from the first:

1. $3d^2 + 4e^2, 2d^2 - e^2$
2. $7k^2 - 3k, -5k^2 - 2k$
3. $3x^2 + 2y + 7, 8x^2 - 3y$
4. $5x + 2y - 3, -3x - 3y - 7$

Basic Operations with Algebraic Expressions

5. $-8a + 3b - 5c$, $-9a + 4b - 6c$
6. $-5m + mn - n$, $3m - n$
7. $7h - 6k$, $-11h + 2k - 7$
8. $5d + 17e$, $-5e + g - 11h$
9. $4x - 3y$, $5x + 8y - 9$
10. $2j - m + 6n$, $j - m + 2n$
11. $-8a^2 + 3b - 4c$, $-11a^2 + 7b - 3c$
12. $5g^2 - 6h^2 + 9$, $3g^2 + 4h^2 - 4$

5-3 Multiplication of Algebraic Expressions

The simplest and most basic case of multiplication is multiplication of monomials. Consider the multiplication $(-2x^2y^3)(3x^4y^5)$. Using the commutative law for multiplication, the rule for multiplication of signed numbers, and the laws of exponents, we have

$$(-2x^2y^3)(3x^4y^5) = -2 \cdot 3 \cdot x^2 \cdot x^4 \cdot y^3 \cdot y^5 \quad \text{(by the Commutative Law of Multiplication)}$$

$$= -6 \cdot x^2 \cdot x^4 \cdot y^3 \cdot y^5 \quad \text{(by the rule for multiplication of signed numbers)}$$

$$= -6x^6y^8 \quad \text{(by the Law 1 of exponents)}$$

It follows from the above example that *to multiply two or more algebraic terms:*

1. *Find the product of their numerical coefficients. This product and its sign is the positive or negative numerical coefficient of the final product.*
2. *Multiply the literal parts of the terms being multiplied, using the laws of exponents. This product is the literal part of the final product.*

EXAMPLE 5-6 Multiply (a) $(3xy^3z)(-7x^3yz^4)$
(b) $(-2ab^2c^3)(-5a^2bc)$
(c) $(-m^3n)(-8k^2mn^4)(-km^3n)$

SOLUTION

(a) $(3xy^3z)(-7x^3yz^4) = -21x^4y^4z^5$
(b) $(-2ab^2c^3)(-5a^2bc) = 10a^3b^3c^4$
(c) $(-m^3n)(-8k^2mn^4)(-km^3n) = -8k^3m^7n^6$

To multiply a polynomial by a monomial, multiply every term of the polynomial by the monomial.

EXAMPLE 5-7 Multiply $a(g + h - k)$.

SOLUTION This is the familiar case of the distributive law. Thus,

$$a(g + h - k) = ag + ah - ak$$

EXAMPLE 5-8 Multiply $2ab^2(a^2b^3 - 3ab^5 + 4a^3b^3)$.

SOLUTION

$$2ab^2(a^2b^3 - 3ab^5 + 4a^3b^3) = 2a^3b^5 - 6a^2b^7 + 8a^4b^5$$

Multiplication of Algebraic Expressions

To multiply a polynomial by a polynomial:

1. *Multiply each of the terms of one polynomial by each of the terms of the other polynomial.*
2. *Combine similar terms of these products.*

If, for instance, we want to multiply the two binomials $5x - 2$ and $3a + b$, we multiply each term of the binomial $3a + b$ by $5x$, the first term of the binomial $5x - 2$. Then we multiply each term of the binomial $3a + b$ by -2, the second term of the binomial $5x - 2$. The procedure is shown below:

$$(5x - 2)(3a + b) = 5x \cdot 3a + 5x \cdot b - 2 \cdot 3a - 2 \cdot b = 15ax + 5bx - 6a - 2b$$

EXAMPLE 5-9   Multiply $(2a + 3c)(7x - 2y)$.

SOLUTION   We first multiply each of the terms of the binomial $(7x - 2y)$ by $2a$. Then we multiply each of the terms of the binomial $(7x - 2y)$ by $3c$, and then we combine the two products. Thus,

$$(2a + 3c)(7x - 2y) = 2a \cdot 7x + 2a(-2y) + 3c \cdot 7x + 3c(-2y)$$
$$= 14ax - 4ay + 21cx - 6cy$$

EXAMPLE 5-10   Multiply $(5d - 2)(3d + 2e - 4)$.

SOLUTION   We multiply each of the terms of the trinomial $(3d + 2e - 4)$ first by $5d$, then by -2, and finally we combine similar terms. Thus,

$$(5d - 2)(3d + 2e - 4) = 5d \cdot 3d + 5d \cdot 2e + 5d(-4) - 2 \cdot 3d - 2 \cdot 2e - 2(-4)$$
$$= 15d^2 + 10de - 20d - 6d - 4e + 8$$
$$= 15d^2 + 10de - 26d - 4e + 8$$

Polynomials can be multiplied using the vertical form of multiplication. Care should be taken in the partial products to line up similar terms in columns, to facilitate the combination of similar terms at the end. Here are two illustrations:

$$
\begin{array}{r}
2x^2 - x + 5 \\
x - 2 \\
\hline
2x^3 - x^2 + 5x \\
- 4x^2 + 2x - 10 \\
\hline
2x^3 - 5x^2 + 7x - 10
\end{array}
\qquad
\begin{array}{r}
7x + 3 \\
7x - 3 \\
\hline
49x^2 + 21x \\
- 21x - 9 \\
\hline
49x^2 - 9
\end{array}
$$

EXERCISE 5-3

Multiply each of the following:

1. $(3xy)(2xy)$
2. $(-2a^2b)(4ab^3)$
3. $3xy^2(x^2y + x)$
4. $2b(c^2d + 5)$
5. $7ab^2(3a - 2b - 3)$
6. $(h^2 - 3h + 2)(h + 3)$

Basic Operations with Algebraic Expressions

7. $(m^2 - mt + t^2)(m^2 + mt + t^2)$
9. $(5x - g)^2$
11. $(2a - b)(3a - 2b)$
13. $(2a - 3)(a + 2)(a - 1)$

8. $(2v + 3w)^2$
10. $(8a + 3b)(8a - 3b)$
12. $(2k - 3m)(3k^2 + 2km + m^2)$
14. $(x + 4)(3x - 1)(5x + 2)$

5-4 Special Products

Multiplications of certain types of polynomials occur frequently enough in algebra for us to give them special attention. The products of such multiplications are called **special products.** The purpose of this section is to make such multiplications easily recognizable, and to perform them without actual multiplication.

One special product is the familiar distributive law. This multiplication can really be performed by inspection. Thus,

$$5(x + 3) = 5 \cdot x + 5 \cdot 3$$
$$= 5x + 15$$

Also

$$a(x + y + 7) = ax + ay + 7a$$

The formula for the first case of the special products is

$$a(x + y + z) = ax + ay + az$$

Let us see now the product of the binomials $(x + 3)(x + 2)$, in which the first term in both binomials is the same variable with 1 as its numerical coefficient. To make our discussion of the binomials and their product easier, we will give to the terms of the binomials, the names shown below.

firsts

$$(x + 3)(x + 2)$$

seconds

By multiplying the binomials $(x + 3)(x + 2)$ with the known method, we have

$$(x + 3)(x + 2) = x \cdot x \quad + 2 \cdot x + 3 \cdot x \quad + 3 \cdot 2$$

$$(2 + 3)x$$

$$\begin{pmatrix} \text{product} \\ \text{of firsts} \end{pmatrix} \qquad \begin{pmatrix} \text{sum of} \\ \text{seconds} \end{pmatrix} \qquad \begin{pmatrix} \text{product of} \\ \text{seconds} \end{pmatrix}$$

$$x^2 \qquad\qquad +5x \qquad\qquad +6$$

We can say then that *the product of two binomials having the first term of each the same variable with 1 as its numerical coefficient is a trinomial. In that trinomial product,*

1. *The first term is the product of the "firsts," a perfect square,*
2. *The numerical coefficient of the second term is the algebraic sum of the "seconds," and*
3. *The third term is the product of the "seconds."*

Special Products

The formula for two such binomials is

$$(x + a)(x + b) = x^2 + (a + b)x + ab$$

EXAMPLE 5-11 Multiply $(x + 5)(x - 7)$.

SOLUTION

$$(x + 5)(x - 7) = x^2 + (5 - 7)x + 5(-7)$$
$$= x^2 - 2x - 35$$

EXAMPLE 5-12 Multiply $(x - 2)(x - 3)$.

SOLUTION

$$(x - 2)(x - 3) = x^2 + (-2 - 3)x + (-2)(-3)$$
$$= x^2 - 5x + 6$$

We might have to multiply two binomials in which the first terms of the binomials as well as the second terms of the binomials are the same. Below are the three possible cases:

(a) $(x + 5)(x + 5)$ or $(x + 5)^2$

(b) $(x - 5)(x - 5)$ or $(x - 5)^2$

(c) $(x + 5)(x - 5)$

If we apply the rule of the previous case, we have

(a) $(x + 5)(x + 5) = x^2 + (5 + 5)x + 5 \cdot 5$
$$= x^2 + 2 \cdot 5 \cdot x + 5^2$$
$$= x^2 + 10x + 25$$

(b) $(x - 5)(x - 5) = x^2 + (-5 - 5)x + (-5)(-5)$
$$= x^2 + 2(-5)x + (-5)^2$$
$$= x^2 - 10x + 25$$

(c) $(x + 5)(x - 5) = x^2 + (-5 + 5)x + 5(-5)$
$$= x^2 + (0)x - 5^2$$
$$= x^2 - 25$$

From the above examples we can generalize as follows:

1. *The square of the sum of two terms is equal to the square of the first term, plus twice the product of the two terms, plus the square of the second term.* The formula is

$$(x + a)^2 = x^2 + 2ax + a^2$$

2. *The square of the difference of two terms is equal to the square of the first term minus twice the product of the two terms, plus the square of the second term.* The

Basic Operations with Algebraic Expressions

formula is

$$(x - a)^2 = x^2 - 2ax + a^2$$

3. *The product of the sum and the difference of two terms is equal to the square of the first term minus the square of the second term.* The formula is

$$(x + a)(x - a) = x^2 - a^2$$

EXAMPLE 5-13 Multiply by inspection:

(a) $(x + 7)^2$ (b) $(x - 4)^2$ (c) $(x + 6)(x - 6)$

SOLUTION

(a) $(x + 7)^2$ is the square of the sum of two terms. Therefore,

$$(x + 7)^2 = x^2 + 2 \cdot 7 \cdot x + 7^2$$
$$= x^2 + 14x + 49$$

(b) $(x - 4)^2$ is the square of the difference of two terms. Therefore,

$$(x - 4) = x^2 - 2 \cdot 4 \cdot x + 4^2$$
$$= x^2 - 8x + 16$$

(c) $(x + 6)(x - 6)$ is the product of the sum and the difference of two terms. Therefore,

$$(x + 6)(x - 6) = x^2 - 6^2$$
$$= x^2 - 36$$

EXAMPLE 5-14 Multiply by inspection:

(a) $(3h - 4)^2$ (b) $(4R + a)^2$ (c) $(3c + d)(3c - d)$

SOLUTION

$$\text{(a)} \quad (3h - 4)^2 = (3h)^2 - 2 \cdot 12 \cdot h + 4^2$$
$$= 9h^2 - 24h + 16$$
$$\text{(b)} \quad (4R + a)^2 = (4R)^2 + 2 \cdot 4 \cdot aR + a^2$$
$$= 16R^2 + 8aR + a^2$$
$$\text{(c)} \quad (3c + d)(3c - d) = (3c)^2 - d^2$$
$$= 9c^2 - d^2$$

EXERCISE 5-4

Find each of the following special products using as little writing as possible.

1. $(m - 2)^2$
2. $(g + 4)^2$
3. $(c + 3)^2$
4. $(d + 8)(d - 8)$
5. $(t + 11)(t - 11)$
6. $(k - 9)^2$
7. $3(v + 5w - x + 2y)$
8. $3a(2d + 6e)$
9. $(a - 12)^2$
10. $(2t - a)^2$
11. $(3 + b)^2$
12. $(5 - 2m)^2$
13. $(x + 7)(x + 3)$
14. $(h + 5)(h + 1)$
15. $(6g + h)(6g - h)$
16. $(3R - S)^2$
17. $(2v + 7)^2$
18. $(x - 6)(x + 2)$

Special Products

99

19. $(N - 3)(N - 8)$ **20.** $(3z + 8)^2$

21. $(2d - 8)^2$ **22.** $R(k^2 + 2mR + 3nR^2)$

23. $(g + R)^2$ **24.** $2xy(2x^3 - 3xy^2 + y^3)$

25. $(h - 3b)^2$ **26.** $(2x + 3)(2x - 3)$

27. $(4x + 5y)(4x - 5y)$ **28.** $(2f - 8g)^2$

29. $(7r + 6t)^2$ **30.** $(3A + 4B)(3A - 4B)$

5-5 More Special Products

Let us now look at another case of the special products. Consider the two binomials $(2x + 5)$ and $(3x + 4)$.

Here all four terms of the binomials are different. Of great importance in this kind of special products are the constants. In this problem, the constants are 2, 5, 3 and 4. To make our discussion of this special product easier, we will give to the different products of these constants the symbols A, B, and C, as shown below. The student is urged to examine carefully the diagram below. Multiplying

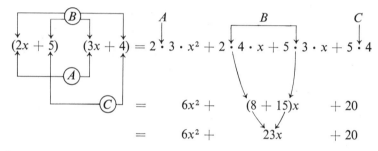

$$= 6x^2 + (8 + 15)x + 20$$
$$= 6x^2 + 23x + 20$$

Notice that

1. The numerical coefficient of x^2, 6, in the first term of the trinomial product is the product A.
2. The numerical coefficient of x, 23, in the second term of the trinomial product is the sum of products B.
3. The last term, 20, of the trinomial product is the product C.

If we substitute the constants of the diagram above, 2, 5, 3, and 4 with four different letters, say a, b, c, and d, we will obtain the general form of this type of special products shown below.

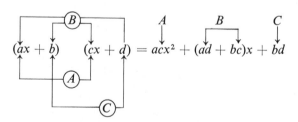

We can say now that *given the binomials $(ax + b)(cx + d)$,*

Basic Operations with Algebraic Expressions

1. The coefficient A, of x² in the first term of the trinomial product, is the product ac,
2. The coefficient B, of x in the second term of the trinomial product, is the sum of the products ad and bc, and
3. The last term C, of the trinomial product, is the product bd.

The formula for this type of special product is

$$(ax + b)(cx + d) = acx^2 + (ad + bc)x + bd$$

The student should have clearly in mind the diagram above while examining the following examples and while working the exercises.

EXAMPLE 5-15 Multiply, with as little writing as possible, the binomials $(3x + 1)$ $(x + 2)$.

SOLUTION In this problem, a, b, c, and d of the formula are 3, 1, 1, and 2 respectively. Therefore,

$$(3x + 1)(x + 2) = 3 \cdot 1x^2 + (3 \cdot 2 + 1 \cdot 1)x + 1 \cdot 2$$
$$= 3x^2 + 7x + 2$$

EXAMPLE 5-16 Multiply $(4k - 1)(3k + 2)$.

SOLUTION Here a, b, c, and d of the formula are 4, −1, 3, and 2, respectively. Therefore,

$$(4k - 1)(3k + 2) = 4 \cdot 3k^2 + [(4 \cdot 2) + (-1)(3)]k + (-1)(2)$$
$$= 12k^2 + 5k - 2$$

EXAMPLE 5-17 Multiply $(3y - 5)(4y - 2)$.

SOLUTION

$$(3y - 5)(4y - 2) = 12y^2 + (-6 - 20)y + 10$$
$$= 12y^2 - 26y + 10$$

For easy reference, the formulas of the six cases of special products we have considered are listed below. The student should know these formulas thoroughly.

1. $a(x + y + z) = ax + ay + az$
2. $(x + a)(x + b) = x^2 + (a + b)x + ab$
3. $(x + a)^2 = x^2 + 2ax + a^2$
4. $(x - a)^2 = x^2 - 2ax + a^2$
5. $(x + a)(x - a) = x^2 - a^2$
6. $(ax + b)(cx + d) = acx^2 + (ad + bc)x + bd$

EXERCISE 5-5

Find each of the following products using as little writing as possible:
1. $(6P + 2)(3P - 7)$
2. $(y - 5)(3y - 8)$
3. $(4z - 3)(2z - 1)$
4. $(8w - 2)(w - 1)$
5. $(2t + 3)(t + 7)$
6. $(b + 4)(2b - 3)$

More Special Products

7. $(9y - 10)(y + 5)$ **8.** $(4v - 5)(v + 7)$
9. $(2v - 1)(3v + 6)$ **10.** $(2v + 3)(3v - 8)$
11. $(3y - 9)(2y + 1)$ **12.** $(7k - 1)(8k - 3)$
13. $(7k - 2)(6k - 3)$ **14.** $(8g + 2)(7g + 3)$
15. $(11t - 3)(4t - 8)$ **16.** $(9h + 1)(h - 2)$
17. $(18d - 5)(21d - 6)$ **18.** $(14A - 13)(17A + 12)$

5-6 Division of Algebraic Expressions

The simplest case is the division of monomials. Consider the following division:

$$-21a^2b^5c^2 \div 7ab^3c^4$$

This division can be expressed as the fraction

$$\frac{-21a^2b^5c^2}{7ab^3c^4}$$

Applying the basic property of fractions (cancellation) and the laws of exponents, we have

$$\frac{-21a^2b^5c^2}{7ab^3c^4} = \frac{-3 \cdot 7}{7} \cdot a^{2-1} \cdot b^{5-3} \cdot c^{2-4}$$

$$= -3ab^2 \cdot \frac{1}{c^2}$$

$$= \frac{-3ab^2}{c^2} \quad \text{or} \quad -\frac{3ab^2}{c^2}$$

It should be evident that this division reduces the fraction denoting the division to its lowest terms by eliminating the common factors from the numerator and the denominator. Therefore, *to divide a monomial by a monomial, express the division as a fraction and determine its sign. This fraction reduced to lowest terms is the quotient.*

    EXAMPLE 5-18   Divide $27x^3y^2c$ by $36x^4y^3c^3$.

    SOLUTION

$$27x^3y^2c \div 36x^4y^3c^3 = \frac{27x^3y^2c}{36x^4y^3c^3}$$

$$= \frac{3}{4xyc^2}$$

To divide a polynomial by a monomial, divide each term of the polynomial by the monomial and add the results. Thus,

$$\frac{15a^2b^3 - 12a^3b + 6a^4b^5}{3a^2b} = \frac{15a^2b^3}{3a^2b} - \frac{12a^3b}{3a^2b} + \frac{6a^4b^5}{3a^2b} = 5b^2 - 4a + 2a^2b^4$$

    EXAMPLE 5-19   Divide $14k^2m^3n - 35k^3m^4n^2 + 21km^2n^4$ by $7km^2n$.

SOLUTION

$$(14k^2m^3n - 35k^3m^4n^2 + 21km^2n^4) \div 7km^2n = \frac{14k^2m^3n - 35k^3m^4n^2 + 21km^2n^4}{7km^2n}$$

$$= 2km - 5k^2m^2n + 3n^3$$

EXAMPLE 5-20 Divide $7a^2bx^3 + 21ab^3x^2 - 3b^5x^2$ by $3ab^2x$.

SOLUTION

$$\frac{7a^2bx^3 + 21ab^3x^2 - 3b^5x^2}{3ab^2x} = \frac{7ax^2}{3b} + 7bx - \frac{b^3x}{a}$$

EXERCISE 5-6

Perform each of the following divisions:

1. $12x^5 \div 4x^2$
2. $-28a^2bc^3 \div 7ab^3c^2$
3. $10k^3m \div 4k^2m^3n$
4. $12h^3k^2m^5 \div 4h^5m^3$
5. $(x^6 + 2x^5) \div 3x^2$
6. $(w^3 - 3w^2 + w + 3) \div 3w$
7. $(35a^5b^3c^5 + 7a^2b^5c^6 + 14a^3b^3c^4) \div 7a^2b^3c^4$
8. $(24d^2e^5g^3 - 9d^5e^3g^3 - 16d^3e^4g^2) \div 3de^3g^2$
9. $(4k^2x^4 - 12km^2x^2 + 9m^3x^2) \div 3k^2mx^2$
10. $(6a^3b^2cw^3 + 17a^2bw^2 - 18aw + 5) \div 6ab^2cw^2$

5-7 Factoring Polynomials

A polynomial that cannot be factored is called a **prime expression**. In this section, we will factor factorable polynomials by reversing the process of the special products. In each case of the special products we are given two factors and we find the product. In factoring, we are given the product and we express it as the indicated multiplication of two simpler factors.

We know from arithmetic that the product of two numbers is divisible by any of the two factors, and that the quotient of the division is the other factor. The same is true for polynomials. Thus,

1. *The product of two polynomials is divisible by any of the two factors.*
2. *When the product of two polynomials is divided by any of its two factors, the quotient is the other factor.*

For example, since

$$(x - 1)(x - 3) = x^2 - 4x + 3,$$

then

$$x^2 - 4x + 3 \div (x - 1) = x - 3$$

and

$$x^2 - 4x + 3 \div (x - 3) = x - 1$$

Suppose we are given the polynomial $mu + mv + mw$ to factor. We notice that all terms of the polynomial have m as a common factor. This means that the polynomial is divisible by m, and that the quotient is the other factor. Thus, $mu + mv + mw \div m = u + v + w$, and $mu + mv + mw = m(u + v + w)$.

To factor $3x^2y^3 + 9xy^2$ completely, we must find the **greatest common factor** of the terms of the polynomial, and divide the polynomial by that factor to obtain the other factor. The greatest common monomial factor is $3xy^2$, and the quotient of $3x^2y^3 + 9xy^2$ divided by $3xy^2$ is $xy + 3$. Thus, $3x^2y^3 + 9xy^2 = 3xy^2(xy + 3)$.

From these two examples it follows that *to factor a polynomial whose terms have a common factor:*

1. *Find the greatest common factor of the terms of the polynomial,*
2. *Divide the polynomial by the common factor to obtain the other factor. The quotient is the other factor.*
3. *Express the polynomial as the indicated multiplication of these two factors.*

The correctness of the factoring can always be checked by performing the indicated multiplication to obtain the original polynomial.

EXAMPLE 5-21   Factor the polynomial $8x^3y^2z + 12x^2y - 24x^2$.

SOLUTION   The greatest common factor is $4x^2$. Then,

$$\frac{8x^3y^2z + 12x^2y - 24x^2}{4x^2} = 2xy^2z + 3y - 6$$

Therefore,

$$8x^3y^2z + 12x^2y - 24x^2 = 4x^2(2xy^2z + 3y - 6)$$

Factoring by grouping is very similar to the factoring of polynomials whose terms have a common factor. Consider the four-term polynomial

$$12aR + 3aS + 4bR + bS$$

We notice first that there is no factor common to all four terms. The two first terms, however, have $3a$ as a common factor, and the two last terms have b as a common factor. **Factoring out** these two common factors from the respective terms, we have,

$$12aR + 3aS + 4bR + bS = 3a(4R + S) + b(4R + S)$$

We notice now that the four-term polynomial has become a two-term expression, and that each term has the binomial $(4R + S)$ as a common factor. Factoring $(4R + S)$ out, and dividing by it each term of $3a(4R + S) + b(4R + S)$ we have

$$3a(4R + S) + b(4R + S) = (4R + S)(3a + b)$$

From the above example, we see that *to factor a four-term polynomial, group together those terms that have common factors, and factor the common factors out. Then, from the two terms of the resulting expression, factor the common binomial factor.*

Basic Operations with Algebraic Expressions

EXAMPLE 5-22 Factor the polynomial $3v^2 + 15v - 2vw - 10w$.

SOLUTION The greatest common factor for the first two terms is $3v$ and for the last two terms is $-2w$. Thus,

$$3v^2 + 15v - 2vw - 10w = 3v(v + 5) - 2w(v + 5)$$
$$= (v + 5)(3v - 2w)$$

Notice that $-2w$ was factored out of the last two terms of the polynomial, not $2w$. If $2w$ had been used, the second binomial factor would have been $(-v - 5)$ instead of $(v + 5)$, making the two binomial factors different and further factoring impossible.

Suppose we are given the polynomial $9x^2 - 16y^2$ to factor. We notice first that $9x^2 - 16y^2$ is the difference of two squares. We recall that the difference of two squares is the product of two binomials, one binomial being the sum and the other the difference of two terms. To form these two binomials, starting with the product $9x^2 - 16y^2$, we first indicate their multiplication by two parentheses, one with the addition and the other with the subtraction sign, as shown below:

$$9x^2 - 16y^2 = (\quad + \quad)(\quad - \quad)$$

Then, the first term of each binomial is the square root of $9x^2$ and the second term of each binomial is the square root of $16y^2$. Thus,

$$9x^2 - 16y^2 = (3x + 4y)(3x - 4y)$$

We see from the above example that to factor the difference of two squares,

1. *Find the square roots of each, and then*
2. *Indicate the multiplication of two binomials, one binomial being the sum and the other being the difference of these two square roots.*

EXAMPLE 5-23 Factor the polynomial $36B^2 - 25C^2$.

SOLUTION This is the difference of two squares. The two square roots are $6B$ and $5C$. Therefore, $36B^2 - 25C^2 = (6B + 5C)(6B - 5C)$.

EXAMPLE 5-24 Factor the polynomial $24g^2 - 150h^2$ completely.

SOLUTION Here, as in any other factoring problem, any common factor should be factored out first. Thus,

$$24g^2 - 150h^2 = 6(4g^2 - 25h^2)$$

Notice now that $4g^2 - 25h^2$ is the difference of two squares and can be factored further. Thus,

$$24g^2 - 150h^2 = 6(2g + 5h)(2g - 5h)$$

EXERCISE 5-7

Factor each of the following polynomials completely:
1. $3x + 12$
2. $5x^2 - x$
3. $2R - 10$
4. $3ax + 9ay$

5. $4e - 2$

6. $az^2 - 6z$

7. $14cv^2 - 7dv - 21$

8. $15x^3y + 10x^2 - 20x$

9. $-4abc + 3ac - 5bc$

10. $12xyz - 18xy + 24yz$

11. $9g^3h^2 - 81gh^2 + 27g^2h$

12. $75 - 25x^3 + 15x^2$

13. $6a^2b + 24ab + 66a$

14. $x^3 + x^2y - xy^2$

15. $7uvw - 3uv + 5vw$

16. $2x^3y^3 - 8x^2y^3 + 6x^3y^2 - 10x^2y^2$

17. $3t^2s^4 + 15t^3s^3 - 18t^4s^2$

18. $k^2 + 2k + 2dk + 4d$

19. $4s^2 + 16s + 12st + 48t$

20. $3gs - 3gt - 5s + 5t$

21. $3h - 6g + 2gh - 9$

22. $a^2 - ab + 2ac - 2bc$

23. $5x^2 - 10x + 3xy - 6y$

24. $3ay - 15a + 4y - 20$

25. $2M^2 + 3dM + 9d + 6M$

26. $x^2 - a^2$

27. $t^2 - 4$

28. $9k^2 - 4m^2$

29. $64v^2 - 16w^2$

30. $81x^2 - 1$

31. $1 - D^2$

32. $121 - 49R^2$

33. $16y^2 - 25z^2$

34. $81u^2 - 144v^2$

35. $e^2 - 1$

36. $25 - H^2$

37. $F^2 - 4$

38. $49x^2 - 64y^2$

39. $169g^2 - 9h^2$

40. $25N^2 - 1$

41. $50 - 2A^2$

42. $3 - 27P^2$

43. $28R^2 - 7$

44. $5t^2 - 20$

45. $27x^2 - 12y^2$

46. $2x^2 - 2y^2$

5-8 Factoring Trinomials

Another case of factoring we will consider is the factoring of trinomials that are perfect squares. We have seen that the square of a binomial is a trinomial that is a perfect square. For instance, we know that

$$(3a + 2)^2 = 9a^2 + 12a + 4$$

It must be evident at this point that the trinomial $9a^2 + 12a + 4$ can be factored into the square of the binomial $(3a + 2)$. Thus,

$$9a^2 + 12a + 4 = (3a + 2)^2$$

Any time a given trinomial is a perfect square, it is the square of a binomial whose first term is the square root of the first term of the trinomial. Its second term is the square root of the last term of the trinomial. The sign of the second term of the binomial is the same as the sign of the second term of the trinomial.

A perfect square trinomial can be factored readily. It is necessary, however, to be able to recognize perfect square trinomials. Let us see what makes a trinomial a perfect square.

Notice in $9a^2 + 12a + 4$, both the first and the last terms are perfect squares. Notice also that the middle term, $12a$ is twice the product of the square root of $9a^2$ and the square root of 4. That is,

$$12a = 2 \cdot 3a \cdot 2$$

Basic Operations with Algebraic Expressions

From the above example, we see that *for a trinomial to be a perfect square, the following two things are necessary:*

1. *Both the first and the third terms of the trinomial must be perfect squares,*
2. *The middle term of the trinomial must be twice the square roots of its first and last terms.*

EXAMPLE 5-25   Factor $16x^2 + 8xy + y^2$ completely.

SOLUTION   The trinomial $16x^2 + 8xy + y^2$ is a perfect square since both the first and the third terms are perfect squares, and the middle term is $2 \cdot 4x \cdot y = 8xy$. Then,

$$16x^2 + 8xy + y^2 = (4x + y)^2$$

EXAMPLE 5-26   Factor $9s^2 - 30st + 25t^2$ completely.

SOLUTION   The trinomial $9s^2 + 30st + 25t^2$ is a perfect square since both the first and the third terms are perfect squares, and the middle term is $2 \cdot 3s \cdot 5t = 30st$. Therefore,

$$9s^2 - 30st + 25t^2 = (3s - 5t)^2$$

EXAMPLE 5-27   Factor $4x^2 + 5xy + y^2$ completely.

SOLUTION   The first and the third terms of the trinomial are perfect squares. The middle term, however, instead of being $2 \cdot 2x \cdot y = 4xy$, is $5xy$. Therefore, $4x^2 + 5xy + y^2$ is not a perfect square and cannot be factored as the square of a binomial. We will see this case of factoring next.

The last case of factoring we will consider is the factoring of trinomials which are not perfect squares. We know from previous sections that

$$(2x + 1)(3x + 4) = 6x^2 + 8x + 3x + 4$$
$$= 6x^2 + (8 + 3)x + 4$$
$$= 6x^2 + 11x + 4$$

Notice that the numerical coefficient 11 of the middle term of the trinomial product is the sum of 8 and 3. Notice also that the product of 8 and 3, 24, is the same as the product of 6, the numerical coefficient of the first term of the trinomial product, and 4, the numerical coefficient of the last term of the trinomial.

Since these facts are not a coincidence with this trinomial only, they can be used to reconstruct the four-term polynomial from which the trinomial to be factored was formed. Once the four-term polynomial is reconstructed the factoring is a very simple matter. Let us trace the reconstruction of the four-term polynomial from which the trinomial $6x^2 + 11x + 4$ was formed when similar terms were combined. In order to do that, the constant 11, the coefficient of the middle term of the trinomial, must be split into two numbers such that, their sum will be 11 and their product will be $6 \cdot 4 = 24$. If we represent these two numbers we are looking for by A and B, then A and B must be such that

$$A + B = 11 \qquad \text{and} \qquad A \cdot B = 6 \cdot 4 = 24$$

Twenty-four is the product of 6, the numerical coefficient of the first term, and 4, the last term of the trinomial. The possible pairs of factors of 24 are

$$1 \times 24$$
$$2 \times 12$$
$$3 \times 8$$
$$4 \times 6$$

Notice that only the pair 3 and 8 will satisfy both the conditions

$$A + B = 11 \qquad \text{and} \qquad A \cdot B = 24$$

These then are the numbers we are looking for. With 3 and 8 we rewrite the trinomial $6x^2 + 11x + 4$ as the four-term polynomial $6x^2 + 3x + 8x + 4$. Then, using factoring by grouping, the complete factoring is as follows:

$$6x^2 + 11x + 4 = 6x^2 + 3x + 8x + 4$$
$$= 3x(2x + 1) + 4(2x + 1)$$
$$= (2x + 1)(3x + 4)$$

From the above illustration, we conclude that to factor a trinomial that is not a perfect square:

1. *Form a four-term polynomial by splitting the numerical coefficient of the middle term of the trinomial into two numbers such that their sum is equal to the coefficient itself, and their product is equal to the first numerical coefficient times the last numerical coefficient of the trinomial, or the last constant of the trinomial.*
2. *Complete the factoring using factoring by grouping.*

EXAMPLE 5-28 Factor the trinomial $2x^2 + 7x + 6$ completely.

SOLUTION Since $2x^2 + 7x + 6$ is not a perfect square trinomial, we will form first a four-term polynomial by splitting 7 into two numbers, A and B, such that

$$A + B = 7 \qquad \text{and} \qquad A \cdot B = 2 \cdot 6 = 12$$

The possible pairs of factors for 12 are

$$1 \times 12$$
$$2 \times 6$$
$$3 \times 4$$

The numbers we are looking for are 3 and 4 since only 3 and 4 satisfy the above conditions. Then,

$$2x^2 + 7x + 6 = 2x^2 + 3x + 4x + 6$$
$$= x(2x + 3) + 2(2x + 3)$$
$$= (2x + 3)(x + 2)$$

EXAMPLE 5-29   Factor the trinomial $k^2 - 2k - 24$ completely.

SOLUTION   Since $k^2 - 2k - 24$ is not a perfect square trinomial, we need two numbers, A and B, such that

$$A + B = -2 \quad \text{and} \quad A \cdot B = 1 \cdot (-24) = -24$$

The possible pairs of factors of -24 are:

$$
\begin{array}{ccc}
1 \times (-24) & & -1 \times (24) \\
2 \times (-12) & & -2 \times (12) \\
& \text{and} & \\
3 \times (-8) & & -3 \times (8) \\
4 \times (-6) & & -4 \times (6)
\end{array}
$$

The only pair which satisfies both the sum and the product conditions is 4 and -6. Then,

$$k^2 - 2k - 24 = k^2 + 4k - 6k - 24$$
$$= k(k + 4) - 6(k + 4)$$
$$= (k + 4)(k - 6)$$

EXAMPLE 5-30   Factor the trinomial $6x^2 - xy - 2y^2$ completely.

SOLUTION   We must determine two numbers, A and B, such that

$$A + B = -1 \quad \text{and} \quad A \cdot B = 6 \cdot (-2) = -12$$

Possible pairs of factors are

$$
\begin{array}{ccc}
1 \times (-12) & & -1 \times (12) \\
2 \times (-6) & \text{and} & -2 \times (6) \\
3 \times (-4) & & -3 \times (4)
\end{array}
$$

The numbers we are looking for are 3 and -4. Therefore,

$$6x^2 - xy - 2y^2 = 6x^2 + 3xy - 4xy - 2y^2$$
$$= 3x(2x + y) - 2y(2x + y)$$
$$= (2x + y)(3x - 2y)$$

EXAMPLE 5-31   Factor the trinomial $d^2 - 15d + 36$ completely.

SOLUTION   We need two numbers, A and B, such that

$$A + B = -15 \quad \text{and} \quad A \cdot B = 1 \cdot 36 = 36$$

Possible pairs of factors are

$$
\begin{array}{ccc}
1 \times 36 & & -1 \times (-36) \\
2 \times 18 & & -2 \times (-18) \\
3 \times 12 & \text{and} & -3 \times (-12) \\
4 \times 9 & & -4 \times (-9) \\
6 \times 6 & & -6 \times (-6)
\end{array}
$$

The two numbers we are looking for are -3 and -12. Therefore,

$$d^2 - 15d + 36 = d^2 - 3d - 12d + 36$$
$$= d(d - 3) - 12(d - 3)$$
$$= (d - 3)(d - 12)$$

EXAMPLE 5-32 Factor the trinomial $12k^2 - 2k - 2$ completely.

SOLUTION The three terms of the trinomial $12k^2 - 2k - 2$ have the factor 2 as a common factor. First 2 is factored out:

$$12k^2 - 2k - 2 = 2(6k^2 - k - 1)$$

Then we proceed as before to factor the trinomial $6k^2 - k - 1$. Here the numbers A and B are 2 and -3. Then,

$$6k^2 - k - 1 = 6k^2 + 2k - 3k - 1$$
$$= 2k(3k + 1) - 1(3k + 1)$$
$$= (3k + 1)(2k - 1)$$

Therefore,

$$12k^2 - 2k - 2 = 2(3k + 1)(2k - 1)$$

EXERCISE 5-8

Factor each of the following trinomials completely:

1. $m^2 + 2mn + n^2$
2. $4A^2 + 12A + 9$
3. $R^2 - 2RS + S^2$
4. $d^2 - 12d + 36$
5. $25L^2 + 20LM + 4M^2$
6. $9x^2 + 24xy + 16y^2$
7. $9g^2 - 6g + 1$
8. $1 - 2w + w^2$
9. $4t^2 + 20ts + 25s^2$
10. $64m^2 - 48mn + 9n^2$
11. $36g^2 + 84gh + 49h^2$
12. $121v^2 - 22vw + w^2$
13. $k^2 - 10km + 25m^2$
14. $4P^2 + 4P + 1$
15. $9x^2 - 12xy + 4y^2$
16. $16d^2 - 24d + 9$
17. $x^2 + 2x - 24$
18. $h^2 + 2h - 3$
19. $x^2 + 4x - 21$
20. $c^2 - 2c - 63$
21. $t^2 + 8t + 7$
22. $8x^2 + 10xy - 3y^2$
23. $14k^2 - 13k + 3$
24. $6D^2 - D - 1$
25. $9x^2 - 12x - 21$
26. $6x^2 + 13x - 5$
27. $6x^2 - 13x + 6$
28. $12t^2 + 14t - 10$
29. $3t^2 - 14t - 5$
30. $6m^2 - 15m - 21$
31. $6d^2 - 8d - 30$
32. $28P^2 - 30P - 18$

5-9 Evaluation of Algebraic Expressions

Evaluation of an algebraic expression is finding its numerical value when the numerical values of the literal factors are given. All it takes to evaluate an expression is to substitute numbers for letters, and then perform the indicated operations.

EXAMPLE 5-33 Find the value of $3x + 2y + 7z$, if $x = 4$, $y = 1$, and $z = 2$.

Basic Operations with Algebraic Expressions

SOLUTION

$$3x + 2y + 7z = 3(4) + 2(1) + 7(2)$$
$$= 12 + 2 + 14$$
$$= 28$$

EXAMPLE 5-34 Find the value of $5v(7 - 8w) - 2(w - 6v)$, if $v = 4$, and $w = -2$.

SOLUTION

$$5v(7 - 8w) - 2(w - 6v) = 35v - 40vw - 2w + 12v$$
$$= 47v - 40vw - 2w$$

Substituting values,

$$47v - 40vw - 2w = 47(4) - 40(4)(-2) - 2(-2)$$
$$= 188 + 320 + 4$$
$$= 512$$

EXERCISE 5-9

Evaluate each of the following expressions for $u = 2$, $v = -1$, $w = 3$, $x = -2$, and $y = 4$:

1. $2v + 5w - 2x + 3y$
2. $7uwy$
3. $\dfrac{x - u}{w}$
4. $7v - 2$
5. $2v^2 x$
6. $4uwy^2$
7. $3w^2 - 2uv$
8. $\dfrac{3y + 2v^2 - 3x - 4}{8}$
9. $3uv - vw^3 + wx - 2xy$
10. $4(u - v)$
11. $(x + y)^2$
12. $7u^3 v^3$
13. $5u^2 - 7v + 3x$
14. $\dfrac{1}{u} + \dfrac{1}{w}$
15. $\dfrac{2}{v} - \dfrac{y}{x}$
16. $7y - 4(2 + 3v)$
17. $3u(x + 2v - 8)$
18. $\dfrac{8x + 2y}{u}$
19. $\dfrac{3y}{x} - w$
20. $\dfrac{5u}{v} + \dfrac{w}{x}$

REVIEW QUESTIONS

1. Is the distributive law a special product?
2. State the rule for the product of two binomials having the first term of each the same variable with 1 as its numerical coefficient.
3. State the rule for the square of the sum of two terms.
4. State the rule for the square of the difference of two terms.
5. State the rule for the product of the sum and the difference of two terms.
6. State the rule for the product of two binomials whose terms are all different.
7. State the rule for factoring a polynomial whose terms have a common factor.
8. State the rule for factoring a four-term polynomial by grouping.
9. State the rule for factoring the difference of two squares.

10. What are the two necessary conditions for a trinomial to be a perfect square?

11. State the rule for the factoring of a perfect square trinomial.

12. State the rule for the factoring of a trinomial that is not a perfect square.

13. What is meant by the evaluation of an algebraic expression?

REVIEW EXERCISES

Add the expressions in each of the following:

1. $3d + 5e - 8g,\ 4d - 3g$ **2.** $5ax^2 - 3x + 2y,\ 8ax^2 + 7x - 3y$

In exercises 3 and 4, subtract the second expression from the first:

3. $8h^2 + 3k - 7,\ 11h^2 + 5k + 8$ **4.** $7x^2 + 11y - 6,\ 3x^2 - 2$

Multiply each of the following:

5. $(4x^2y)(-3xy^2)$ **6.** $3c(5x - 2y - 8)$

7. $(t^2 + mt + m^2)(2t^2 - 3mt + 4m^2)$

Find each of the following products using as little writing as possible:

8. $(p + 11)^2$ **9.** $(r - q)^2$ **10.** $7a(p^2 - 3q + rt^3)$

11. $(M - 7)(M - 8)$ **12.** $(Q + 3)(Q - 1)$ **13.** $(2A + 5B)(3A - 5B)$

14. $(3s - 2t)^2$ **15.** $(7h - 1)(3h + 5)$ **16.** $(B + 2)(3B - 1)$

Perform each of the following divisions:

17. $24a^3bc^2 \div 6ab^2c^3$ **18.** $(6R^3 + 3R^2 - 2R - 3) \div 3R$

Factor each of the following:

19. $7x^2 - 21x - 2tx + 6t$ **20.** $3y^2 + 6y + 5dy + 10d$

21. $64x^2 - 16xy + y^2$ **22.** $4p^2 - 12pq + 9q^2$

23. $2x^2 + x - 3$ **24.** $20d^2 - 7d - 6$

Evaluate each of the following expressions for $a = 3,\ b = -1,\ x = -2,$ and $y = 2$:

25. $\dfrac{ax^2 - by^3}{b}$ **26.** $\dfrac{2a^2x + 3by^2}{4ax - 5b^2y^2}$

Basic Operations with Algebraic Expressions

Algebraic Fractions

6-1 Reducing Algebraic Fractions to Lowest Terms

In an algebraic fraction, the numerator can be an integer, a monomial, or a polynomial. The same is true of the denominator. Thus,

$$\frac{3}{y}, \qquad \frac{-2x}{x+5}, \qquad \frac{x-y}{2x+y}, \qquad \text{and} \qquad \frac{-x(x+3)}{y(x-7)}$$

are algebraic fractions. Notice in these fractions that everything above the fraction bar is the numerator, and everything below the fraction bar is the denominator of the fraction. The numerators of the fractions are 3, $-2x$, $x-y$, and $-x(x+3)$ and y, $x+5$, $2x+y$, and $y(x-7)$ are the denominators.

An algebraic fraction is said to be in its lowest terms if there is no algebraic expression which will divide evenly both the numerator and the denominator of the fraction. This is the same thing as saying that *an algebraic fraction is in its lowest terms if the numerator and the denominator have no common factor.* This is not always easy to determine, and, as in arithmetic fractions, the safest way is to use complete factorization. Let us see some examples:

The fraction

$$\frac{a^2}{c^3}$$

is in its lowest terms because a^2 and c^3 have no common factor.

However, the fraction

$$\frac{ab^2}{b^3c}$$

is not in lowest terms. Having simplified expressions of this kind in the section for exponents, the student can see that there are pairs of factors common to both the numerator and the denominator of this fraction. We can reduce $\dfrac{ab^2}{b^3c}$ to its lowest terms by using the fundamental principle of fractions, that is by dividing out (cancelling) these pairs of factors in exactly the same way we did with arithmetic fractions. Thus,

$$\frac{ab^2}{b^3c} = \frac{a \cdot b \cdot b}{b \cdot b \cdot b \cdot c} \quad \text{(by factoring completely)}$$

$$= \frac{a \cdot \cancel{b} \cdot \cancel{b}}{\cancel{b} \cdot \cancel{b} \cdot b \cdot c} \quad \text{(by cancelling pairs of common factors)}$$

$$= \frac{a}{bc} \quad \text{(lowest terms)}$$

NOTE: Complete factorization was used here for illustrative purposes. The experienced student will use the laws of exponents in similar cases.

In the same way, the fraction

$$\frac{x+1}{x-3}$$

is in its lowest terms, because both its numerator and denominator are prime expressions (cannot be factored).

However, the fraction

$$\frac{2x^2 + x}{6x + 3}$$

is not in its lowest terms because,

$$\frac{2x^2 + x}{6x + 3} = \frac{x(2x+1)}{3(2x+1)} \quad \text{(by factoring completely)}$$

$$= \frac{x\cancel{(2x+1)}}{3\cancel{(2x+1)}} \quad \text{(by cancelling pairs of common factors)}$$

$$= \frac{x}{3} \quad \text{(lowest terms)}$$

These examples show that in algebraic fractions, as in arithmetic fractions, *to reduce a fraction to its lowest terms, first factor both its terms completely. Then cross out (cancel) all pairs of factors common to both numerator and denominator in the factored terms of the fraction.*

EXAMPLE 6-1   Reduce the fraction $\dfrac{14a^2b^2}{21ab^3}$.

SOLUTION

$$\frac{14a^2b^2}{21ab^3} = \frac{2 \cdot 7 \cdot a \cdot a \cdot b \cdot b}{3 \cdot 7 \cdot a \cdot b \cdot b \cdot b} \quad \text{(by factoring completely)}$$

$$= \frac{2 \cdot \cancel{7} \cdot \cancel{a} \cdot a \cdot \cancel{b} \cdot \cancel{b}}{3 \cdot \cancel{7} \cdot \cancel{a} \cdot \cancel{b} \cdot \cancel{b} \cdot b} \quad \text{(by cancelling pairs of common factors)}$$

$$= \frac{2a}{3b} \quad \text{(lowest terms)}$$

EXAMPLE 6-2 Reduce the fraction $\dfrac{(x+5)}{x^2+10x+25}$.

SOLUTION

$$\frac{(x+5)}{x^2+10x+25}=\frac{(x+5)}{(x+5)(x+5)} \qquad \text{(by factoring complete-} \\ \text{ly)}$$

$$=\frac{\cancel{(x+5)}}{\cancel{(x+5)}(x+5)} \qquad \text{(by cancelling pairs of common factors)}$$

$$=\frac{1}{(x+5)} \qquad \text{(lowest terms)}$$

EXAMPLE 6-3 Reduce the fraction $\dfrac{3x^2-27}{4x^2+20x+24}$.

SOLUTION

$$\frac{3x^2-27}{4x^2+20x+24}=\frac{3(x^2-9)}{4(x^2+5x+6)}$$

$$=\frac{3\cancel{(x+3)}(x-3)}{4\cancel{(x+3)}(x+2)}$$

$$=\frac{3(x-3)}{4(x+2)}$$

EXERCISE 6-1

Reduce each of the following fractions to its lowest terms:

1. $\dfrac{3x}{18y}$

2. $\dfrac{8x}{6xy}$

3. $\dfrac{7h}{8h}$

4. $\dfrac{36xy}{18y}$

5. $\dfrac{3v^2}{5v^3}$

6. $\dfrac{RT^5}{ST^4}$

7. $\dfrac{27m^2k^3}{12m^3k}$

8. $\dfrac{26w^5x^4y}{39w^2x^3y^3}$

9. $\dfrac{24ACD^4}{32BD^3E}$

10. $\dfrac{a^2-a}{a}$

11. $\dfrac{b}{3b^2+b}$

12. $\dfrac{b+1}{3b+3}$

13. $\dfrac{2x^3-6x^2}{5x-15}$

14. $\dfrac{3+12w}{1-16w^2}$

15. $\dfrac{2x-y}{4x^2-4xy+y^2}$

16. $\dfrac{h+3m}{h^2+6hm+9m^2}$

17. $\dfrac{6t^2-13t-5}{2t-5}$

18. $\dfrac{3d^2+4d-4}{ad+2a}$

19. $\dfrac{7k^2-k-8}{k^2-1}$

20. $\dfrac{3z^2-10z-8}{2z^2-32}$

21. $\dfrac{6y^2-7y+2}{2y^2+5y-3}$

Reducing Algebraic Fractions to Lowest Terms

6-2 Finding the Least Common Multiple (LCM) of Algebraic Expressions

The LCM of two or more expressions is the smallest expression which is divisible by each of the given expressions. To find the LCM of algebraic expressions, use the same two-step procedure used in arithmetic:

1. *Factor each of the given expressions completely.*
2. *Indicate the product of all the different prime factors, each taken the largest number of times it occurs in any of the given expressions.*

EXAMPLE 6-4   Find the LCM of $12ab^2$ and $15a^2b^3$.

SOLUTION

$$12ab^2 = 2 \cdot 2 \cdot 3 \cdot a \cdot b \cdot b$$
$$15a^2b^3 = 3 \cdot 5 \cdot a \cdot a \cdot b \cdot b \cdot b$$
$$\text{LCM} = 2 \cdot 2 \cdot 3 \cdot 5 \cdot a \cdot a \cdot b \cdot b \cdot b$$
$$= 60a^2b^3$$

Notice that $60a^2b^3$ is divisible by both $12ab^2$ and $15a^2b^3$.

EXAMPLE 6-5   Find the LCM of $4h^2 + 12h + 9$ and $6h - 9$.

SOLUTION

$$4h^2 + 12h + 9 = (2h + 3)(2h + 3)$$
$$6h - 9 = 3(2h - 3)$$
$$\text{LCM} = 3(2h - 3)(2h + 3)(2h + 3)$$

EXAMPLE 6-6   Find the LCM of $d^2 - d - 2$ and $d^2 - 4$.

SOLUTION

$$d^2 - d - 2 = (d - 2)(d + 1)$$
$$d^2 - 4 = (d + 2)(d - 2)$$
$$\text{LCM} = (d - 2)(d + 1)(d + 2)$$

Notice that the LCM is divisible by both $d^2 - d - 2$ and $d^2 - 4$. Thus,

$$\frac{(d - 2)(d + 1)(d + 2)}{(d - 2)(d + 1)} = \frac{\cancel{(d - 2)}\cancel{(d + 1)}(d + 2)}{\cancel{(d - 2)}\cancel{(d + 1)}}$$
$$= d + 2$$

and

$$\frac{(d - 2)(d + 1)(d + 2)}{(d + 2)(d - 2)} = \frac{\cancel{(d - 2)}(d + 1)\cancel{(d + 2)}}{\cancel{(d + 2)}\cancel{(d - 2)}}$$
$$= d + 1$$

EXAMPLE 6-7   Find the LCM of $3x + 4$, $3x^2 + x - 4$, and $9x^2 + 24x + 16$.

SOLUTION

$$3x + 4 = 3x + 4$$
$$3x^2 + x - 4 = (3x + 4)(x - 1)$$

$$9x^2 + 24x + 16 = (3x + 4)(3x + 4)$$
$$\text{LCM} = (3x + 4)(3x + 4)(x - 1)$$

Find the LCM in each of the following exercises. Check in each case to see if the LCM is divisible by every expression in the exercise.

1. $12xy^2z^3$, $6x^2yz^2$
2. $4k^2m^3$, $9km^2$
3. $10u^2vw^3$, $6u^3x$
4. $3s^2t$, r^3s^5
5. $6x^2y^3$, $4x^3y^2$, $8x^5y$
6. $27d^2h^5k^3$, $6h^3k^2$, $9d^3h^4$
7. $6x - 3y^3$, $2x - y^3$
8. $7d - 21g$, $4d - 12g$
9. $3b + 6c$, $8b + 16c$, $6b + 12c$
10. $3a^2 + a^2x$, $3a^3 + a^3x$, $3k + kx$
11. $x^2 - y^2$, $x^2 + 2xy + y^2$
12. $4x^2 - 12x + 9$, $2x^2 + x - 6$
13. $4m^2 - 25$, $4m^2 - 20m + 25$, $2km - 5k$
14. $21t^2 + 20t + 4$, $24t + 16$, $12t + 8$
15. $h^2 - 2hk + k^2$, $6h - 6k$, $4h - 4k$
16. $h^2 + 6h + 9$, $h^2 - 6h + 9$, $h^2 - 9$

6-3 Addition and Subtraction of Algebraic Fractions

To add or subtract algebraic fractions, follow this four-step procedure:

1. *Find the LCM of all the denominators of the given fractions. This LCM is the least common denominator (LCD) of the fractions.*
2. *Change each fraction to an equivalent fraction having the LCD as its denominator by multiplying both its terms by the number of times its denominator will go into the LCD.*
3. *Place the numerator of each of these equivalent fractions over a single LCD. Because the fraction bar is a grouping symbol, each polynomial numerator should be enclosed in parentheses with the sign of its fraction in front of it.*
4. *Remove the parentheses and combine similar terms.*

EXAMPLE 6-8 Add $\dfrac{5}{3a} + \dfrac{7}{2a}$.

SOLUTION

$$1. \qquad \text{LCD} = 6a$$

$$2.\ \frac{5}{3a} = \frac{5 \cdot 2}{3a \cdot 2} = \frac{10}{6a}$$

(by multiplying both terms of the first fraction by 2, the number of times its denominator 3a will go into the LCD, 6a)

Addition and Subtraction of Algebraic Fractions

$$\frac{7}{2a} = \frac{7 \cdot 3}{2a \cdot 3} = \frac{21}{6a}$$

(*by multiplying both terms of the second fraction by 3, the number of times its denominator 2a will go into the LCD, 6a*)

Thus,

$$\frac{5}{3a} + \frac{7}{2a} = \frac{10}{6a} + \frac{21}{6a}$$

3. $$= \frac{10 + 21}{6a}$$

(*by placing the numerators of the equivalent fractions over a single LCD. Each numerator is preceded by the sign of its fraction*)

4. $$= \frac{31}{6a}$$

(*by combining similar terms*)

EXAMPLE 6-9   Add $\dfrac{a}{a + 1} + \dfrac{a}{a - 1}$.

SOLUTION

$$\text{LCD} = (a + 1)(a - 1)$$

$$\frac{a}{a + 1} + \frac{a}{a - 1} = \frac{a(a - 1)}{(a + 1)(a - 1)}$$

(*by multiplying both terms of each fraction by the number of times its denominator will go into the LCD*)

$$+ \frac{a(a + 1)}{(a + 1)(a - 1)}$$

$$= \frac{a(a - 1) + a(a + 1)}{(a + 1)(a - 1)}$$

(*by placing the numerators of the equivalent fractions over a single LCD, each numerator preceded by the sign of its fraction*)

$$= \frac{a^2 - a + a^2 + a}{(a + 1)(a - 1)}$$

(*by removing parentheses*)

$$= \frac{2a^2}{(a + 1)(a - 1)}$$

(*by combining similar terms*)

EXAMPLE 6-10   Perform the subtraction $\dfrac{5}{2x} - \dfrac{3}{x + 2}$.

SOLUTION

$$\text{LCD} = 2x(x + 2)$$

$$\frac{5}{2x} - \frac{3}{x + 2} = \frac{5(x + 2)}{2x(x + 2)} - \frac{3 \cdot 2x}{(x + 2)2x}$$

$$= \frac{5(x + 2) - 3 \cdot 2x}{2x(x + 2)}$$

(*notice that each numerator has the sign of its fraction in front of it*)

$$= \frac{5x + 10 - 6x}{2x(x + 2)}$$

$$= \frac{10 - x}{2x(x + 2)}$$

EXAMPLE 6-11 Combine $\dfrac{3x^2 - 2}{3x} - \dfrac{x - 3}{4x} - \dfrac{7}{12}$.

SOLUTION

$$\text{LCD} = 12x$$

$$\frac{3x^2 - 2}{3x} - \frac{x - 3}{4x} - \frac{7}{12} = \frac{4(3x^2 - 2)}{12x} - \frac{3(x - 3)}{12x} - \frac{7 \cdot x}{12x}$$

$$= \frac{4(3x^2 - 2) - 3(x - 3) - 7x}{12x}$$

$$= \frac{12x^2 - 8 - 3x + 9 - 7x}{12x}$$

$$= \frac{12x^2 - 10x + 1}{12x}$$

EXAMPLE 6-12 Perform the subtraction $\dfrac{-3}{4x^2} - \dfrac{5}{6xy}$.

SOLUTION

$$\text{LCD} = 12x^2y$$

$$\frac{-3}{4x^2} - \frac{5}{6xy} = \frac{-3 \cdot 3y}{12x^2y} - \frac{5 \cdot 2x}{12x^2y}$$

$$= \frac{-9y - 10x}{12x^2y}$$

EXAMPLE 6-13 Perform the subtraction $1 - \dfrac{1}{d - 2}$.

SOLUTION The student might be helped if the 1 is converted into the fraction $\dfrac{1}{1}$.
Then, clearly, the LCD is $d - 2$, and we have,

$$1 - \frac{1}{d - 2} = \frac{1}{1} - \frac{1}{d - 2}$$

$$= \frac{1(d - 2)}{d - 2} - \frac{1}{d - 2}$$

$$= \frac{d - 2 - 1}{d - 2}$$

$$= \frac{d - 3}{d - 2}$$

EXERCISE 6-3

In each of the following, perform the indicated operations.

1. $\dfrac{8v}{3} + \dfrac{5v}{12}$

2. $\dfrac{5b}{6} - \dfrac{b}{8}$

3. $\dfrac{3}{4} - \dfrac{2x - 3}{5x}$

4. $\dfrac{t}{t - 2} + \dfrac{t}{t + 2}$

5. $\dfrac{7}{3c} - \dfrac{5}{c + 3}$

6. $\dfrac{-2}{5d} - \dfrac{3}{d - 2}$

7. $\dfrac{9m}{3t} - \dfrac{4m}{5t}$

8. $\dfrac{-4}{5x} + \dfrac{7}{6x^2}$

Addition and Subtraction of Algebraic Fractions

9. $\dfrac{h}{x^2} - \dfrac{1}{w}$

10. $\dfrac{1}{7} + \dfrac{3u - 2}{4u}$

11. $\dfrac{5g - 4}{6} + \dfrac{3g + 2}{2g}$

12. $1 + \dfrac{2}{5 - x}$

13. $\dfrac{3}{k - 1} - 1$

14. $\dfrac{1}{3x - 2} - \dfrac{1}{3x + 2}$

15. $\dfrac{5}{x - 3} + \dfrac{2}{x^2 - 6x + 9}$

16. $\dfrac{-8}{5w^2} - \dfrac{1}{3wx}$

17. $\dfrac{k}{mt} - \dfrac{S}{t}$

18. $\dfrac{3d^2 + 4}{5d} + \dfrac{d - 2}{3d} - 1$

19. $x + \dfrac{5}{x - 3}$

20. $\dfrac{7}{x + 2} + x$

21. $x - \dfrac{3x}{x - 1} + \dfrac{x}{x^2 - 1}$

22. $\dfrac{2}{a} - \dfrac{3}{b} + \dfrac{4}{c}$

23. $\dfrac{a + b}{x} + \dfrac{a - b}{3x} - \dfrac{3a + 2b}{7x}$

24. $\dfrac{k}{2 - k} + \dfrac{2}{2 - k} - \dfrac{5}{1 + k}$

25. $\dfrac{3x}{x + 3} - \dfrac{2x}{x - 1} + \dfrac{2x + 5}{x^2 + 6x + 9}$

26. $\dfrac{1}{x + y} + \dfrac{2y}{x(x - y)} - \dfrac{3x}{x^2 - y^2}$

27. $\dfrac{6}{x + 2} - \dfrac{5}{x^2 + 4x + 4} - \dfrac{7}{x^2 - 4}$

28. $\dfrac{2g}{h + k} - \dfrac{5g}{8(h + k)} + \dfrac{g}{3(h + k)}$

29. $\dfrac{a}{a^2 - 9} - a + \dfrac{2a}{a + 3}$

30. $\dfrac{5x^2}{a^2 - x^2} + \dfrac{4a}{a + x} - \dfrac{3x}{a + x}$

6-4 Multiplication and Division of Algebraic Fractions

The rule for the multiplication of two or more algebraic fractions is the same as the rule for common fractions. Here again, it is highly recommended to factor completely and cancel before multiplying. If this is done, the result will be in its lowest terms.

EXAMPLE 6-14 Multiply $\dfrac{5z^2}{12x^3y^2} \cdot \dfrac{24x^2y}{5z^3} \cdot \dfrac{9xz^2}{16y^3}$.

SOLUTION

$$\dfrac{5z^2}{12x^3y^2} \cdot \dfrac{24x^2y}{5z^3} \cdot \dfrac{9xz^2}{16y^3} = \dfrac{5 \cdot 24 \cdot 9 \cdot z^2 \cdot x^2 \cdot y \cdot x \cdot z^2}{12 \cdot 5 \cdot 16 \cdot x^3 \cdot y^2 \cdot z^3 \cdot y^3}$$

$$= \dfrac{9z}{8y^4}$$

EXAMPLE 6-15 Multiply $\dfrac{2b}{3a^2} \cdot \dfrac{9(a + b)}{4b^2(a - b)} \cdot \dfrac{a^2b^2}{a^2 - b^2}$.

SOLUTION

$$\dfrac{2b}{3a^2} \cdot \dfrac{9(a + b)}{4b^2(a - b)} \cdot \dfrac{a^2b^2}{a^2 - b^2} = \dfrac{2 \cdot 9 \cdot b \cdot (a + b) \cdot a^2 \cdot b^2}{3 \cdot 4 \cdot a^2 \cdot b^2 \cdot (a - b)(a + b)(a - b)}$$

$$= \dfrac{3b}{2(a - b)^2}$$

EXAMPLE 6-16 Multiply $\dfrac{x^2 - 9}{2x - 10} \cdot \dfrac{ax - 5a}{x^2 - 3x}$.

SOLUTION

$$\frac{x^2 - 9}{2x - 10} \cdot \frac{ax - 5a}{x^2 - 3x} = \frac{(x + 3)(x - 3)}{2(x - 5)} \cdot \frac{a(x - 5)}{x(x - 3)}$$

$$= \frac{(x + 3)(x - 3)}{2(x - 5)} \cdot \frac{a(x - 5)}{x(x - 3)}$$

$$= \frac{a(x + 3)}{2x}$$

The rule for the division of one algebraic fraction by another is the same as the rule for the division of common fractions. Once more, it is best to factor completely and cancel after the division has been converted into multiplication.

EXAMPLE 6-17 Divide $\dfrac{a^2 - 3a^2b}{a^2 - 4b^2} \div \dfrac{ab - 3ab^2}{5a - 10b}$.

SOLUTION

$$\frac{a^2 - 3a^2b}{a^2 - 4b^2} \div \frac{ab - 3ab^2}{5a - 10b} = \frac{a^2 - 3a^2b}{a^2 - 4b^2} \cdot \frac{5a - 10b}{ab - 3ab^2}$$

$$= \frac{a^2(1 - 3b)}{(a + 2b)(a - 2b)} \cdot \frac{5(a - 2b)}{ab(1 - 3b)}$$

$$= \frac{a^2(1 - 3b)}{(a + 2b)(a - 2b)} \cdot \frac{5(a - 2b)}{ab(1 - 3b)}$$

$$= \frac{5a}{b(a + 2b)}$$

EXERCISE 6-4

Perform each of the following indicated operations. Use factoring and cancellation so that the result will be in its lowest terms:

1. $\dfrac{2}{c + d} \cdot \dfrac{3}{c + d}$

2. $\dfrac{5}{x + y} \cdot \dfrac{3}{x - y}$

3. $\dfrac{3x + 6}{2x - 3} \cdot \dfrac{10x - 15}{x + 2}$

4. $\dfrac{x^2 - 6x + 9}{3x - y} \cdot \dfrac{6x - 2y}{x^2 - 9}$

5. $\dfrac{x^2 + 2x - 3}{x^2 - 2x - 8} \cdot \dfrac{x^2 - 3x - 4}{x^2 + x - 6}$

6. $\dfrac{x + 2}{2y + 2a} \cdot \dfrac{y^2 - a^2}{x^2 + 3x + 2}$

7. $\dfrac{s + t}{s^2 + 5s + 6} \cdot \dfrac{5s + 10}{3s + 3t}$

8. $\dfrac{x + 11}{x^3 - 2x^2} \cdot (x^2 - 2x)$

9. $(1 - 2x) \cdot \dfrac{3x + 8}{3 - 5x - 2x^2}$

10. $\dfrac{1 - 4a^2}{9x^2 + 12x + 4} \cdot \dfrac{9x^2 - 4}{1 - 2a}$

11. $\dfrac{3x}{7a^2c} \div \dfrac{2x}{14c}$

12. $\dfrac{8a^2x}{3av^3} \div \dfrac{4by^2}{9a^2v}$

13. $\dfrac{25d^3}{12g^2h} \div \dfrac{15d^2}{4h^2}$

14. $\dfrac{21s^2t}{10ab^2} \div \dfrac{14st^3}{15abc}$

15. $\dfrac{3ab + 9ac}{16ac + 8bc} \div \dfrac{5bd + 15cd}{6a + 3b}$

16. $\dfrac{2kx + 3ky}{2st - 2su} \div \dfrac{2mx + 3my}{3tv - 3uv}$

17. $(a - x) \div \dfrac{a^2 - 2ax + x^2}{5x}$

18. $\dfrac{s^2 - 4t^2}{7x + 2} \div (s - 2t)$

19. $\dfrac{2x^2 + 7x - 15}{3xyz} \div \dfrac{2x^2 - x - 3}{2xyz}$

20. $\dfrac{k^2 + k - 12}{x^2 - 3x} \div \dfrac{k^2 + 2k - 15}{x^2 - 5x + 6}$

6-5 Complex Algebraic Fractions

In algebra also, as in arithmetic, a complex fraction is the fraction that has one or more fractions in its numerator or denominator, or both. *Complex fractions are simplified by reducing the numerator and the denominator into a single fraction and performing the indicated division.*

In a complex fraction, the fraction bar between the main numerator and the denominator should be made longer. Thus, in the complex fraction,

$$\frac{y + \dfrac{5}{8}}{2 - \dfrac{y}{y + 3}}$$

$y + \dfrac{5}{8}$ is the main numerator, and $2 - \dfrac{y}{y + 3}$ is the main denominator.

EXAMPLE 6-18 Simplify the complex fraction:

$$\frac{\dfrac{1}{x + 1} + 5}{1 - \dfrac{3}{x - 1}}$$

SOLUTION The LCD of the main numerator is $x + 1$, and the LCD of the main denominator is $x - 1$. Thus,

$$\frac{\dfrac{1}{x + 1} + 5}{1 - \dfrac{3}{x - 1}} = \frac{\dfrac{1}{x + 1} + \dfrac{5(x + 1)}{(x + 1)}}{\dfrac{x - 1}{x - 1} - \dfrac{3}{x - 1}}$$

$$= \frac{\dfrac{1 + 5(x + 1)}{x + 1}}{\dfrac{(x - 1) - 3}{x - 1}}$$

$$= \frac{\dfrac{5x + 6}{x + 1}}{\dfrac{x - 4}{x - 1}}$$

Dividing now the main numerator by the main denominator, we have

$$\frac{5x + 6}{x + 1} \div \frac{x - 4}{x - 1} = \frac{5x + 6}{x + 1} \cdot \frac{x - 1}{x - 4}$$

$$= \frac{5x^2 + x - 6}{x^2 - 3x - 4}$$

EXERCISE 6-5

Simplify each of the following complex fractions:

1. $\dfrac{\dfrac{1}{x}}{\dfrac{2}{3x}}$

2. $\dfrac{\dfrac{x + 3}{x - 2}}{\dfrac{x - 1}{x + 5}}$

3. $\dfrac{1 + \dfrac{3}{4}}{3 - \dfrac{1}{t}}$

4. $\dfrac{\dfrac{1}{a} - \dfrac{2}{b}}{\dfrac{1}{a}}$

5. $\dfrac{\dfrac{g}{h}}{\dfrac{3}{g} + \dfrac{5}{h}}$

6. $\dfrac{m + \dfrac{1}{m^2}}{\dfrac{1}{m^2} - 1}$

7. $\dfrac{\dfrac{2}{x} - \dfrac{3}{x^2}}{\dfrac{5}{x^2} - \dfrac{1}{x}}$

8. $\dfrac{\dfrac{2 + c}{3 + c}}{\dfrac{c - 5}{c + 1}}$

9. $\dfrac{1 + \dfrac{xy}{x - y}}{\dfrac{xy}{x^2 - y^2} - 1}$

10. $\dfrac{x - \dfrac{1}{x + 2}}{x + \dfrac{1}{x^2 - 4}}$

REVIEW QUESTIONS

1. When is an algebraic fraction in its lowest terms?

2. How do we reduce an algebraic fraction to its lowest terms?

3. State the two-step procedure for finding the LCM of algebraic fractions.

4. State the four-step procedure for the addition or subtraction of algebraic fractions.

5. State the rules for the multiplication and division of algebraic fractions.

REVIEW EXERCISES

Reduce each of the fractions in exercises 1–8 to its lowest terms:

1. $\dfrac{7d}{28h}$

2. $\dfrac{21x}{3xy}$

3. $\dfrac{40kx^2}{32x}$

4. $\dfrac{BC^2}{CD^2}$

5. $\dfrac{m - 1}{7m - 7}$

6. $\dfrac{3d + e}{21d^2 + de - 2e^2}$

7. $\dfrac{2m^2 + 5m + 3}{1 - m^2}$

8. $\dfrac{6x^2 + 3x - 3}{14x^2 - 15x + 4}$

Find the LCM of the expressions in each of exercises 9–11:

9. $3t - r^2,\ 21t - 7r^2$

10. $6y - 6z,\ 8y - 4z,\ 18y - 9z$

11. $x^2 - 16,\ 3x - 12,\ x^2 + 8x + 16$

In each of the exercises 12–21, perform the indicated operations.

12. $\dfrac{7c}{4} + \dfrac{c}{6}$

13. $\dfrac{7d}{3} - \dfrac{d}{8}$

14. $\dfrac{5}{2x} + \dfrac{2}{x - 2}$

15. $\dfrac{7k}{3b} - \dfrac{2k}{8b}$

16. $\dfrac{A}{B^2} - \dfrac{1}{C}$

17. $1 + \dfrac{6}{7 - t}$

18. $\dfrac{A}{x - 1} - 1$

19. $\dfrac{5}{3 - x} + \dfrac{3}{9 - 6x + x^2}$

20. $\dfrac{2k}{2x + 4} + \dfrac{k}{x + 2} - \dfrac{k}{3x + 6}$

21. $\dfrac{7m}{3(x + y)} - \dfrac{2m}{6(x - y)} + \dfrac{6m}{8(x - y)}$

In each of the exercises 22–28, perform the indicated operations:

22. $\dfrac{h - 3}{k + 7} \cdot \dfrac{k^2 + 14k + 49}{2h - 6}$

23. $(1 - 6d) \cdot \dfrac{b - c}{1 - 12d + 36d^2}$

24. $\dfrac{1 - 9x^2}{4x^2 + 12x + 9} \cdot \dfrac{4x^2 - 9}{1 + 3x}$

25. $\dfrac{5x}{18k^2m} \div \dfrac{7x}{3km^2}$

26. $\dfrac{s - u}{d^2 + 4d + 4} \div \dfrac{s^2 - u^2}{d^2 - 4}$

27. $\dfrac{7A}{36x^2 - 60x + 25} \div \dfrac{1}{6x - 5}$

28. $\dfrac{7g - 21h}{D + 2H} \div (g - 3h)$

Simplify each of the following complex fractions:

29. $\dfrac{\dfrac{3}{R}}{\dfrac{5}{6R}}$

30. $\dfrac{\dfrac{x - 2}{x + 5}}{\dfrac{x - 2}{x - 5}}$

31. $\dfrac{\dfrac{7 + d}{c - 5}}{\dfrac{c + 3}{c}}$

32. $\dfrac{\dfrac{1}{m} - \dfrac{2}{k}}{\dfrac{3}{k}}$

33. $\dfrac{d + \dfrac{2}{c^2}}{\dfrac{1}{c^2} + 1}$

Simple Equations

So far we have been concerned only with introductory topics of algebra. We come now to the very important case in which two algebraic expressions are said to be equal, as in the example

$$3x + 2 = 5x - 6$$

The above is a mathematical statement. It says that the expression on the left side of the symbol (=) represents the same number as the expression to the right of it. Such a mathematical statement is called an **equation**.

When there is only one kind of variable in an equation, and that variable is of the first power, the equation is called a **first degree equation in one variable,** *or a* **linear equation in one variable.**

The expression on the left side of the equality symbol is called the **left side of the equation,** and the expression on the right side of the symbol of equality is called the **right side of the equation.**

Consider now once more the equation

$$3x + 2 = 5x - 6$$

Upon substituting 4 for x, both sides of the equation become numerically equal. Thus,

$$3(4) + 2 = 5(4) - 6$$

and

$$14 = 14$$

We say then that 4 is the **root,** or **solution,** of the equation $3x + 2 = 5x - 6$, or that 4 **satisfies** the equation $3x + 2 = 5x - 6$. In general, *the specific number that can replace the variable of an equation and prove the equation a true statement is the root or the solution of that equation.*

When two or more equations have a common root, they are called **equivalent equations**. Thus, the two equations $2x + 3 = 15$ and $3x = 18$ are equivalent equations; their common solution is 6.

7-2 Solving Linear Equations in One Variable

The procedure for finding the root of an equation is known as solving the equation. This procedure consists of changing the given equation into successive equivalent equations. In order to do this, the following four simple and self-evident facts are used:

1. *When the same number is added to two equal quantities, the results are equal.*
2. *When the same number is subtracted from two equal quantities, the results are equal.*
3. *When two equal quantities are multiplied by the same nonzero number, the products are equal.*
4. *When two equal quantities are divided by the same nonzero number, the quotients are equal.*

These operations are simple, but they can be made still simpler. We know that subtraction can be expressed as addition using **opposites**, and that division can be expressed as multiplication using **reciprocals**. Therefore, in solving equations, we can use only addition in the place of the first two of the above four operations, and only multiplication in the place of the last two. Our two new operations from now on will be called **addition principle** and **multiplication principle**, respectively.

Using the addition and the multiplication principles, a given equation can be changed into successively simpler equivalent equations, until the last of these equivalent equations is the solution.

Suppose, for instance, that we are given the equation

$$5x - 3 = 17$$

to solve. An equation is solved when the variable x of the equation stands alone on one side, and everything else in the equation is on the other side. This means that the equation takes the form $x = N$, where N is a number, the solution. Notice that on the left side of the given equation, there are two numbers on the same side with x, the negative number -3 and the numerical coefficient of x, 5. Since x must be left alone on the left side of the equation, these two numbers must be eliminated from that side. This can be done using the addition and the multiplication principles. The solution is as follows:

$$5x - 3 = 17 \qquad \text{(given equation)}$$
$$5x - 3 + 3 = 17 + 3 \qquad \text{[adding } +3, \text{ the opposite of } -3, \text{ to both sides of the equation (addition principle)]}$$
$$5x = 20 \qquad \text{(collecting similar terms)}$$

$$5x \cdot \frac{1}{5} = 20 \cdot \frac{1}{5} \qquad \textit{[multiplying both sides of the equation by 1/5, the reciprocal of 5 (multiplication principle)]}$$

$$x = 4 \qquad \textit{(simplifying)}$$

Notice that every time one of the two principles is used, the resulting equivalent equation is simpler than the one before, until at the end, the last equivalent equation is the solution.

EXAMPLE 7-1 Solve the linear equation $x + 7 = 18$ and check your solution.

SOLUTION To arrive at the equivalent equation $x = N$, only the term 7 has to be eliminated from the left side. Thus,

$$x + 7 = 18 \qquad \textit{(given equation)}$$

$$x + 7 - 7 = 18 - 7 \qquad \textit{[adding -7 to both sides of the equation (addition principle)]}$$

$$x = 11 \qquad \textit{(collecting similar terms)}$$

In checking, we substitute 11 for x in the original form of the equation. Thus,

$$11 + 7 \overset{?}{=} 18 \qquad \text{and} \qquad 18 = 18 \checkmark$$

EXAMPLE 7-2 Solve the linear equation $3x = 24$ and check your solution.

SOLUTION To arrive at the equivalent equation $x = N$, only the numerical coefficient of x, the number 3, must be changed to 1. Thus,

$$3x = 24 \qquad \textit{(given equation)}$$

$$3x \cdot \frac{1}{3} = 24 \cdot \frac{1}{3} \qquad \textit{[multiplying both sides of the equation by the reciprocal of 3 (multiplication principle)]}$$

$$x = 8 \qquad \textit{(simplifying)}$$

Checking:

$$3 \cdot 8 \overset{?}{=} 24 \qquad \text{and} \qquad 24 = 24 \checkmark$$

EXAMPLE 7-3 Solve the equation $\frac{x}{4} = 3$ and check your solution.

SOLUTION To arrive at the equivalent equation $x = N$, the numerical coefficient of x, the fraction $\frac{1}{4}$, must be changed to 1. Thus,

$$\frac{x}{4} = 3 \qquad \textit{(given equation)}$$

$$\frac{x}{4} \cdot \frac{4}{1} = 3 \cdot \frac{4}{1} \qquad \textit{[multiplying both sides of the equation by the reciprocal of 1/4 (multiplication principle)]}$$

$$x = 12 \qquad \textit{(simplifying)}$$

Solving Linear Equations in One Variable

Checking:

$$\frac{12}{4} \overset{?}{=} 3 \quad \text{and} \quad 3 = 3 \checkmark$$

EXAMPLE 7-4 Solve the equation $\frac{2}{3}x + 5 = 9$ and check your solution.

SOLUTION To arrive at the equivalent equation $x = N$, first the term 5 must be eliminated from the left side, and then the numerical coefficient of x must be changed to 1. Thus,

$$\frac{2}{3}x + 5 = 9$$

$$\frac{2}{3}x + 5 - 5 = 9 - 5$$

$$\frac{2}{3}x = 4$$

$$\frac{2}{3}x \cdot \frac{3}{2} = 4 \cdot \frac{3}{2}$$

$$x = 6$$

Checking:

$$\frac{2}{3}(6) + 5 \overset{?}{=} 9 \quad \text{and} \quad 9 = 9 \checkmark$$

The student should notice in the above examples that *the addition principle is used to eliminate terms only,* and that *the multiplication principle is used only to change the numerical coefficient of the variable to 1.*

EXERCISE 7-1

Solve each of the following equations and check the root.

1. $x + 3 = 7$
2. $x + 7 = 12$
3. $x + 12 = 17$
4. $2x = 8$
5. $7x = 21$
6. $16x = 48$
7. $8x = 32$
8. $\frac{x}{2} = 16$
9. $\frac{x}{3} = 9$
10. $\frac{x}{14} = 5$
11. $\frac{x}{21} = 3$
12. $\frac{x}{4} = 7$
13. $\frac{3}{7}x = 3$
14. $\frac{3}{4}x = 6$
15. $\frac{3}{7}x = 6$
16. $\frac{1}{3}x + 11 = 13$
17. $\frac{3}{5}x - 2 = 7$
18. $\frac{1}{8}x + 7 = 10$
19. $\frac{7}{11}x - 9 = 5$
20. $\frac{5}{6}x - 8 = 7$
21. $\frac{3}{2}x + 3 = 27$
22. $\frac{11}{7}x - 13 = 20$
23. $\frac{1}{3}x + 3 = 6$
24. $\frac{1}{8}x - 37 = 3$

Simple Equations

7-3 More About Linear Equations in One Variable

The student must have noticed that in the equations solved so far *when the addition principle is used, and a term is eliminated from one side of the equation, the same term appears on the other side with its sign changed.* We can take advantage of this fact and use a shortcut in solving equations. This shortcut is the process known as **transposition**. *Using transposition, a term is moved from one side of an equation to the other by just changing its sign.* The student should remember, however, that *only terms can be transposed. A coefficient can never be transposed.*

There are, of course, linear equations which are more involved than the equations we have solved so far. Following, however, the four-step procedure given below, any linear equation can be given, step-by-step, one of the forms we have already seen, and then solved readily.

1. *Remove all grouping symbols, and combine similar terms on both sides of the equation.*
2. *Using transposition, move to the left side all terms involving the variable, and to the right side all other terms. Again combine similar terms.*
3. *Multiply both sides of the equation by the reciprocal of the numerical coefficient of the variable.*
4. *Check the solution with the original form of the equation.*

NOTE: It is not necessary to transpose all terms involving the variable to the left side; we can move such terms to the right and all other terms to the left if we want. If we do this, however, the solution will be in the form $N = x$, which is obviously equivalent to $x = N$.

EXAMPLE 7-5 Solve the equation $5x - 2 = 3x + 6$ and check your solution.

SOLUTION

$$5x - 2 = 3x + 6$$
$$5x - 3x = 6 + 2 \quad \text{(transposing all terms involving } x \text{ to the left side, and all other terms to the right)}$$
$$2x = 8 \quad \text{(combining similar terms)}$$
$$x = 4$$

Checking:

$$5(4) - 2 \stackrel{?}{=} 3(4) + 6$$
$$20 - 2 \stackrel{?}{=} 12 + 6 \quad \text{and} \quad 18 = 18 \checkmark$$

EXAMPLE 7-6 Solve the equation $4x - 2(x + 1) = x + 2(x - 4)$ and check your solution.

SOLUTION

$$4x - 2(x + 1) = x + 2(x - 4)$$
$$4x - 2x - 2 = x + 2x - 8 \quad \text{(removing parentheses)}$$
$$2x - 2 = 3x - 8 \quad \text{(combining similar terms)}$$

More About Linear Equations in One Variable 129

$$2x - 3x = -8 + 2 \qquad \text{(using transposition)}$$
$$-x = -6 \qquad \text{(combining similar terms)}$$
$$x = 6 \qquad \text{(multiplying both sides by } -1)$$

Checking:

$$4(6) - 2(6 + 1) \stackrel{?}{=} (6) + 2(6 - 4)$$
$$24 - 14 \stackrel{?}{=} 6 + 4 \qquad \text{and} \qquad 10 = 10\checkmark$$

EXERCISE 7-2

Solve each of the following equations for the variable and check the root.

1. $6x - 2x + 3 = x + 12$
2. $7x - 3x + 3 = 5x - 5$
3. $3x - 5x + 6 = 8x - 84$
4. $11 + 9x = 44 - 2x$
5. $17 - 13x = 7x - 23$
6. $11x + 42 = 18x - 43 - 2x$
7. $8x - 2x - 173 = x + 17$
8. $1208 - 41x = 14x + 53$
9. $38 - 3(x - 2) - 2x = 10 - (x + 1) + x$
10. $7 + (5x - 12) = 72 - 12x - 2(x + 10)$
11. $5x - (x + 3) - 7 = 2x + 36$
12. $3t + 5(t - 8) + 7 = 100 - (2t - 27)$
13. $7(2x - 5) + 71 - 3(x + 2) = 585 - 4(x + 15)$
14. $191 - 3(2x - 5) + 8x = 486 - 7(2x + 8)$
15. $82 - 4(3k - 6) + 5(4k + 2) - 8 = 165 - 11(2k - 3)$
16. $32 + 7(x - 7) - 2(x + 1) = 43 - (2x - 1)$
17. $3(8y + 7) - (3y + 6) + 42 = 9(6y + 7) - 6(5y + 3)$
18. $2(3v - 7) + (7v - 6) + 4 = 17v - 40$
19. $583 - 8(2x - 7) + (3x + 24) = 112 + 5(2x + 64) - (9x + 35)$
20. $17r - 2(2r + 7) + 70 = 2(6r + 18) + 2r - 9$

7-4 Solving Problems Using Linear Equations in One Variable

The solution of "word" problems involves translating the verbal problem into an equation. Since there are a great variety of such problems, no general formula can be given for their solution. This is more so with technical problems which usually require technical background. The solution, therefore, is a matter of background, experience, and judgment, rather than following a number of rules. Experience will come to the student who is careful and who persists in spite of difficulty at first.

The following are not specific rules for the solution of word problems, but rather a very broad general procedure:

1. *Read the problem carefully more than once, until you can say the problem to yourself. Notice carefully the relationship that the various numbers or quantities of the problem have with each other.*

2. *Represent the unknown with an appropriate letter and make a list of the different quantities of the problem.*
3. *Make a rough diagram in the case of a technical problem. Such a sketch showing relations of quantities is extremely helpful in most cases.*
4. *Write an equation using two expressions which are equal.*
5. *Solve the equation and check to see if the solution satisfies the requirements of the problem.*

Common technical problems solved using linear equations in one variable are *motion problems, mixture problems,* and *moment, or lever problems.*

EXAMPLE 7-7 If seven times a number is decreased by eight, the result is 55. Find the number.

SOLUTION In this problem the relation is simple. Representing the unknown by x, we have

$$7x - 8 = 55$$

Solving the equation for x, we find the number to be 9. This result checks with the requirements of the problem since $7(9) - 8$ is 55.

EXAMPLE 7-8 A father is 21 years older than his son. Six years ago, the father was twice as old as the son was. What is the present age of each?

SOLUTION Reading the problem carefully, we notice that the father's age is given in terms of the son's age. Representing the son's age by x, we have the following list of related quantities:

—son's present age: x
—father's present age: $x + 21$
—son's age six years ago: $x - 6$
—father's age six years ago: $(x + 21) - 6$

Since six years ago, the father's age was twice that of the son, we write the equation

$$(x + 21) - 6 = 2(x - 6)$$

Solving the equation, the son is 27 years old, and the father is $27 + 21$, or 48 years old. This result checks with the requirements of the problem, since $48 - 6$, or 42, is twice $27 - 6$, or 21.

EXAMPLE 7-9 A metal alloy contains 20% copper. Another metal alloy contains 30% copper. How many pounds of each alloy should be mixed to make 300 pounds of a third alloy which will contain 26% copper? (Recall that a percent is changed into a decimal by dropping the percent symbol and moving the decimal point two places to the left.)

SOLUTION Reading the problem carefully we realize that if we represent the number of pounds needed from the first alloy with x, then all the other quantities in the problem can be expressed in terms of x. Thus, we have the following list:

—first alloy: x lb
—second alloy $(300 - x)$ lb
—copper in x pounds of first alloy: $0.20(x)$ lb
—copper in $(300 - x)$ pounds of second alloy: $0.30(300 - x)$ lb
—copper in 300 pounds of mixture: $0.26(300)$ lb

Solving Problems Using Linear Equations in One Variable

Since the total amount of copper in the mixture is equal to the sum of the copper in the two alloys contributing to the mixture, we can write the equation:

$$0.20x + 0.30(300 - x) = 0.26(300)$$

Solving the equation, we have

$$0.20x + 90 - 0.30x = 78$$
$$-0.10x = -12$$
$$x = 120$$

Therefore, 120 lb of the first alloy are needed, and $300 - 120$, or 180, lb of the second alloy. Checking, we find that this answer is correct because

$$0.20(120) + 0.30(180) = 0.26(300)$$

EXAMPLE 7-10 The length of a rectangular field is 60 yd longer than twice its width. If the perimeter of the field is 1800-yd long, how long and how wide is the field?

SOLUTION A careful reading of the problem shows that the length of the field is given in terms of the width. Representing the width with the letter w, we have the following quantities of the problem:

—width: w yd
—length: $(2w + 60)$ yd
—perimeter: 1800 yd

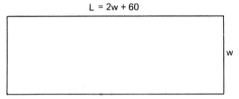

FIG. 7-1

Figure 7-1 shows a sketch of the field with its length and width in terms of w. Since the perimeter is the sum of all four sides, we write the equation

$$(2w + 60) + (2w + 60) + w + w = 1800$$
$$2(2w + 60) + 2w = 1800$$

Solving the equation, $w = 280$. Therefore, the width is 280 yd, and the length is 620 yd. This result is in agreement with the requirements of the problem, since

$$2(620) + 2(280) = 1800$$

EXAMPLE 7-11 A pilot flies his plane to an airfield at A and back to A in 7 hr. A 20-mph wind is a head wind on the way to airfield B, and a tail wind on the way to airfield A. If the plane can go 140 mph when there is no wind, find the distance between the two airfields.

SOLUTION This is a motion problem. Motion of this kind is described by the formula,

$$\text{distance} = \text{speed} \times \text{time}$$

Notice that we can express the distance in this problem in terms of time, since the speed is given. Representing time with the letter t, we have the following list of different quantities of the problem:

—time to fly from A to B: t hr
—time to fly from B to A: $(7 - t)$ hr

—speed of plane flying from A to B: $(140 - 20)$ mph
—speed of plane flying from B to A: $(140 + 20)$ mph
—distance from A to B: $t(140 - 20)$ mi
—distance from B to A: $(7 - t)(140 + 20)$ mi

A diagram of these relations is shown in Fig. 7-2.

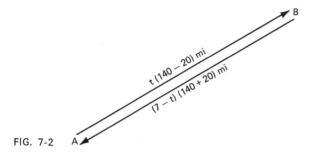

FIG. 7-2

Since the distance from A to B is the same as the distance from B to A, we write the equation

$$t(140 - 20) = (7 - t)(140 + 20)$$
$$120t = 1120 - 160t$$
$$t = 4$$

Thus, the distance between the airfields, A and B is $4(120)$, or $3(160)$, or 480 mi.

EXAMPLE 7-12 One angle of a triangle is 5 degrees less than twice the smallest angle. The third angle of the triangle is 15 degrees larger than the second angle. Find the size of each angle of the triangle.

SOLUTION We notice that two of the angles of the triangle are described in terms of the smallest angle. Representing the size of the smallest angle by x, we have the following list of the three angles in terms of x:

—first angle: x degrees
—second angle: $(2x - 5)$ degrees
—third angle: $(2x - 5) + 15$ degrees

Since the sum of the three angles of any triangle is 180 degrees, we write the equation

$$x + (2x - 5) + (2x - 5) + 15 = 180$$

Solving the equation, $x = 35$ degrees. Therefore, the second angle is 65 degrees, and the third is 80 degrees.

Moment problems require some background on moments, and this is what the next two paragraphs are about.

When two boys, A and B, of different weights sit on a see-saw at different distances from the pivot point F (see Fig. 7-3), the weight W_1 of boy A tends to turn the see-saw counterclockwise around point F, and the weight W_2 of boy B tends to turn the see-saw clockwise around point F.

The product $d_1 W_1$, of the distance d_1 and the weight W_1, is called a **moment**, and so is the product $d_2 W_2$. Moment $d_1 W_1$ is a **counterclockwise moment**, and the moment $d_2 W_2$ is a **clockwise moment**. The law of the lever says that *a bar will be in*

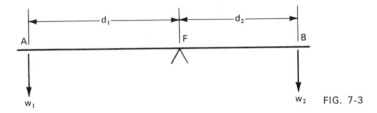

FIG. 7-3

equilibrium when the sum of all the clockwise moments is equal to the sum of all the counterclockwise moments.

EXAMPLE 7-13 A uniform steel rod is resting on point F, and weights of 22.0 kg, 5.00 kg, and 9.00 kg are hanging from the rod as shown in Fig. 7-4. Determine the distance, x, of the point F from the 22.0-kg weight necessary to keep the lever in equilibrium. Neglect the weight of the lever.

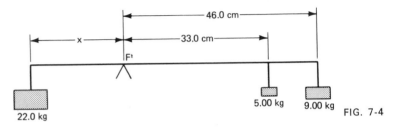

FIG. 7-4

SOLUTION Notice that the 5.00-kg and the 9.00-kg weights with their respective distances produce clockwise moments. According to the law of the lever, the sum of these two moments should be equal to the counterclockwise moment, $22.0x$. Thus, we write the equation,

$$22.0x = 33.0(5.00) + 46.0(9.00)$$

Solving the equation,

$$x = 26.3 \text{ cm}$$

EXERCISE 7-3

Solve each of the following problems.
1. Subtracting 2 from three times a number gives the same result as adding 6 to 2 times that number. Find the number.
2. When 8 is added to three times a certain number, the result is equal to 4 times the number reduced by 13. Find the number.
3. The sum of two numbers is 68, and one of the numbers is three times the other. Find the numbers.
4. A father is 20 years older than his son. In three years, he will be three times as old as his son. Find the age of each.
5. There are 28 students in a computer programming class. If there are 6 more men than women in that class, what is the number of each?
6. In a pile of coins, there are 5 more dimes than nickels, and 4 fewer quarters than dimes.

Simple Equations

The total value of the coins in the pile is $3.15. How many of each kind of coin are there in the pile?

7. The length of a rectangular garden is 1 m less than three times its width. If the perimeter of the garden is 38 m, find the length and the width of the garden.

8. One angle of a triangle is 2 degrees less than three times the smaller angle, and the third angle is 37 degrees more than the second angle. Find the size of each angle.

9. The bill for a redwood sundeck that a carpenter and his assistant built is for $952.00. It was agreed that the carpenter would be paid $8.00 an hour and his assistant $6.00 an hour. If the cost of the materials is $430.00, and the assistant worked 11 hours less than the carpenter, how many hours did each work?

10. A secretary has 31¢ stamps and 15¢ stamps in her desk drawer. If the total value of the stamps is $11.95, and there are 3 more 15¢ stamps than 31¢ stamps, how many stamps of each kind are there in her drawer?

11. Hematite, the best quality of iron ore, contains 70% iron. Taconite, an inferior kind of iron ore contains 30% iron. How many of each kind should be mixed to give 120 t of ore containing 45% iron?

12. In China, there are many more bicycles than cars. In a certain small town, there are 12,114 bicycles, and a total of 26,272 tires (ignoring spare tires) of cars and bicycles in circulation. How many cars are there in that town?

13. A spice wholesaler wants to mix two kinds of tea to make a blend of 200.0 kg worth $11.25 per kilogram. The first of the two kinds of tea he is mixing sells for $10.00 a kilogram, and the second $12.00 a kilogram. How many kilograms of each kind should he mix?

14. If a boy weighing 120 lb sits on a see-saw 5 ft from the pivot point, where should his father sit to balance the see-saw if he weighs 150 lb?

15. Two boys weighing 140 lb and 105 lb, respectively, balance each other on a see-saw sitting 7 ft apart. What is the distance of each from the pivot point?

16. If a father, weighing 180 lb, and his older son, weighing 140 lb, sit on a see-saw 6 ft each from the pivot point, where should his younger son sit to balance the see-saw, if the boy's weight is 80 lb?

17. Weights of 2 kg, 7 kg, and 3 kg are hanging from a rigid, uniform steel bar, supported at point F, as shown in Fig. 7-5. Ignoring the weight of the steel bar, what should be the weight of the fourth weight, 50 cm to the left of point F, in order to keep the steel bar in equilibrium?

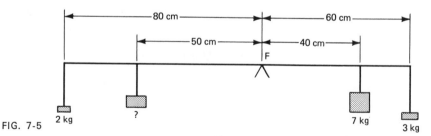

FIG. 7-5

18. Weights of 30 kg and 38 kg, respectively, are hung from the two ends of a rigid rod 1-m long. The rod is supported in a horizontal position by a pivot. Neglecting the weight of the rod, what should the distance of the pivot be from the 30-kg weight?

19. A motor boat can run 25 mph in still water. The boat starts at 7:00 AM going upstream in a river whose current runs at 5 mph. How far up the river can the boat go and still be back at 12:00 noon?

Solving Problems Using Linear Equations in One Variable

20. A crew can row at the rate of 12 mph in still water. They find that it takes them twice as long to go up a river as to go down. Find the rate of the flow of the river.

21. Two freight trains leave from the same station at 10 AM going in opposite directions. At what time will the trains be 120 mi apart if the speed of the first train is 50 mph and the speed of the second train is 30 mph?

22. A truck leaves town A at 7:00 AM going to town B at the speed of 35 mph. A tractor leaves town B at the same time going to town A at the speed of 15 mph. The distance from A to B is 200 mi. At what time will the truck and the tractor meet? How far will they be from town A when they meet?

23. Two cars leave town A at the same time going in the same direction. The speed of one car is 40 mph, and the speed of the other car is 55 mph. In how many hours will the cars be 45 mi apart?

24. An airplane flew from airport A to airport B in 2 hr with a 30 kph tail wind. Because of the head wind, the return flight took $2\frac{1}{2}$ hr. What is the distance from A to B? What is the speed of the plane in still air?

25. A car leaves town A going to town B, and at the same time another car leaves town B going to town A. If the distance between A and B is 210 mi, and the rates of the cars are 40 mph and 50 mph, respectively, when will the cars meet?

REVIEW QUESTIONS

1. When is an equation called a linear equation in one variable?
2. What is the root or solution of an equation?
3. When are two or more equations called equivalent equations?
4. Which of the two principles, the addition principle or the multiplication principle, will eliminate a term?
5. Which of the two principles, the addition principle or the multiplication principle, will change the numerical coefficient of the variable to 1?
6. What do we do to both sides of the equation in using the addition principle?
7. What do we do to both sides of the equation in using the multiplication principle?
8. How is a term transposed from one side of an equation to the other?
9. Can a coefficient be transposed?
10. State the four-step procedure for solving any type of linear equation in one variable.

REVIEW EXERCISES

Solve each of equations 1–6.

1. $13x - 2(4x - 2) = 3(2x - 1) - (3x - 25)$
2. $925 - (6h - 17) = 3(3h - 8) - 24$
3. $3(x - 3) + 5(x + 2) = 4(x - 10) + 3(x + 51) - 41$
4. $2(3x + 4) - 3(x + 5) = (4x + 10) - (2x + 6)$
5. $7x + 2(81 - 3x) = 3(x + 8) + 5(50 - 2x)$
6. $3(5x - 21) - 2(45 - x) = (3 + 3x) + 2(4x + 6)$
7. Six more than four times a number is 74. Find the number.
8. Eleven less than eight times a number is 157. Find the number.

9. A collection of quarters, dimes, and nickels totals $5.10. There are three times as many nickels as there are quarters, and 2 more dimes than nickels. How many coins of each kind are there in the collection?

10. A landscape contractor asked for $390.00 for a job he completed. He charges $9.00 per hour for himself and $6.00 per hour for his assistant. How many hours did each work if the contractor worked 9 hours more than his assistant, and the cost of the materials was $84.00?

11. The large cantilever bridge shown in Fig. 7-6 has three sections. Each section extending from the banks of the river is 95 m longer than the central section. If the total length of the bridge is 535 m, what is the length of each of the two outer arms, and the length of the central span?

FIG. 7-6

12. Bauxite, an aluminum ore, is considered of good quality if it contains 45% or more aluminum oxide. A poor quality bauxite containing 36% aluminum oxide, was delivered to a mill. To improve its quality, the ore will be mixed with another ore containing 52% aluminum oxide. How many tons of each should be mixed to produce 75 tons of bauxite containing 42% aluminum oxide?

13. At what distance from the fulcrum should an 105-lb boy sit on a see-saw to balance a 70-lb boy sitting 6 ft from the fulcrum?

14. A boat can go 30 mph in still water. How far downstream can the boat go in a river with a 6 mph current, if the boat must be back to its starting point in 5 hr?

8

Fractional Equations.
Formulas

8-1 Fractional Equations

In this book we will call an equation a **fractional equation** if one or more terms are fractions. Thus,

$$\frac{1}{x} + \frac{4x}{3} = \frac{7}{3}, \qquad \frac{3x}{4} + 2x = 3x - 2, \qquad \text{and} \qquad \frac{1}{x} = \frac{1}{8 - x}$$

are called fractional equations.

It is very easy to remove the fractions from a fractional equation, and thus change it to the kind of equation which we already know how to solve. Once this is done, the four-step procedure of the previous chapter can be used in this case also. *To clear a fractional equation of fractions, multiply all terms of both sides of the equation by the LCM of all the denominators in the equation.*

EXAMPLE 8-1 Solve the fractional equation $\dfrac{3}{2x} + \dfrac{5}{4} = \dfrac{4}{x}$.

SOLUTION First we multiply all three terms of the equation by $4x$, the LCM of x, $2x$, and 4. Thus,

$$4x \cdot \frac{3}{2x} + 4x \cdot \frac{5}{4} = 4x \cdot \frac{4}{x}$$

Simplifying, we have

$$6 + 5x = 16$$

and solving,

$$5x = 10$$

$$x = 2$$

Checking:

$$\frac{3}{2 \cdot 2} + \frac{5}{4} \overset{?}{=} \frac{4}{2} \qquad \text{and} \qquad \frac{8}{4} = 2\checkmark$$

We can make the procedure of clearing an equation of fractions simpler by multiplying the numerator of each fraction by the number of times its denominator will go into the LCM.

EXAMPLE 8-2 Solve the equation $\dfrac{5}{x} - \dfrac{3}{2x} = \dfrac{7}{6}$.

SOLUTION

$$\frac{5}{x} - \frac{3}{2x} = \frac{7}{6}$$

$$5 \cdot 6 - 3 \cdot 3 = 7 \cdot x \qquad \text{\textit{(by multiplying each numera-}}$$
$$\text{\textit{tor by the number of times its}}$$
$$\text{\textit{denominator goes into 6x, the}}$$
$$\text{\textit{LCM)}}$$

$$21 = 7x$$

$$x = 3$$

Checking:

$$\frac{5}{3} - \frac{3}{6} \overset{?}{=} \frac{7}{6} \qquad \text{and} \qquad \frac{7}{6} = \frac{7}{6} \checkmark$$

EXAMPLE 8-3 Solve the equation $\dfrac{2x}{x+1} - 4 = \dfrac{7-3x}{x+1}$.

SOLUTION

$$\frac{2x}{x+1} - 4 = \frac{7-3x}{x+1}$$

$$2x - 4(x+1) = 7 - 3x \qquad \text{\textit{(by multiplying each numer-}}$$
$$\text{\textit{ator by the number of times}}$$
$$\text{\textit{its denominator goes into}}$$
$$\text{\textit{x + 1, the LCM)}}$$

$$2x - 4x - 4 = 7 - 3x$$

$$x = 11$$

Checking:

$$\frac{2 \cdot 11}{11+1} - 4 \overset{?}{=} \frac{7 - 3 \cdot 11}{11+1}$$

$$\frac{22}{12} - \frac{48}{12} \overset{?}{=} \frac{7 - 33}{12}$$

$$-\frac{26}{12} = -\frac{26}{12} \checkmark$$

EXAMPLE 8-4 Solve the equation $\dfrac{2}{x-2} = \dfrac{x}{x-2} - 2$.

SOLUTION

$$\frac{2}{x-2} = \frac{x}{x-2} - 2$$

$$2 = x - 2(x-2) \qquad \text{\textit{[by multiplying each numera-}}$$
$$\text{\textit{tor by the number of times}}$$
$$\text{\textit{its denominator goes into}}$$
$$\text{\textit{(x - 2), the LCM]}}$$

Fractional Equations

$$2 = x - 2x + 4$$
$$-2 = -x$$
$$x = 2$$

Checking:

$$\frac{2}{2-2} \overset{?}{=} \frac{2}{2-2} - 2 \quad \text{and} \quad \frac{2}{0} \overset{?}{=} \frac{2}{0} - 2$$

Since division by zero is meaningless, the root $x = 2$ is not valid.

NOTE: Why do we say that division by zero is meaningless? Suppose $a \div 0 = x$. Then, $0 \cdot x$ should equal a. But this is impossible since zero times any number equals zero. Therefore, division by zero is meaningless.

Since the solution $x = 2$ does not satisfy the original equation, the equation has no solution. And in general, *any time the check of a solution causes a division by zero, that root is not a solution.* The root $x = 2$ is not a real solution. It "slipped in" when we used the method of multiplying all terms of the original equation by the LCM. Such solutions are called **extraneous solutions,** or "solutions from outside." This should make the student see the value of checking solutions in the original equation.

EXAMPLE 8-5 Solve the equation $\dfrac{5}{x} + \dfrac{3}{x-1} = \dfrac{2}{x^2 - x}$.

SOLUTION

$$\frac{5}{x} + \frac{3}{x-1} = \frac{2}{x^2 - x}$$

$$5(x - 1) + 3x = 2 \qquad \textit{[by multiplying each numerator by the number of times its denominator goes into } x(x-1), \textit{ the LCM]}$$

$$5x - 5 + 3x = 2$$
$$8x = 7$$
$$x = \frac{7}{8}$$

It is not simple to check this solution in the original equation. A quick check, however, will show that no zero denominator will result if we substitute $\dfrac{7}{8}$ for x in the original equation. This suggests that the solution $\dfrac{7}{8}$ is real if there is no computational error. In the case of a fractional root, it is often easier to rework the problem than to do a regular check.

When both sides of an equation are simple fractions, instead of multiplying both sides by the LCM to clear the equation of fractions, we can multiply the numerator of each side by the denominator of the other side. This process is called **cross-multiplication.** In such cases, multiplying by the LCM and cross-multiplying give the same result.

Thus, given the equation, $\dfrac{6}{x} = \dfrac{4}{2}$

multiplying by the LCM   cross-multiplying

$$\frac{6}{x} = \frac{4}{2} \qquad\qquad \frac{6}{x} \;\; \frac{4}{2}$$

$$2 \cdot 6 = 4x \qquad\qquad 4x = 6 \cdot 2$$

$$12 = 4x \qquad\qquad 4x = 12$$

Fractional Equations. Formulas

EXAMPLE 8-6 Solve the equation $\dfrac{2}{x+2} = \dfrac{3}{4}$.

SOLUTION

$$\frac{2}{x+2} = \frac{3}{4}$$

$$2 \cdot 4 = 3(x+2) \quad \textit{(by cross-multiplying)}$$

$$8 = 3x + 6$$

$$x = \frac{2}{3}$$

Checking:

$$\frac{2}{\frac{2}{3}+2} \overset{?}{=} \frac{3}{4}$$

$$\frac{2}{\frac{8}{3}} \overset{?}{=} \frac{3}{4}$$

$$\frac{3}{4} = \frac{3}{4} \checkmark$$

Solve each of the fractional equations 1–30 and check your answer.

1. $\dfrac{x}{3} + \dfrac{x}{6} = 2$

2. $\dfrac{2}{x} - \dfrac{1}{3} = \dfrac{1}{x}$

3. $\dfrac{3}{x} - \dfrac{1}{x} = \dfrac{2}{5}$

4. $\dfrac{3}{x} + \dfrac{1}{4} = \dfrac{5}{8}$

5. $\dfrac{2}{x} - \dfrac{1}{3} = 2$

6. $\dfrac{x}{2} + \dfrac{x}{3} = 1$

7. $\dfrac{x}{8} + \dfrac{11}{24} = \dfrac{x+1}{6}$

8. $\dfrac{x-3}{3} = \dfrac{x}{7} - \dfrac{5}{21}$

9. $\dfrac{2x+2}{6} + 1 = \dfrac{7x-5}{10}$

10. $\dfrac{5-x}{9} - \dfrac{x+2}{16} = \dfrac{1}{12}$

11. $\dfrac{3g-1}{4} - \dfrac{2g+1}{7} = 1$

12. $\dfrac{x}{x-5} = \dfrac{5}{x-5} + 3$

13. $\dfrac{x+2}{6} + \dfrac{x-7}{12} = 2$

14. $\dfrac{3x-1}{4} - \dfrac{x+5}{3} = 1$

15. $\dfrac{2x-4}{3} - 5 = \dfrac{x+1}{5}$

16. $\dfrac{7x-6}{11} + 1 = \dfrac{x+8}{4}$

17. $2 + \dfrac{1}{x-1} = \dfrac{x}{x-1}$

18. $\dfrac{5x-1}{9} - 3 = \dfrac{2x-1}{7}$

19. $\dfrac{5}{x} - \dfrac{3}{x-2} = \dfrac{6}{x^2-2x}$

20. $\dfrac{x}{x-4} + 3 = \dfrac{4}{x-4}$

21. $\dfrac{3x}{x-3} - 4 = \dfrac{x}{x-3}$

22. $\dfrac{2x-1}{3} + \dfrac{x}{2} = \dfrac{3x}{4}$

23. $\dfrac{3h}{4} - \dfrac{h}{2} = \dfrac{h-1}{2} - \dfrac{h}{8}$

24. $\dfrac{x}{2} - \dfrac{2x+1}{6} = \dfrac{x}{4} - \dfrac{3x-8}{12}$

25. $\dfrac{1}{3} = \dfrac{x}{21}$

26. $\dfrac{x}{5} = \dfrac{1}{9}$

Fractional Equations

27. $\dfrac{21}{24} = \dfrac{x}{32}$ **28.** $\dfrac{15}{18} = \dfrac{35}{h}$

29. $\dfrac{14}{4} = \dfrac{21}{x}$ **30.** $\dfrac{2d}{7} = \dfrac{3d-1}{8}$

31. One third of a man's age 12 years ago equals one fourth the age he will be in 8 years. What is his present age?

32. Divide the number 46 into two parts, such that, if one part is divided by 7, and the other part is divided by 3, the sum of the two quotients will be 10.

33. A sum of money is to be divided among three persons. The first person is to get $25 less than one half, the second person $20 less than one third, and the third person $9 more than one fourth. How much is the sum and how much does each person receive?

8-2 Formulas

A formula is a mathematical or a scientific rule or law in the form of an equation. A very common physics rule says, for instance, that the distance a car will travel at a given speed in a given number of hours is equal to the product of the speed and the time. This physics rule in the form of a formula is

$$d = rt$$

where d is the distance, r is the rate of motion, and t is the time. In mathematics, science, engineering, business, and in other fields, many formulas are used.

In many cases, it becomes necessary to change the form of a formula because the letter for which the formula is solved is known, and another letter in the formula is unknown. Suppose the distance a car has traveled, and the time it took the car to travel that distance are both known. To find what to do to calculate the speed of the car, we should solve the formula $d = rt$ for r. We have

$$r = \frac{d}{t}$$

Our original formula, now solved for r, becomes a rule for finding the speed. To solve a formula for any of its letters is nothing more than solving an equation for one of its letters. Manipulating a formula, that is, solving the formula for a given letter, is highly important and is one of the immediate applications of algebra in technology. Here is one more illustration. A very familiar formula from electricity is

$$E = IR$$

Using the formula in this form, we can find the voltage E in an electric circuit by multiplying the current I by the resistance R. If, however, the voltage E, and the current I are known, to find the rule for finding the resistance, all we have to do is to solve the formula $E = IR$ for R. In the same way, we can find the rule for finding the current by solving the formula $E = IR$ for I.

EXAMPLE 8-7 Solve the formula $E = IR$ for R.

SOLUTION

$$E = IR$$

$$\frac{E}{I} = R$$

or

$$R = \frac{E}{I}$$

EXAMPLE 8-8 Solve the formula $E = IR$ for I.

SOLUTION

$$E = IR$$

$$\frac{E}{R} = I$$

or

$$I = \frac{E}{R}$$

The student is reminded that in a product of several factors, the coefficient of any of the factors is the product of all the other factors. Thus, in abc, the coefficient of a is bc, the coefficient of b is ac, and the coefficient of c is ab. Also, in $ax(b + c)$, the coefficient of a is $x(b + c)$, the coefficient of x is $a(b + c)$, and the coefficient of $(b + c)$ is ax.

EXAMPLE 8-9 The formula for finding the Celsius scale temperature corresponding to a Fahrenheit scale temperature is $C = \frac{5}{9}(F - 32)$. Solve this formula for F; that is, find the formula for changing a Celsius scale temperature to the corresponding Fahrenheit scale temperature.

SOLUTION

$$C = \frac{5}{9}(F - 32)$$

$$\frac{9}{5}C = F - 32 \qquad \text{[\textit{multiplying both sides by the reciprocal of the coefficient of} $(F - 32)$]}$$

$$\frac{9}{5}C + 32 = F$$

or

$$F = \frac{9}{5}C + 32$$

EXAMPLE 8-10 The formula $\frac{1}{f} = (n - 1)\left(\frac{1}{R_1} + \frac{1}{R_2}\right)$ is known as the lensmaker's equation. Solve this formula for R_2. (R_1 and R_2 are different variables and should be treated as such.)

SOLUTION

$$\frac{1}{f} = (n - 1)\left(\frac{1}{R_1} + \frac{1}{R_2}\right)$$

$$\frac{1}{f(n - 1)} = \frac{1}{R_1} + \frac{1}{R_2} \qquad \left[\textit{multiplying both sides by the reciprocal of the coefficient of} \left(\frac{1}{R_1} + \frac{1}{R_2}\right)\right]$$

Formulas

$$R_1 R_2 = f R_2(n-1) + f R_1(n-1)$$ *(multiplying all terms by the LCM)*

$$R_1 R_2 - f R_2(n-1) = f R_1(n-1)$$ *(by transposition)*

$$R_2[R_1 - f(n-1)] = f R_1(n-1)$$ *(by factoring)*

$$R_2 = \frac{f R_1(n-1)}{R_1 - f(n-1)}$$ *(multiplying both sides by the reciprocal of the coefficient of R_2)*

EXERCISE 8-2

Solve each of the following formulas for the letter indicated:

1. $I = PRT$ $P = ?$

2. $R = R_1 + R_2$ $R_1 = ?$

3. $S = v_0 t + \frac{1}{2} a t^2$ $v_0 = ?$

4. $v = v_0 + at$ $t = ?$

5. $f = \frac{R}{2}$ $R = ?$

6. $R = \frac{I}{PT}$ $T = ?$

7. $I = \frac{E - e}{R}$ $e = ?$

8. $y = mx + b$ $m = ?$

9. $D = \frac{W}{V}$ $V = ?$

10. $\frac{P_1}{P_2} = \frac{T_1}{T_2}$ $T_2 = ?$

11. $\frac{V_1}{V_2} = \frac{T_1}{T_2}$ $V_2 = ?$

12. $\frac{P_1 V_1}{P_2 V_2} = \frac{T_1}{T_2}$ $V_2 = ?$

13. $\frac{V_1}{V_2} = \frac{P_1}{P_2}$ $P_2 = ?$

14. $a = \frac{V_2 - V_1}{t}$ $V_1 = ?$

15. $r = \frac{FR}{W}$ $R = ?$

16. $F = \frac{1}{2} W$ $W = ?$

17. $F = \frac{1}{n} W$ $n = ?$

18. $F = \frac{Wl}{L}$ $l = ?$

19. $P = \frac{Fs}{t}$ $s = ?$

20. $Ft = M(V_2 - V_1)$ $V_2 = ?$

21. $\frac{1}{R} = \frac{1}{R_1} + \frac{1}{R_2}$ $R_2 = ?$

22. $I = \dfrac{E}{r + \dfrac{R}{n}}$ $R = ?$

REVIEW QUESTIONS

1. What is a fractional equation?
2. How is a fractional equation cleared of fractions?
3. Why is division by zero meaningless?
4. If, when checking the solution of a fractional equation, a zero denominator results, what do we say about that solution?
5. What is an extraneous solution?
6. In which case of fractional equations can we use cross-multiplication?
7. What is a formula?

REVIEW EXERCISES

Solve each of fractional equations 1–6 and check your answer.

1. $\dfrac{3}{2x - 8} - \dfrac{1}{x - 4} = \dfrac{1}{x - 2}$

2. $\dfrac{5n + 2}{3n + 2} = \dfrac{4}{3n + 2} + 1$

Fractional Equations. Formulas

3. $\dfrac{x-4}{x-7} - 2 = \dfrac{3}{x-7}$

4. $\dfrac{8}{3u+7} = \dfrac{10}{5u}$

5. $\dfrac{3k+6}{2} = \dfrac{5k+7}{3}$

6. $\dfrac{6}{2d-4} = \dfrac{1}{d-6}$

7. The numerator of a fraction is one less than its denominator. If the numerator is multiplied by 5, and the denominator by 3, the result will be $\dfrac{5}{4}$. Find the fraction.

8. Divide the number 66 into two parts, such that, if one part is divided by 8, and the other part is divided by 13, the sum of the two quotients will be 7.

9. Three persons drive a car from St. Louis, MO to New York City. The first person drove 10 mi more than one third of the distance, the second person drove 5 mi more than one fourth of the distance, and the third person drove 93 mi less than one half of the distance. What is the distance from St. Louis to New York City?

Solve each of formulas 10–16 for the letter indicated.

10. $Q = mc(t_2 - t_1)$ $t_2 = ?$

11. $\dfrac{1}{p} + \dfrac{1}{q} = \dfrac{1}{f}$ $q = ?$

12. $I = \dfrac{E}{R_1 + R_2 + R_3}$ $R_1 = ?$

13. $n = \dfrac{NT_1T_2}{t_1t_2}$ $t_1 = ?$

14. $v = v_0 + at$ $t = ?$

15. $W = \dfrac{2PR}{R-r}$ $r = ?$

16. $s = \dfrac{W}{Z}(d - x)$ $x = ?$

9

Dimensional Analysis

9-1 Basic Ideas About Dimension

There are three dimensions in geometry: **length, width,** and **height.** From this point of view, the world we live in is three dimensional. The student should notice at this point that although we say that there are three geometrical dimensions, all are of the same basic kind: they are length.

There is more to the concept of dimension from the physics point of view. In physics there are three kinds of basic quantities. The three fundamental quantities, for the gravitational system of physics, are **length** (L), **force** (F), and **time** (T). (In the absolute system of physics, the fundamental quantities are length, mass, and time.) To measure these fundamental quantities, **fundamental units** are used, which are different for different systems. In the British engineering system of units, the fundamental units are the **foot**, the **pound-force**, and the **second**. This system of units is used in the United States in engineering, and essentially in daily life. The fundamental units for the mks (meter, kilogram, second) system of units are the **meter**, the **newton**, and the **second**. All other units for the measurement of other physical quantities are expressed in terms of length, force, and time, and are derived from the fundamental units. Units so derived are called **derived units**. For instance, the unit of area is the square foot, and it is a derived unit because it is the product of a length times a length (L^2). The unit of velocity is the foot per second, and it is a derived unit since it is the quotient of a length divided by time (L/T). Similarly, the unit for density is pounds per cubic foot, and it is a derived unit since it is the quotient of a force divided by a length raised to the third power (F/L^3). In these three examples, we say that L^2, L/T, and F/L^3 are the **dimensions** of area, average velocity, and density, respectively.

Technical problems and science and engineering formulas involve physical quantities and dimensions. When a new formula is developed in engineering or physics, it is subjected to a test for its correctness. The technique of such a test is known as **dimensional analysis**.

We do not have to check the soundness of formulas in this book, but rather use formulas correctly. We will also have to convert from one unit to another and to work technical problems, often involving several units. To do this, a method is needed to make the procedure simple and safe from mistakes as much as possible. This method is based on dimensional analysis, and it will be introduced and illustrated promptly.

9-2 Denominate Numbers

Numbers representing a physical quantity are called **denominate numbers**. *Denominate numbers have both a magnitude and a dimension.* The dimension of a denominate number is indicated by the unit in which the physical quantity is measured. Thus, 5 m, 3 lb, 2 hr, and 11 cm are denominate numbers.

Two denominate numbers may be added, subtracted, multiplied, or divided only if both their dimensions and their units are of the same kind. In all these basic operations, the unit part of a denominate number should be treated like a literal factor. For example:

1. $5 \text{ cm} + 8 \text{ cm} = 13 \text{ cm}$
2. $9 \text{ hr} - 7 \text{ hr} = 2 \text{ hr}$
3. $(3 \text{ ft})(5 \text{ ft}) = 15 \text{ ft}^2$
4. $12 \text{ in.}^2 \div 4 \text{ in.} = 3 \text{ in.}$
5. $(4 \text{ m})^2 = 16 \text{ m}^2$

Notice that all these operations can be completed because the above basic requirement, that both the dimensions and the units of the two denominate numbers must be of the same kind, is satisfied.

It is not so with the following operations which cannot be completed:

6. $3 \text{ hr} + 8 \text{ ft} = ?$
7. $12 \text{ mi} - 3 \text{ yd} = ?$
8. $(3 \text{ ft})(4 \text{ in.}) = ?$
9. $18 \text{ m}^2 - 2 \text{ cm}^2 = ?$

Notice that the operations (6), (7), (8), and (9) cannot be completed because the above basic requirement, that both the dimensions and the units of the two denominate numbers must be of the same kind, is not satisfied. The operations in (7), (8), and (9) are not really impossible; an appropriate conversion in each case is all that is needed for those operations to be completed. We will take up the subject of conversions again in the next section.

There are also denominate numbers in which a dimension of the same kind is measured in two or three units of different kind. Such denominate numbers are called **compound denominate numbers**. Thus, 2 lb 3 oz, 5 ft 7 in., and 4 hr 41 min 12 sec are compound denominate numbers. *To add or subtract two compound denominate numbers, the numbers must be arranged in columns of the same unit.*

EXAMPLE 9-1 Add 3 hr 28 min 11 sec and 4 hr 17 min 26 sec.

SOLUTION

3 hr	28 min	11 sec
4 hr	17 min	26 sec
7 hr	45 min	37 sec

EXAMPLE 9-2   Add 12 lb 14 oz and 5 lb 7 oz.

SOLUTION

$$
\begin{array}{r r}
12\ \text{lb} & 14\ \text{oz} \\
5\ \text{lb} & 7\ \text{oz} \\
\hline
17\ \text{lb} & 21\ \text{oz}
\end{array}
$$

Since 21 oz is equal to 1 lb 5 oz, the final answer is 18 lb 5 oz.

EXAMPLE 9-3   Subtract 3 ft 7 in. from 12 ft 5 in.

SOLUTION

$$
\begin{array}{r r}
12\ \text{ft} & 5\ \text{in.} \\
3\ \text{ft} & 7\ \text{in.}
\end{array}
$$

Since 7 in. is more than 5 in., we borrow 1 ft from 12 ft and add it to the 5 in. We rewrite and complete the subtraction as follows:

$$
\begin{array}{r r}
11\ \text{ft} & 17\ \text{in.} \\
3\ \text{ft} & 7\ \text{in.} \\
\hline
8\ \text{ft} & 10\ \text{in.}
\end{array}
$$

Compound denominate numbers may be multiplied or divided by a constant if each part of the compound denominate number is multiplied or divided by the constant.

EXAMPLE 9-4   Multiply 7 lb 11 oz by 6.

SOLUTION

$$
\begin{array}{r r}
7\ \text{lb} & 11\ \text{oz} \\
& 6 \\
\hline
42\ \text{lb} & 66\ \text{oz}
\end{array}
$$

Simplifying this product, the final answer is 46 lb 2 oz.

EXAMPLE 9-5   Divide 53 hr 7 min 45 sec by 3.

SOLUTION   To divide 53 hr 7 min 45 sec by 3, we have to perform three divisions. We start with the hours.

$$
\begin{array}{r}
17\ \text{hr} \\
3\,\overline{)53\ \text{hr}} \\
\underline{3} \\
23 \\
\underline{21} \\
2\ \text{hr} \qquad 2\ \text{hr} = 120\ \text{min}
\end{array}
$$

The remaining 120 min are added to the 7 min, and the second division becomes

$$
\begin{array}{r}
42\ \text{min} \\
3\,\overline{)127\ \text{min}} \\
\underline{12} \\
07 \\
\underline{6} \\
1\ \text{min} \qquad 1\ \text{min} = 60\ \text{sec}
\end{array}
$$

Dimensional Analysis

The remaining 60 sec are added to the 45 sec, and the division becomes

$$
\begin{array}{r}
35 \text{ sec} \\
3\overline{\smash)105 \text{ sec}} \\
\underline{9} \\
15 \\
\underline{15} \\
0
\end{array}
$$

The final quotient then is 17 hr 42 min 35 sec.

EXERCISE 9-1

Perform each of the indicated operations in exercises 1–13. In each case where you conclude that an operation cannot be completed (in the form in which it is given,) explain your conclusion.

1. $7\ell + 2\ell$ 2. $15 \text{ gal} - 3 \text{ hr}$ 3. $9 \text{ m} - 7 \text{ m}$
4. $21 \text{ kg} + 54 \text{ g}$ 5. $17 \text{ lb} + 21 \text{ lb}$ 6. $23 \text{ mi} - 7 \text{ mi}$
7. $9 \text{ ft} + 4 \text{ gal}$ 8. $(8 \text{ ft})(3 \text{ ft})$ 9. $21 \text{ in.}^3 \div 7 \text{ in.}$
10. $24 \text{ ft}^2 \div 6 \text{ in.}$ 11. $(6 \text{ m})(4 \text{ cm})$ 12. $(3 \text{ m}^2)(2 \text{ m})$
13. $28 \text{ in.}^3 \div 7 \text{ in.}^2$

In exercises 14–18, perform the stated operations of compound denominate numbers. Arrange the numbers in columns of the same kind of units, and wherever applicable, simplify the result.

14. Add 8 yd 2 ft 2 in. and 7 yd 1 ft 5 in.
15. Subtract 2 hr 11 min 17 sec from 5 hr 8 min 3 sec.
16. Add 3 gal 3 qt 1 pt and 4 gal 2 qt.
17. Add 9 tons 618 lb and 3 tons 1817 lb.
18. Subtract 2 yd 9 in. from 9 yd 2 ft 7 in.

Perform each operation in exercises 19–29. Whenever applicable simplify the result.

19. Multiply 18.0 lb 7.0 oz by 8.
20. Multiply 5.0 gal 3.0 qt 1.0 pt by 5.
21. Multiply 2.0 tons 1267.0 lb by 2.
22. Multiply 3.0 mi 2786.0 ft by 4.
23. Multiply 4.0 yd 1.0 ft 7.0 in. by 8.
24. Multiply 23.0 ft 11.0 in. by 9.
25. Divide 19 lb 14 oz by 6.
26. Divide 39 gal 1 qt 1 pt by 5.
27. Divide 13 tons 448 lb by 3.
28. Divide 19 mi 1002 ft by 2.
29. Divide 26 hr 18 min 32 sec by 4.

9-3 Conversion of Units

In this section we will see a method of conversions different from the one we used in Chapter 2. Here we will use in calculations the units as well as the numbers of the denominate numbers. The units will be treated like literal factors, obeying the usual laws of algebra.

We start the method with a simple conversion. Suppose we want to convert

3.28 m to centimeters. The conversion factor for this case is $1\text{ m} = 100\text{ cm}$. We express this conversion factor as the fraction $100\text{ cm}/1\text{ m}$ (read "100 centimeters per meter"). The fractional conversion factor $100\text{ cm}/1\text{ m}$ is equivalent to 1. Therefore, when a denominate number is multiplied by it, the value of the denominate number is not really changed. We indicate now the multiplication of the given denominate number and the fractional conversion factor as follows:

$$3.28\text{ m} = \frac{3.28\text{ m}}{1} \cdot \frac{100\text{ cm}}{1\text{ m}}$$

The only thing left to do now is to perform the indicated operations, remembering to treat the units as literal factors. Thus,

$$3.28\text{ m} = \frac{3.28\ \cancel{\text{m}}}{1} \cdot \frac{100\text{ cm}}{1\ \cancel{\text{m}}}$$

$$= 328\text{ cm}$$

Notice that of the two ways we could have written the conversion factor as a fraction, $\dfrac{100\text{ cm}}{1\text{ m}}$ or $\dfrac{1\text{ m}}{100\text{ cm}}$, we chose the one with the unit m in the denominator. We did so because the unit m that is to be converted is in the numerator of the denominate number.

The following is a three-step procedure for this method of conversion. To convert a denominate number from one unit of measure to another:

1. *Write the given denominate number as a fraction.*
2. *Choose from the table of conversion factors the correct factor for the case, and write it as a fraction.*
   a. *Have the unit to be converted in the numerator of the conversion factor if it is in the denominator of the denominate number.*
   b. *Have the unit to be converted in the denominator of the conversion factor if it is in the numerator of the denominate number.*
3. *Multiply the given denominate number by the conversion factor using cancellation of numbers and units like in any other case of multiplication of algebraic fractions.*

EXAMPLE 9-6   Convert 28 in. to ft.

SOLUTION   We first write the denominate number 28 in. as the fraction $\dfrac{28\text{ in.}}{1}$. Then, we write the conversion factor $1\text{ ft} = 12\text{ in.}$ as the fraction $\dfrac{1\text{ ft}}{12\text{ in.}}$. Thus, we have

$$28\text{ in.} = \frac{28\ \cancel{\text{in.}}}{1} \cdot \frac{1\text{ ft}}{12\ \cancel{\text{in.}}}$$

$$= \frac{28}{12}\text{ ft}$$

$$= 2\frac{1}{3}\text{ ft}$$

EXAMPLE 9-7   Change 2.34 km to mi.

SOLUTION   The conversion factor to be used here is $\dfrac{1 \text{ mi}}{1.609 \text{ km}}$. Thus,

$$2.34 \text{ km} = \frac{2.34 \text{ km}}{1} \cdot \frac{1 \text{ mi}}{1.609 \text{ km}}$$

$$= \frac{2.34}{1.609} \text{ mi}$$

$$= 1.45 \text{ mi}$$

For more involved conversions of units more than one conversion factor is needed. The procedure, however, is the same.

EXAMPLE 9-8   Convert 1.500 days to minutes.

SOLUTION   The conversion factors to use here are $\dfrac{24 \text{ hr}}{1 \text{ day}}$ and $\dfrac{60 \text{ min}}{1 \text{ hr}}$. Then,

$$1.500 \text{ day} = \frac{1.500 \text{ day}}{1} \cdot \frac{24 \text{ hr}}{1 \text{ day}} \cdot \frac{60 \text{ min}}{1 \text{ hr}}$$

$$= (1.500)(24)(60) \text{ min}$$

$$= 2160 \text{ min}$$

EXAMPLE 9-9   Change 36.0 mi/hr to ft/sec.

SOLUTION   The conversion factors to be used here are $\dfrac{1 \text{ hr}}{60 \text{ min}}$ and $\dfrac{1 \text{ min}}{60 \text{ sec}}$, to change hr to sec, and $\dfrac{5280 \text{ ft}}{1 \text{ mi}}$, to change mi to ft. Then,

$$\frac{36.0 \text{ mi}}{\text{hr}} = \frac{36.0 \text{ mi}}{\text{hr}} \cdot \frac{1 \text{ hr}}{60 \text{ min}} \cdot \frac{1 \text{ min}}{60 \text{ sec}} \cdot \frac{5280 \text{ ft}}{1 \text{ mi}}$$

$$= \frac{(36.0)(5280) \text{ ft}}{(60)(60) \text{ sec}}$$

$$= 52.8 \frac{\text{ft}}{\text{sec}}$$

EXAMPLE 9-10   The density of iron is 490 lb/ft³. Find the density of iron in g/cm³.

SOLUTION   The conversion factors to be used here are $\dfrac{16 \text{ oz}}{1 \text{ lb}}$, $\dfrac{1 \text{ g}}{0.03527 \text{ oz}}$, $\dfrac{1 \text{ ft}^3}{1728 \text{ in.}^3}$, and $\dfrac{1 \text{ in.}^3}{16.4 \text{ cm}^3}$. Thus,

$$\frac{490 \text{ lb}}{\text{ft}^3} = \frac{490 \text{ lb}}{\text{ft}^3} \cdot \frac{16 \text{ oz}}{1 \text{ lb}} \cdot \frac{1 \text{ g}}{0.03527 \text{ oz}} \cdot \frac{1 \text{ ft}^3}{1728 \text{ in.}^3} \cdot \frac{1 \text{ in.}^3}{16.4 \text{ cm}^3}$$

$$= \frac{(490)(16) \text{ g}}{(0.03527)(1728)(16.4) \text{ cm}^3}$$

$$= 7.8 \frac{\text{g}}{\text{cm}^3}$$

EXERCISE 9-2

In the following exercises use the tables of conversion factors found inside the back cover to convert the given measurements as indicated.

1. 3.70 yd to in.
2. 482.0 g to oz
3. 28.7 kg to lb
4. 23.2 gal to ℓ

5. 7.10 km to mi	**6.** 8.20 cm to in.
7. 7230 ft to mi	**8.** 273 m to ft
9. 14.0 lb to kg	**10.** 7.8 oz to g
11. 8.20 ℓ to gal	**12.** 218.0 mi to km
13. 18.25 in. to cm	**14.** 2.70 mi to ft
15. 326.4 ft to m	**16.** 2.2 ft^2 to in.^{2}
17. 2.8 gal to in.^{3}	**18.** 172.9 ft^3 to yd^3
19. 341 in.^{2} to ft^2	**20.** 783 in.^{3} to gal
21. 817 cm^3 to in.^{3}	**22.** 423 ft^2 to m^2
23. 724 ft^3 to m^3	**24.** 11.3 yd^3 to ft^3
25. 14$\bar{0}$0 in.^{2} to ft^2	**26.** 233.4 in.^{3} to cm^3
27. 3.6 yd^3 to in.^{3}	**28.** 3.12 ft^3 to gal
29. 6$\bar{0}$00 cm^3 to gal	**30.** 72.0 in./sec to ft/min
31. 98$\bar{0}$ cm/sec to ft/min	**32.** 425 ft/min to in./sec
33. 32.0 lb/in.^{2} to kg/cm^2	**34.** 35.0 ft/sec to mi/hr
35. 4$\bar{0}$ mi/hr to ft/sec	**36.** 45.0 mi/hr to ft/sec and to m/sec
37. 18.3 gal/min to ℓ/sec	**38.** 211 lb/ft^3 to tons/yd^3
39. 45.0 mi/hr to km/hr	**40.** 10$\bar{0}$ km/hr to mi/hr

41. The density of lead is 705 lb/ft^3. Convert this density to g/cm^3.

42. A Volkswagen Rabbit was bought in Dresden, Germany, by a tourist for 10,588 marks. If the price of the mark that day was $0.5383, what was the price of the car in dollars?

43. If you are driving a foreign car with a speedometer showing kph (kilometers per hour), what should the approximate indication of the speedometer be to know that you are not exceeding the 55 mph limit?

44. The density of gold is 19.30 g/cm^3. Find the density of gold in lb/ft^3.

45. A water pump is pumping 2350 gal/min. What is the pumping rate in ℓ/sec?

46. The tires of a car are inflated to the pressure of 30 lb/in.^{2} What is this pressure in kg/cm^2?

9-4 Evaluation of Formulas

When evaluating a formula, both the magnitude and the unit of the denominate number must be substituted in the formula. As the indicated operations are performed after the substitution, both the numerical part and the unit of the denominate number are treated like algebraic numbers by the usual algebraic laws, as we have already seen. In an involved technical, or scientific formula with unfamiliar units, it is not easy to predict the dimensions of the result. To substitute both magnitude and units is very important in evaluating such formulas because the units in a result are as important as the number.

In substituting denominate numbers in a formula, one must *make sure that the units expressing the dimensions of a given physical quantity are all of the same kind.* For example, length cannot be substituted in the same formula in feet and in inches. *If the values were given in different units, they must be converted before substitution.*

EXAMPLE 9-11 Evaluate the formula $A = LW$, when $L = 2.15$ m and $W = 76$ cm.

SOLUTION We can express both lengths in either (a) meters or (b) centimeters.

a. Converting 76 cm to m

$$76 \text{ cm} = \frac{76 \text{ cm}}{1} \cdot \frac{1 \text{ m}}{100 \text{ cm}}$$

$$= 0.76 \text{ m}$$

Substituting in the formula

$$A = (2.15 \text{ m})(0.76 \text{ m})$$

$$= 1.634 \text{ m}^2$$

$$= 1.6 \text{ m}^2$$

b. Converting 2.15 m to cm

$$2.15 \text{ m} = \frac{2.15 \text{ m}}{1} \cdot \frac{100 \text{ cm}}{1 \text{ m}}$$

$$= 215 \text{ cm}$$

Substituting in the formula

$$A = (215 \text{ cm})(76 \text{ cm})$$

$$= 16\,340 \text{ cm}^2$$

$$= 16\,000 \text{ cm}^2$$

The student should realize that the two areas, 1.6 m² and 16 000 cm², are the same.

EXAMPLE 9-12   The braking time, the time it takes for a car to stop from a given velocity, if its brakes can give it a certain negative acceleration, is given by the formula

$$t = \frac{v}{a}$$

where t is the time, v is the velocity, and a is the acceleration. Evaluate the formula for $v = 55.0$ mph, and $a = 30.0$ ft/sec².

SOLUTION   Before substituting into the formula, we will convert the velocity from mi/hr to ft/sec so that the units for time in velocity and acceleration will be the same. Thus,

$$\frac{55.0 \text{ mi}}{\text{hr}} = \frac{55.0 \text{ mi}}{\text{hr}} \cdot \frac{5280 \text{ ft}}{1 \text{ mi}} \cdot \frac{1 \text{ hr}}{60 \text{ min}} \cdot \frac{1 \text{ min}}{60 \text{ sec}}$$

$$= \frac{(55)(5280) \text{ ft}}{(60)(60) \text{ sec}}$$

$$= 80.66 \text{ ft/sec}$$

$$= 80.7 \text{ ft/sec}$$

Substituting now into the formula, we have

$$t = \frac{\dfrac{80.7 \text{ ft}}{\text{sec}}}{\dfrac{30.0 \text{ ft}}{\text{sec}^2}}$$

$$= \frac{80.7 \text{ ft sec}^2}{30.0 \text{ ft sec}}$$

$$= 2.69 \text{ sec}$$

Evaluation of Formulas   153

EXAMPLE 9-13 Evaluate the formula

$$I = \frac{E - e}{R}$$

when $E = 125$ V, $e = 5$ V, and $R = 750\ \Omega$ (ohms).

SOLUTION Since the quantities to be subtracted are both of the same unit (volts), nothing has to be converted, and we proceed with the substitution.

$$I = \frac{125\ V - 5\ V}{750\ \Omega}$$

$$= \frac{120\ V}{750\ \Omega}$$

$$= 0.16\ V/\Omega$$

NOTE: Volts/ohms is amperes, the unit for current. It will be enough for the student to arrive at the correct dimension in the following exercises; he will learn in physics about derived units like newtons, joules, farads, etc.

EXERCISE 9-3

Evaluate each of the following formulas using the values given for each case. Be sure to convert wherever the units are different. It is not necessary that the student be familiar with these formulas and units to be able to work these exercises. The answer will still have the correct dimensions if what has been said about dimensions and denominate numbers is followed.

1. $V = e^3$, when $e = 3.2$ m.
2. $A = \frac{1}{2}bh$, when $b = 2.8$ ft, and $h = 0.25$ ft.
3. $V = LWH$, when $L = 3.0$ m, $W = 1.2$ m, and $H = 83$ cm.
4. $A = \frac{1}{2}(a + b)h$, when $a = 8.0$ ft, $b = 7.0$ ft, and $h = 4.0$ ft.
5. $W = \dfrac{V}{LH}$, when $V = 3.0\ m^3$, $L = 2.8$ m, and $H = 1.3$ m.
6. $L = \dfrac{V}{WH}$, when $V = 490$ in.^{3}, $W = 8.5$ in., and $H = 18$ in.
7. $IMA = \dfrac{\text{input distance}}{\text{output distance}}$, when input distance $= 0.03$ m, and output distance $= 0.003$ m.
8. $S = \frac{1}{2}gt^2$, when $g = 980$ cm/sec^2, and $t = 3.0$ sec.
9. $W = FS$, when $F = 23.2$ lb, and $S = 7.8$ ft.
10. $F = ma$, when $m = 72.0$ kg, and $a = 15.0$ m/sec^2.
11. $P = hD$, when $h = 3$ ft 9 in., and $D = 62.4$ lb/ft^3.
12. $R = \dfrac{E}{I}$, when $E = 1.50$ V, and $I = 0.150$ A.
13. $v_2 = v_1 + at$, when $v_1 = 4\overline{0}$ ft/sec, $a = 32.2$ ft/sec^2, and $t = 4.0$ sec.
14. $R_x = \dfrac{R_1 R_3}{R_2}$, when $R_1 = 75\ \Omega$, $R_2 = 25\ \Omega$, and $R_3 = 45\ \Omega$.
15. $S = v_0 t + \frac{1}{2}at^2$, when $v_0 = 3\overline{0}$ m/sec, $a = 425$ cm/sec^2, and $t = 8.0$ sec.
16. $S = vt$, when $v = 25$ mi/hr, and $t = 45$ min.
17. $P = \dfrac{F}{A}$, when $F = 180$ lb, and $A = 4.5$ in.^{2}.
18. $V_2 = \dfrac{V_1 P_1}{P_2}$, when $V_1 = 40\overline{0}$ in.^{3}, $P_1 = 3\overline{0}$ lb/in.^{2}, and $P_2 = 10\overline{0}$ lb/in.^{2}.

19. $E_k = \frac{1}{2}mv^2$, when $m = 52$ kg, and $v = 3.0$ m/sec.

20. $a = \dfrac{v_2 - v_1}{t}$, when $v_2 = 98.0$ km/hr, $v_1 = 33.0$ km/hr, and $t = 6.70$ sec.

REVIEW QUESTIONS

1. How many geometric dimensions are there?
2. Of what basic kind are all geometric dimensions?
3. Which system of fundamental units is used in the United States in engineering and in daily life?
4. What is a denominate number?
5. What are the parts of a denominate number?
6. What is the basic requirement which must be satisfied before two denominate numbers may be added, subtracted, multiplied, or divided?
7. In algebraic operations with denominate numbers, how is the unit part of a denominate number treated by the laws of algebra?
8. What are compound denominate numbers?
9. How are compound denominate numbers added or subtracted?
10. How are compound denominate numbers multiplied or divided by a constant?
11. State the three-step procedure for converting a given denominate number from one unit of measure to another.
12. If the denominate numbers to be substituted in a formula are of different units, what must be done before substitution?
13. In evaluating a formula or solving a technical problem, is the unit part of the answer important?

REVIEW EXERCISES

Perform each of the indicated operations in exercises 1–12. In case you decide that an operation cannot be performed (in the form in which it is given) justify your decision.

1. (7 mm)(8 cm)	**2.** 3 oz + 4 oz	**3.** 35 mi² ÷ 7 mi
4. 18 gal − 7 ℓ	**5.** 23 ℓ − 9 ℓ	**6.** 12 km + 8 mi
7. (35 ft)(8.0 ft²)	**8.** (6 m²)(5 m)	**9.** 42 m² ÷ 7 m²
10. 32 sec + 14 sec	**11.** 11 lb − 4 lb	**12.** 8 hr ÷ 2 ft

Perform each of operations 13–22, and simplify the result wherever applicable.

13. Add 7 yd 4 ft 3 in. and 8 yd 9 ft 11 in.
14. Subtract 7 lb 3 oz from 11 lb 1 oz.
15. Subtract 7 gal 3 qt 2 pt from 15 gal 2 qt 1 pt.
16. Add 2 tons 812 lb and 3 tons 1216 lb.
17. Subtract 2 hr 52 sec from 7 hr 18 min 7 sec.
18. Multiply 10 ft 2 in. by 3.
19. Multiply 12 lb 14 oz by 7.
20. Multiply 3 hr 21 min by 3.
21. Divide 19 tons 211 lb by 3.

22. Divide 5 mi 420 ft by 4.

In each of the exercises 23–28, convert as indicated.

23. 85 in./sec to ft/min

24. 31.0 ft/sec to mi/hr

25. 42.0 lb/in.² to kg/cm²

26. 112 lb/ft³ to ton/yd³

27. 50.0 mi/hr to km/hr

28. 87.0 km/hr to mi/hr

29. An advertisement claims that a gallon of a certain paint will cover $41\bar{0}$ ft². How much paint is needed to cover 9265 ft²?

30. Crude oil is flowing through a pipe at the rate of 18.7 gal/min. What is the rate in ℓ/sec?

31. The density of aluminum is 169 lb/ft³. Find the density of aluminum in g/cm³.

32. A drum is filled with water to the height of 4.0 ft. The pressure at the bottom of the drum is $25\bar{0}$ lb/ft². Find the pressure in kg/cm².

33. The density of concrete is 2.30 g/cm³. Find the density of concrete in lb/ft³.

Evaluate each of the following formulas.

34. $MA = \dfrac{\text{output force}}{\text{input force}}$, when input force = 8.2 kg, and output force = 73.8 kg.

35. $P = \dfrac{F}{A}$, when $F = 42.3$ kg, and $A = 6.3$ cm².

36. $P = hD$, when $h = 5.7$ ft, and $D = 62.4$ lb/ft³.

37. $S = vt$, when $v = 38$ km/hr, and $t = 38$ min.

38. $R = \dfrac{E}{I}$, when $E = 1.65$ V, and $I = 0.180$ A.

39. $P_2 = \dfrac{V_1 P_1}{V_2}$, when $V_1 = 32\bar{0}$ in.³, $P_1 = 35$ lb/in.², and $V_2 = 16\bar{0}$ in.³.

40. $a = \dfrac{v_2 - v_1}{t}$, when $v_2 = 55.0$ mi/hr, $v_1 = 15.0$ mi/hr, and $t = 10.0$ sec.

10

Ratio, Proportion, and Variation

We can compare two quantities either by subtraction (by finding their difference), or by division (by finding the number of times one quantity is larger than the other). *Ratio is comparison by division.* Consider the two sticks of Fig. 10-1. We may say that the larger stick is 4 ft longer than the smaller stick (by subtraction), because $6 - 2 = 4$, or we may say that the larger stick is three times as long as the smaller stick (by division), because $6 \div 2 = 3$. In short, the ratio of one number to another is the quotient of the division of the first number by the second. Since division can be expressed with a fraction, the ratio of the large stick to the small stick is $\frac{6}{2}$ or 3. We can compare the small stick to the large, if we please, and say that the ratio of the small stick to the large stick is $\frac{2}{6}$ or $\frac{1}{3}$. *The terms of a fraction expressing a ratio are called the terms of the ratio.* The student should notice in the above examples that *a fraction expressing a ratio is reduced to its lowest terms.*

A ratio is often written with a colon (:). For instance, the above ratio of the 2-ft stick to the 6-ft stick which is $\frac{2}{6} = \frac{1}{3}$ can be written

$$1 : 3 \text{ (read "1 to 3")}$$

Also, the ratio of the 6-ft stick to the 2-ft stick which is $\frac{6}{2} = \frac{3}{1}$ can be written

$$3 : 1 \text{ (read "3 to 1")}$$

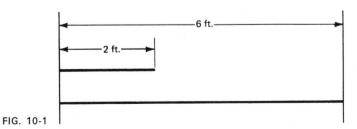

FIG. 10-1

The student should be careful about the following important point: since we can compare only quantities of the same kind, *for two quantities to be a ratio, they must be of the same dimension, and the dimension must be expressed in the same unit of measurement.*

EXAMPLE 10-1   Find the ratio of 3 ft to 8 in.

SOLUTION   We first form the fraction $\frac{3 \text{ ft}}{8 \text{ in.}}$. Notice that the terms of the ratio are of the same dimension, length, but not of the same unit. Therefore, strictly speaking, $\frac{3 \text{ ft}}{8 \text{ in.}}$ is not a **proper ratio**, since only one of the two necessary conditions for a comparison to be a ratio is met. If, however, we convert 3 ft into 36 in., we have

$$\frac{3 \text{ ft}}{8 \text{ in.}} = \frac{36 \text{ in.}}{8 \text{ in.}} = \frac{9}{2}$$

and $\frac{9}{2}$ is a true ratio.

EXAMPLE 10-2   What is the ratio of the circumference of the circle of Fig. 10-2, to its diameter?

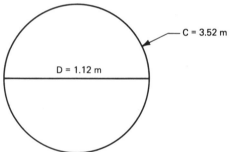

FIG. 10-2

SOLUTION   First we express the ratio as a fraction and then we divide. Thus,

$$\frac{C}{D} = \frac{3.52 \text{ m}}{1.12 \text{ m}} = 3.14$$

The student is familiar with the constant 3.14. This number is represented by the symbol π (read "pi") and is the ratio of the circumference of any circle to its diameter. The constant π means that the circumference of any circle is approximately 3.14 times as large as its diameter. Notice that *π is a pure number and has no units. Like π, all proper ratios are constants. They are not denominate numbers.* In technical mathematics constants of this kind are of particular importance. The trigonometric ratios are constants of this kind, as we will see later.

The comparison $\frac{26 \text{ mi}}{\text{gal}}$ (read "twenty-six miles per gallon") is not a proper ratio, since neither the dimensions nor the units are of the same kind. Other comparisons of this kind are $\frac{32 \text{ ft}}{\text{sec}^2}$, $\frac{680 \text{ rotations}}{\text{min}}$, $\frac{\$2.30}{\text{sq ft}}$, etc.

It is proper to call such comparisons "rates" rather than ratios. We will see in

Ratio, Proportion, and Variation

the next section that "rates" as well as ratios can be used in solutions of problems, as long as "rates" are not treated as ratios.

EXAMPLE 10-3   Is $\dfrac{980 \text{ cm}}{\text{sec}^2}$ a ratio?

SOLUTION   No, it is not a proper ratio because neither the dimensions nor the units are of the same kind. 980 cm/sec² is a rate.

For each comparison in exercises 1–5, state whether it is a ratio or a rate.
1. 3 ft to 2 sec
2. 14 oz to 18 oz
3. 16 kg to 11 kg
4. 74 cm² to 39 cm²
5. $\dfrac{15 \text{ dollars}}{2 \text{ hours}}$

Express each of exercises 6–13 as a ratio reduced to its lowest terms and with no units.
6. 10 in. to 5 ft
7. 3 hr to 40 min
8. $8.00 to $15.00
9. 1215 ft to 3 mi
10. 48 g to 2 kg
11. 21 sq in. to 3 sq ft
12. 18 mm² to 2 cm²
13. 24 cm² to 1 m²
14. The ratio of two distances is 3:1. The smaller distance is how many times the larger distance?
15. An oil tank has a capacity of 300 gal. If there are 60 gal of oil in the tank, what is the ratio of the number of gallons of oil in the tank to the capacity of the tank?
16. The smaller of two gears makes 180 revolutions per minute, and the larger one makes 45 revolutions per minute. Find the ratio of the speed of the smaller gear to the speed of the larger gear.
17. Permalloy, a special alloy for very strong artificial magnets, is made of 80 parts of nickel and 20 parts of iron. What is the ratio of iron to nickel in permalloy?
18. An express train is traveling at the rate of 60 mph, and a plane is flying at the rate of 280 mph. What is the ratio (a) of the speed of the train to the speed of the plane? (b) of the speed of the plane to the speed of the train?
19. Carpenters describe the amount of the slope of a roof using the term **pitch**. The pitch of a rafter is the ratio between the rise and the run of the rafter (see Fig. 10-3). Thus, pitch = rise/run. What is the pitch of a roof with a rise of 6 ft and a run of 15 ft?

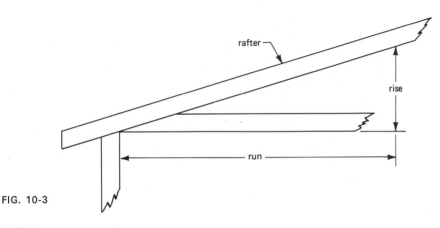

rafter

rise

run

FIG. 10-3

10-2 Proportion

The equality of two ratios is called a proportion. For instance,

$$\frac{9}{12} = \frac{3}{4}, \quad \frac{18}{20} = \frac{9}{10} \quad \text{and} \quad \frac{40 \text{ mi}}{120 \text{ mi}} = \frac{2 \text{ gal}}{6 \text{ gal}}$$

are proportions. The first of these proportions is read "9 is to 12 as 3 is to 4." The last of the above proportions is read "40 miles is to 120 miles as 2 gallons is to 6 gallons."

Another way to express a proportion is to write it using colons. For instance, the proportion

$$\frac{a}{b} = \frac{c}{d}$$

can be written

$$a : b = c : d$$

Both forms of the above proportion are read "*a* is to *b* as *c* is to *d*."

The four quantities of a proportion are called its **terms**. Thus, the terms of the proportion $a:b = c:d$ are *a*, *b*, *c*, and *d*. Of the four terms of a proportion, the first and the last terms are called the **extremes**, and the second and the third terms are called the **means**. Thus, for the proportion $a:b = c:d$, *a* and *d* are the extremes, and *b* and *c* are the means.

Proportions have some outstanding and very useful properties, which will be used in making the solution of many types of problems simple. The most important of these properties in terms of the proportion $\frac{a}{b} = \frac{c}{d}$ are:

1. *The product of the extremes is equal to the product of the means.* Thus, if

$$\frac{a}{b} = \frac{c}{d} \quad \text{then} \quad ad = bc$$

This property is known as the cross-multiplication property.

2. *Reciprocals of equals are equal.* Thus if

$$\frac{a}{b} = \frac{c}{d} \quad \text{then} \quad \frac{b}{a} = \frac{d}{c}$$

3. *The terms of a proportion are proportional when taken alternately.* Thus, if

$$\frac{a}{b} = \frac{c}{d} \quad \text{then} \quad \frac{a}{c} = \frac{b}{d} \quad \text{and} \quad \frac{d}{b} = \frac{c}{a}$$

Ratio, Proportion, and Variation

EXAMPLE 10-4 Is $\frac{6}{7} = \frac{18}{21}$ a true proportion? Why?

SOLUTION Yes, $\frac{6}{7} = \frac{18}{21}$ is a true proportion because by the cross-multiplication property of a proportion the product of the extremes 6 and 21 is 126, and the product of the means 7 and 18 is also 126.

EXAMPLE 10-5 Rewrite the proportion $\frac{3}{x} = \frac{y}{5}$, using the second property of proportions.

SOLUTION Since the reciprocals of equals are equal, $\frac{3}{x} = \frac{y}{5}$ can be written as $\frac{x}{3} = \frac{5}{y}$.

From the fact that a proportion is a simple equation it must be evident that if any three of the four terms of a proportion are known, the value of the fourth term can readily be found.

EXAMPLE 10-6 In the proportion $\frac{2}{7} = \frac{x}{42}$, solve for x.

SOLUTION Assuming that $\frac{2}{7} = \frac{x}{42}$ is a true proportion, we cross-multiply to get

$$2 \cdot 42 = 7x$$
$$x = \frac{84}{7}$$
$$= 12$$

EXAMPLE 10-7 Solve the expression below for x.

$$\frac{x}{d + x} = \frac{c}{d}$$

SOLUTION By cross-multiplication we have

$$dx = c(d + x)$$
$$dx = cd + cx$$
$$dx - cx = cd$$
$$x(d - c) = cd$$
$$x = \frac{cd}{d - c}$$

EXERCISE 10-2

Decide whether problems 1–12 are true proportions:

1. $\frac{2}{3} = \frac{18}{27}$
2. $\frac{5}{8} = \frac{3}{4}$
3. $\frac{21}{45} = \frac{7}{15}$
4. $\frac{6}{7} = \frac{12}{15}$
5. $\frac{31}{21} = \frac{155}{105}$
6. $\frac{2k}{5k} = \frac{18}{45}$
7. $\frac{36}{18} = \frac{18m}{9m}$
8. $\frac{3e}{5d} = \frac{33e}{55d}$
9. $\frac{7P}{2R} = \frac{14R}{4P}$
10. $\frac{4v}{3w} = \frac{12w}{9v}$
11. $\frac{6x}{7y} = \frac{48x}{56y}$
12. $\frac{12x}{4w} = \frac{3x}{y}$

Proportion

Assuming that each of the following equations are true proportions, solve for the variable.

13. $\dfrac{x}{5} = \dfrac{24}{40}$

14. $\dfrac{3}{x} = \dfrac{15}{25}$

15. $\dfrac{2x}{7} = \dfrac{8}{14}$

16. $\dfrac{3}{v+4} = \dfrac{5}{2v+3}$

17. $\dfrac{15}{18} = \dfrac{y-1}{y}$

18. $\dfrac{30}{126} = \dfrac{5}{3x}$

19. $\dfrac{x-2}{7} = \dfrac{2x+2}{28}$

20. $\dfrac{3}{x+1} = \dfrac{18}{9x-3}$

21. $\dfrac{3x+6}{35} = \dfrac{2x-18}{5}$

22. $\dfrac{k}{k+4} = \dfrac{k+3}{k+10}$

23. $\dfrac{d-3}{d} = \dfrac{d+1}{d+8}$

24. $\dfrac{m-1}{m-4} = \dfrac{m+7}{m}$

25. $\dfrac{P+1}{3-P} = \dfrac{1}{2}$

10-3 Proportion Applications

We will show now that proportion problems can be set up either as an equality of ratios, or an equality of "rates." Let us take, for instance, the following problem:

> If a car's mileage is 31 miles per gallon and the car is driven 248 mi, how much gas did it use?

The proportion for this problem is

$$\frac{31 \text{ mi}}{248 \text{ mi}} = \frac{1 \text{ gal}}{x \text{ (gal)}}$$

Using cross-multiplication and proper treatment of denominate numbers, we have

$$x(31 \text{ mi}) = (248 \text{ mi})(1 \text{ gal})$$

$$x = \frac{(248 \text{ m\!i})(1 \text{ gal})}{(31 \text{ m\!i})}$$

$$= 8 \text{ gal}$$

If in the original proportion

$$\frac{31 \text{ mi}}{248 \text{ mi}} = \frac{1 \text{ gal}}{x \text{ (gal)}}$$

we use property 3, and alternate the term 248 mi with the term 1 gal, the proportion will take the form

$$\frac{31 \text{ mi}}{1 \text{ gal}} = \frac{248 \text{ mi}}{x \text{ (gal)}}$$

Notice that this is an equation of "rates." This fact, however, will not make any difference for the solution, since the result will be the same as before. Thus, by cross-multiplying

$$x(31 \text{ mi}) = (1 \text{ gal})(248 \text{ mi})$$

$$x = \frac{(1 \text{ gal})(248 \text{ m\!i})}{(31 \text{ m\!i})}$$

$$= 8 \text{ gal}$$

Ratio, Proportion, and Variation

From the above example it follows that proportion problems can be set up with either proper ratios or with rates. When "rates" are used, the student should make sure that the "unit rates" are (horizontally) the same on both sides of the proportion. Diagram A shows the correct set up of the "unit rates" while diagram B shows the wrong set up.

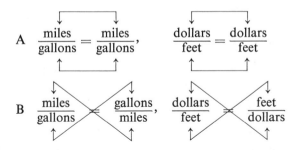

EXAMPLE 10-8 In a factory, the rate of men to women is 7 to 3, and there are 1498 men. What is the number of women in that factory?

SOLUTION Let x represent the number of women. Then,

$$\frac{7 \text{ men}}{3 \text{ women}} = \frac{1498 \text{ men}}{x}$$

$$x = \frac{(3 \text{ women})(1498 \text{ men})}{(7 \text{ men})}$$

$$= 642 \text{ women}$$

EXAMPLE 10-9 Five tickets to a game cost $59.75. What is the cost of eight tickets?

SOLUTION Let x represent the cost of eight tickets. Then,

$$\frac{5 \text{ tickets}}{\$59.75} = \frac{8 \text{ tickets}}{x}$$

$$x = \frac{(\$59.75)(8 \text{ tickets})}{(5 \text{ tickets})}$$

$$= \frac{(\$59.75)(8)}{5}$$

$$= \$95.60$$

EXAMPLE 10-10 The cost of 35 ft of seamless copper tube for boiler feed lines is $26.25. What is the cost of 11 ft of that pipe?

SOLUTION Let us use a proper ratio set up in this problem. We let x represent the cost of 11 ft of pipe. Then,

$$\frac{35 \text{ ft}}{11 \text{ ft}} = \frac{\$26.25}{x}$$

$$x = \frac{(11 \text{ ft})(\$26.25)}{(35 \text{ ft})}$$

$$= \$8.25$$

Proportion Applications

EXERCISE 10-3

1. A drug costs 25¢ per 3 g. How much will 21 g cost?
2. Bill Stone had a yield of 1155 bushels of corn last year from a 35-acre field. If he plants corn in a 42-acre field this year, how many bushels of corn should he expect?
3. A car made 283.5 mi on 21 gal. How much gas is needed for 810 mi?
4. The strength of a drug is 3.5 mg per mℓ. If 17.5 mg are required, how much should be administered?
5. A worker can machine 3 pieces of tapered work in 1 hr 18 min. How long will it take him to machine 7 pieces?
6. A man casts a shadow 8 ft 9 in. long, and a pole, 9 ft 4 in. high, next to him casts a shadow of 14 ft. What is the height of the man?
7. A 683-ft long 15 gauge copper electric wire weighs 6.73 lb. What is the weight of a 1213-ft long piece of that wire?
8. A 7-ft long piece of standard steel tube with an outer diameter of $5\frac{1}{2}$ in. and a thickness of $\frac{9}{32}$ in. weighs 109.8 lb. What is the weight of a 3.8-ft piece of that tube?
9. A 1375-ft long 8 gauge copper electric wire weighs 68.72 lb. What is the length of 39.3 lb of that wire?
10. A 3.4-ft long piece of standard steel tube, with an outer diameter of $6\frac{5}{8}$ in., weighs 64.8 lb. What is the length of a piece of that tube weighing 142.9 lb?
11. A man is 5 ft 7 in. tall and weighs 149 lb. Another man of about the same build is 160 lb. What do you expect his height to be?
12. A man is 1.82 m tall and weighs 84 kg. Another man of about the same build is 1.73 m tall. What do you expect his weight to be in kg?
13. In a scale drawing of a 112-ft by 88-ft rectangular building, the width is 11 in. What is the length of the building in the drawing?
14. In an electric circuit the current increases with voltage. If the current is 1.8 A when the voltage is 18 V, what is the voltage when the current is 5.4 A?
15. Ten dozen resistors cost $4.80. How much do 28 resistors cost?
16. In a fish survey, 1000 fish were tagged and released. Among 154 fish caught in the same pond later, 14 were tagged. Approximately how many fish were there in that pond?
17. An adult dose (assume the adult to weigh 160 lb) is 6 g. What should the dose be for a 60 lb child?
18. A train travels 364.8 km in 3.0 hr. How many km will the train travel in $4\frac{1}{2}$ hr?

10-4 Variation

Many laws of physics use the concept of **variation** and involve quantities with a constant ratio when stated mathematically. Such is the case, for instance, of **Hooke's Law**. This law says that within certain limits, the deformation produced on a body by a force, *varies directly as the force*. Let us assume that the spring of Fig. 10-4 will stretch 3 mm for the first kilogram of force. Then, experiment will show that the spring will stretch an additional 3 mm for every additional kilogram of force. Thus, if the lower point of the spring is at a certain position with a 9-kg weight, it will be 3 mm lower with a 10-kg weight (see Fig. 10-4). A table of different values of the force x, and of corresponding values of elongation, y, is also shown in Fig. 10-4.

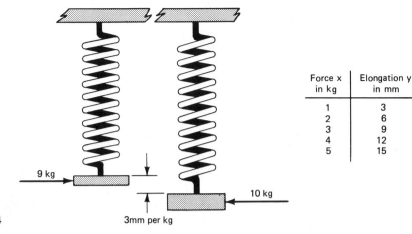

Force x in kg	Elongation y in mm
1	3
2	6
3	9
4	12
5	15

FIG. 10-4   3mm per kg

Notice that the ratio $\frac{y}{x}$ of any pair of corresponding values from the table is equal to the constant 3. Thus,

$$\frac{y}{x} = \frac{6}{2} = \frac{9}{3} = \frac{3}{1} = 3$$

This important fact can be stated in more general terms as follows:

$$\frac{y}{x} = k, \quad \text{and therefore,} \quad y = kx$$

The equation $y = kx$ is the first of four types of variation. In all four types of variation, a constant k is involved and is called **the constant of proportionality**.

The four types of variation are the following:

1. **Direct variation**. When two variables, x and y, are related by the variation equation

$$y = kx$$

we say that *y varies directly as x*, or *y is proportional to x*.

2. **Inverse variation**. When two variables, x and y, are related by the variation equation

$$y = \frac{k}{x}$$

we say that *y varies inversely as x*, or *y is inversely proportional to x*.

3. **Joint variation**. When y is related to two or more variables by the variation equation

$$y = kwxz$$

we say that *y varies jointly as w, x, and z.*

4. **Combined variation**. When y is related to two or more variables by the variation equation

$$y = \frac{kuv}{wx}$$

we say that y *varies jointly as u and v, and inversely as w and x.*

It is important for the student to keep in mind that in direct variation, y increases as x increases, and decreases as x decreases. In inverse variation, however, y increases as x decreases, and decreases as x increases.

EXAMPLE 10-11   Write as a variation equation the statement: t varies directly as s.

SOLUTION   This is direct variation. Therefore,

$$t = ks$$

EXAMPLE 10-12   Write as a variation equation the statement: s is proportional to d, and inversely proportional to t.

SOLUTION   This is a combined variation. Therefore,

$$s = \frac{kd}{t}$$

EXAMPLE 10-13   Write as a variation equation the statement: V varies jointly as L, W, and H.

SOLUTION   This is a joint variation. Therefore,

$$V = kLWH$$

EXAMPLE 10-14   Express as a sentence the equation $T = \frac{kdm}{s}$.

SOLUTION   This is a combined variation. Therefore, T varies jointly as d and m, and inversely as s.

EXAMPLE 10-15   Express as a sentence the equation $P = \frac{k}{R}$.

SOLUTION   This is an inverse variation. Therefore, P varies inversely as R.

EXAMPLE 10-16   Express as a sentence the equation $W = kBT$.

SOLUTION   This is a joint variation. Therefore, W varies jointly as B and T.

Suppose now we are told that the variables x, y, and z are related as follows: y is directly proportional to x and inversely proportional to the square root of z. We are also told that the following data is known about two separate instances of this relationship:

1. $y = 4$ when $x = 8$ and $z = 36$.
2. $y = ?$ when $x = 5$ and $z = 16$.

We can find the value of y in the second instance as follows: Expressing the relationship algebraically we obtain the variation equation

$$y = \frac{kx}{\sqrt{z}}$$

Then substituting in this equation 4 for y, 8 for x and 36 for z (from the first set of data) we solve for k. Thus,

$$4 = \frac{k \cdot 8}{\sqrt{36}} \quad \text{and } k = 3$$

Substituting 3 for k in the variation equation, we get the **general variation formula** for y:

$$y = \frac{3x}{\sqrt{z}}$$

Substituting 5 for x and 16 for z (from the second set of data) we can find the value of y. Thus,

$$y = \frac{3 \cdot 5}{\sqrt{16}}$$
$$= \frac{15}{4}$$

EXAMPLE 10-17   The distance a car will travel at a constant speed varies directly as the time. Find the general variation formula for the distance for a certain car if it is known that the car traveled 220 mi in 4 hr at a constant speed. How many miles will the car travel in 7 hr at that speed?

SOLUTION   The variation equation is

$$d = kt$$

Then substituting 220 mi for d, and 4 hr for t, we solve for k. Thus,

$$220 \text{ mi} = k(4 \text{ hr})$$
$$k = \frac{55 \text{ mi}}{\text{hr}}$$

Notice that the units for the constant of proportionality, k, are mi/hr. Substituting $\frac{55 \text{ mi}}{\text{hr}}$ for k in the variation equation we obtain the general variation formula for d.

$$d = 55t\left(\frac{\text{mi}}{\text{hr}}\right)$$

Then, when $t = 7$ hr

$$d = 55(7 \text{ hr})\left(\frac{\text{mi}}{\text{hr}}\right)$$
$$= 385 \text{ mi}$$

The last two examples suggest the following steps for the solution of variation problems:

1. *Write the variation equation expressing the stated relationship. Make sure to include a k.*
2. *Substitute values from given data in the variation equation and solve for k.*
3. *Obtain the general variation formula by replacing k in the variation equation with the value obtained for k in step 2.*

4. *Solve for the missing value of any variable in the general formula using data from additional information.*

EXAMPLE 10-18 The resistance R of a wire to the flow of electricity varies directly as the length L of the wire. If $R = 7.2\ \Omega$ when $L = 6$ ft, find R when $L = 9.5$ ft.

SOLUTION The variation equation is

$$R = kL$$

Substituting values and solving for k we have

$$7.2\ \Omega = k(6\ \text{ft})$$

$$k = \frac{7.2\ \Omega}{6\ \text{ft}}$$

$$= \frac{1.2\ \Omega}{\text{ft}}$$

Then the general variation formula is

$$R = 1.2\frac{\Omega}{\text{ft}}(L)$$

Substituting 9.5 for L in the general variation formula, we have

$$R = 1.2\frac{\Omega}{\cancel{\text{ft}}}(9.5\ \cancel{\text{ft}})$$

$$= 11.4\ \Omega$$

EXAMPLE 10-19 The weight of a rectangular piece of wood varies jointly as the length, the width, and the thickness of the wood. If the weight of an oak rafter 3 m by 20 cm by 10 cm is 4.8 kg, what is the weight of another rafter of the same kind of wood 4 m by 25 cm by 10 cm?

SOLUTION The variation equation is

$$w = kLWT$$

Substituting values and solving for k we have

$$4.8\ \text{kg} = k(3\ \text{m})(20\ \text{cm})(10\ \text{cm})$$

$$= k(3\ \text{m})(0.2\ \text{m})(0.1\ \text{m})$$

$$= k(0.06\ \text{m}^3)$$

$$k = \frac{4.8\ \text{kg}}{0.06\ \text{m}^3}$$

$$= \frac{80\ \text{kg}}{\text{m}^3}$$

Then the general variation formula is

$$w = 80\frac{\text{kg}}{\text{m}^3}L \cdot W \cdot T$$

The weight of the second rafter is

$$w = 80\frac{\text{kg}}{\text{m}^3}(4\ \text{m})(25\ \text{cm})(10\ \text{cm})$$

$$= 80\frac{\text{kg}}{\text{m}^3}(4\ \text{m})(0.25\ \text{m})(0.1\ \text{m})$$

$$= 80 \frac{\text{kg}}{\text{m}^3} (0.1 \text{ m}^3)$$

$$= 8 \text{ kg}$$

EXAMPLE 10-20 The time needed to fill a tank with a certain liquid varies inversely as the square of the diameter of the pipe used. If it takes 4 hr to fill a railroad tank car with crude oil when a 3-in. pipe is used, how long will it take if a 4-in. pipe is used?

SOLUTION

$$t = \frac{k}{d^2}$$

Then,

$$4 \text{ hr} = \frac{k}{(3 \text{ in.})^2}$$

$$k = (4 \text{ hr})(9 \text{ in.}^2)$$

$$= 36 \text{ hr-in.}^2$$

The general variation formula is

$$t = \frac{36 \text{ hr-in.}^2}{d^2}$$

Now when $d = 4$ in.

$$t = \frac{36 \text{ hr-in.}^2}{(4 \text{ in.})^2}$$

$$= \frac{36 \text{ hr-in.}^2}{16 \text{ in.}^2}$$

$$= 2\tfrac{1}{4} \text{ hr}$$

EXERCISE 10-4

Express each of statements 1–15 as a variation equation.
1. d varies directly as m.
2. P varies jointly as W and Y.
3. T varies inversely as B.
4. N varies directly as the square of R and inversely as s.
5. F varies jointly as Q and U and inversely as L.
6. G varies directly as h.
7. H varies inversely as v.
8. S varies jointly as V and the square of z.
9. C varies directly as E and inversely as the square of R.
10. V varies inversely as the square root of g.
11. M varies jointly as N^2 and R.
12. The velocity of a falling object varies directly as the time of the fall.
13. The distance through which a freely falling object falls varies directly as the square of the time of the fall.
14. The resistance, R, of a wire varies as the length, L, of the wire, and inversely as the square of its diameter, D.
15. In an electric circuit, the current, I, varies directly as the voltage, E.

Variation

In each of exercises 16–25, express the statement as a variation equation, find the value of the constant of proportionality, and the value of the indicated unknown.

16. y varies directly as x. When $y = 24$, $x = 8$. Find y when $x = 5$.

17. P varies directly as the square of R. When $P = 28$, $R = 2$. Find P when $R = 3$.

18. t varies inversely as s. When $t = 3$, $s = 20$. Find t when $s = 30$.

19. B varies jointly as C and the square of E. When $B = 96$, $C = 3$, and $E = 4$. Find B when $C = 5$ and $E = 1$.

20. v varies jointly as r^2 and h. When $r = 2$, and $h = 9$, $v = 113$. Find v when $r = 3$ and $h = 8$.

21. d varies directly as H^2 and inversely as m. When $d = 4$, $H = 2$ and $m = 4$. Find d when $H = 5$ and $m = 8$.

22. t varies directly as B and inversely as the square of P. When $t = 7.5$, $B = 6$ and $P = 2$. Find t when $B = 63$ and $P = 3$.

23. Q varies directly as r and inversely as the product $s \cdot t$. When $Q = 4$, $r = 6$, $s = 3$ and $t = 1$. Find Q when $r = 14$, $s = 1$, and $t = 6$.

24. R varies directly as L and inversely as d^2. When $R = 3$, $L = 8$ and $d = 4$. Find R when $L = 5$ and $d = 1$.

25. t varies directly as g^2 and inversely as m. When $t = 24.3$, $g = 9$ and $m = 10$. Find t when $g = 5$ and $m = 6$.

26. The area of the circle varies directly as the square of its radius. If the area of a circle is 78.5 cm² when its radius is 5.00 cm, find the area of a circle when the radius is 7.00 cm.

27. The weight of a metal rod varies directly as its length. If a 1-in. diameter metal rod 20-in. long weighs 4 lb, find the weight of an 8-in. rod.

28. The velocity of a freely falling body varies directly as the time of the fall. If the velocity at the end of 4.00 sec is 128.8 ft per sec, find the velocity at the end of 6.00 sec.

29. The weight of a block of granite varies jointly as the length, the width and the height of the block. If the weight of a block of granite is 1360 lb when $L = 4.0$ ft, $W = 2.0$ ft and $H = 1.0$ ft, find the weight of a block of granite when $L = 3.0$ ft, $W = 0.50$ ft, and $H = 0.80$ ft.

30. The weight of 36 identical bevel gears is 9.00 oz. How many of these bevel gears will there be in 6.00 lb?

31. The time required to fill a tank varies inversely as the square of the diameter of the pipe used. If it takes 20 hr to fill a given tank with a 2-in. pipe, what should be the diameter of the pipe which can fill the same tank in 5 hr?

32. In an electric circuit the current, I, is proportional to the voltage, E. If $I = 4.0$ A when $E = 24$ V, what is the voltage when $I = 5.5$ A?

33. The distance traveled by a ball rolling down an inclined plane varies directly as the square of the time. If the inclination of the plane is such that the ball traveled 27 ft at the end of 3.0 sec, what is the distance traveled at the end of 2.0 sec?

34. The surface of a sphere varies as the square of its radius. If the surface of a sphere is 16π ft² when the radius is 2 ft, find the surface of the sphere when the radius is 5 ft.

35. The weight of a body above the surface of the earth varies inversely as the square of the distance of the body from the center of the earth (the center of gravity of the earth). If an astronaut weighs $18\bar{0}$ lb on the surface of the earth, how much would he weigh $50\bar{0}$ mi above the surface of the earth? Assume the radius of the earth to be $400\bar{0}$ mi.

36. The kinetic energy of a moving car varies as the square of its velocity. If the kinetic energy of a car is 58,500 ft-lb when it moves at 25.0 mi/hr, what is its kinetic energy when its velocity is 55.0 mi/hr?

37. The cost of labor varies jointly as the number of workers and the number of working days. If 12 men working 11 days each, cost $2772, how much will 15 workers working 16 days each cost?

REVIEW QUESTIONS

1. Write the ratio $a:3$ in its fractional form.
2. What are the two necessary conditions for two quantities to form a ratio?
3. Is π a variable or a constant?
4. Is $\dfrac{38 \text{ mi}}{2 \text{ gal}}$ a proper ratio?
5. What is the difference between ratio and "rate"?
6. What is a proportion?
7. Read the proportion $\dfrac{d}{t} = \dfrac{3}{v}$.
8. Read the proportion $k:m = r:t$.
9. Name the extremes and the means of the proportion $3:5 = 15:x$.
10. State the three properties of a proportion.
11. How can you check if the equality of two given ratios is a true proportion?
12. Can we set up a proportion problem as an equality of rates?
13. Is the set up of the "unit rates" $\dfrac{\text{lb}}{\text{ft}^3} = \dfrac{\text{ft}^3}{\text{lb}}$ correct?
14. Is the set up of the "unit rates" $\dfrac{\text{kg}}{\text{cm}^2} = \dfrac{\text{kg}}{\text{cm}^2}$ correct?
15. What is the constant k called in all types of variation?
16. How many types of variation are there?
17. In direct variation, what happens to y as x decreases?
18. In inverse variation, what happens to y as x increases?

REVIEW EXERCISES

1. Is 5 m to 12 m a ratio or a rate?
2. Is $\dfrac{\$1.20}{3 \text{ ft}^2}$ a ratio or a rate?
3. Express 25 cm to 5 m as a ratio reduced to its lowest terms.
4. In a physics class at a technical institute, there are 24 male students and 3 female students. What is the rate of male to female students in that class?
5. The smaller of two gears makes 240 revolutions per minute, and the larger makes 40 revolutions per minute. What is the ratio of the speed of the larger gear to the speed of the smaller gear?
6. What is the pitch of a roof with a rise of 8 ft and a run of 18 ft?

Solve each of proportions 7–12 for the variable.

7. $\dfrac{5 - x}{8} = \dfrac{19 + x}{56}$

8. $\dfrac{B}{B - 2} = \dfrac{B + 7}{B + 3}$

9. $\dfrac{5}{k+5} = \dfrac{15}{8k}$

10. $\dfrac{Q-2}{14-Q} = \dfrac{1}{2}$

11. $\dfrac{d+5}{d+3} = \dfrac{d+11}{7+d}$

12. $\dfrac{5}{4} = \dfrac{x+2}{x}$

13. The strength of a drug is 2.5 mg per mℓ. If 22.5 mg are required, how much should be administered?

14. In a new factory building, 27 fire alarm devices were installed at first, and then it was decided to install 7 more. If 27 fire alarm devices cost \$1159.65, how much will the additional 7 devices cost?

15. A hydraulic lift which is actuated by a lever lifts a weight of 1500 lb when a force of $1\overline{0}$ lb is applied to the lever. What force should be applied to the lever if a weight of 2700 lb is to be lifted?

16. An effort of 15 lb is applied to the handle of a screw jack to lift a weight of 7800 lb. What force should be applied to the handle if a weight of 22,360 lb is to be lifted?

17. At 200°F, cast iron will expand 1.5 in. per $10\overline{0}$ ft. How much will a piece of $3\overline{0}$ ft expand at that temperature?

18. The weight of 26 ft of plumbing lead pipe of 1.5 in. outside diameter and 0.125 in. thickness is 65 lb. What is the weight of 38 ft of that pipe?

19. The days required to assemble machines is proportional to the number of machines, and inversely proportional to the number of men used. If 12 men take 25 days to assemble and check 20 machines, how many days will it take 30 men to assemble and check 70 machines?

20. The distance a body falls from rest, neglecting air resistance, varies directly with the square of the time. If a body falls 64 ft in the first 2 sec, how far will it fall in 5 sec?

21. The expansion of a metal varies jointly as the temperature and the length of the metal. A brass rod expands 2.70 in. when its length is $10\overline{0}$ ft, and the temperature is 225°F. Find the expansion when the length of the brass rod is $4\overline{0}$ ft, and the temperature is 150°F.

Relations and Functions. Graphs of Equations

11-1 Basic Ideas About Relations and Functions

In science and in technology there is often a relation between two quantities such that when we know the value of the one quantity, we can readily find the value of the other. We have seen in the previous chapter that such a relation exists between the elongation of a spring and the weight (force) which produces it. We expressed this weight–elongation relation by pairs of corresponding numbers, and by the equation $y = kx$, where k, the proportionality constant, was 3 for that particular spring. For easy reference, the table of these pairs of numbers is shown in Table 11-1. Notice how readily we can find any y value of the table using the equation $y = 3x$. Just substitute any x value in the equation and obtain the corresponding value of y. Thus, when $x = 1$, $y = 3$, and when $x = 4$, $y = 12$.

Equations like $y = 3x$, or $y = 2x + 5$, in which both the variables, x and y, are in the first power, are called **first degree equations in two variables**, or **linear equations in two variables**.

The general form of the first degree equation in two variables is $y = mx + b$, where m and b are constants.

Equations like $y = 3x^2$, or $y = 2x^2 + 3x + 1$, in which the variable x appears in the second power, or both in the second and in the first power, are called **second degree equations in two variables**, or **quadratic equations in two variables**. The general form of this particular quadratic equation is $y = ax^2 + bx + c$, where $a \neq 0$, and a,

TABLE 11-1

Force x, in kg	Elongation y, in mm
1	3
2	6
3	9
4	12
5	15

b, and *c* are constants. Because in such equations the value of *y* depends on the value of *x*, as we have already seen in the case of the spring, *y* is called the **dependent variable**, and *x* the **independent variable**.

A pair of corresponding numbers of a relation, like the pair 1 and 3, or 4 and 12, of the Table 11-1 is called an **ordered pair** and it is written in parentheses with the *x* value first, the *y* value second, and a comma between them. Thus, (1, 3) and (4, 12) are two ordered pairs. The order in which an ordered pair is written (*x* first, *y* second) is very important. Thus, the two ordered pairs (1, 3) and (3, 1) are not the same ordered pair.

Mathematical relations are similar to family relations: when two variables of an equation are related so that given the value of one variable the value of the other variable can be determined, we say that *the equation is the relation*. Thus, the equation *y* = 3*x* is a relation. The equation bonds together all the ordered pairs generated by it, and these ordered pairs form a family of associated members.

An equation in two variables which is a relation is also called a **function**. Thus, the equation *y* = 3*x*, which is a relation, is also a function.

The expression *Jim is the "son of" John* shows how these two members of a family are associated. Similarly, the expression *the dependent variable is a "function of" the independent variable* shows how the two variables of an equation which is a function are related. Thus, for the equation *y* = 3*x*, we say that *y* is a "function of" *x*.

NOTE: Usually a more formal and more detailed definition of a function is given, and a distinction is made between a relation and a function. For the scope of this book, however, the above introduction to relations and functions is adequate.

11-2 The Rectangular Coordinate System

It is possible to draw the graph of a function, that is, the graph of an equation. To draw the graph of an equation, a **rectangular coordinate system** is used.

A rectangular coordinate system is shown in Fig. 11-1. Two perpendicular lines intersect at a point marked 0. This point is called the **origin** of the coordinate system. Starting from the origin, points are marked at equal distances using an arbitrary unit of length. Numbers are assigned to these points, as in the case of the number line. The numbers are called the coordinates of these points. The horizontal line of this arrangement is called the **x-axis**. To the right of the origin, on the *x*-axis, is positive, and to left, negative. The vertical line is called the **y-axis**. Up from the origin, on the *y*-axis, is positive, and down is negative. The two axes divide the plane of the coordinate system into four parts called **quadrants**. The four quadrants are numbered counterclockwise with Roman numerals, I, II, III, and IV.

With the points of the plane of the coordinate system, we will associate ordered pairs. Point *A*, of Fig. 11-2, for instance, is associated with the ordered pair (3, 5), because it is 3 units to the right of the origin along the *x*-axis, and 5 units up from the origin, along the *y*-axis. The two numbers of the ordered pair (3, 5) are called the **coordinates** of point *A*. The *x*-coordinate is 3, and the *y*-coordinate is 5.

To find the coordinates of point *B* of Fig. 11-2, we have to specify its distance from the two axes of the coordinate system. To do that we pass one horizontal and

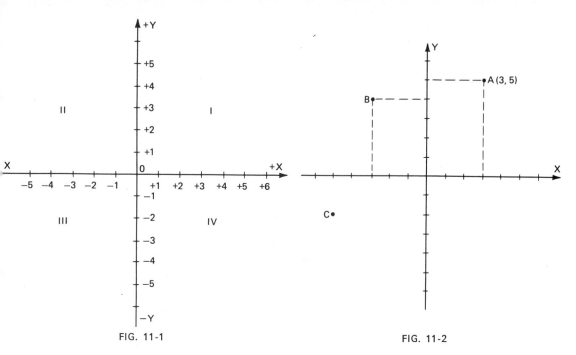

FIG. 11-1   FIG. 11-2

one vertical line through point B. Notice that the vertical dotted line intersects with the x-axis at its coordinate -3, and the horizontal dotted line intersects with the y-axis at its coordinate 4. Therefore, the coordinates of point B are $(-3, 4)$.

To locate or plot point C $(-5, -2)$, we proceed as follows: we first count five units from the origin in the direction of the negative x-axis, and then 2 units down from the origin in the direction of the negative y-axis. Point C is shown plotted in Fig. 11-2.

Finding coordinates of points, or plotting points, is more easily done on graph paper, since horizontal and vertical lines pass through the coordinates of the two axes.

EXAMPLE 11-1 Find and write the coordinates of points A, B, C, and D of Fig. 11-3.

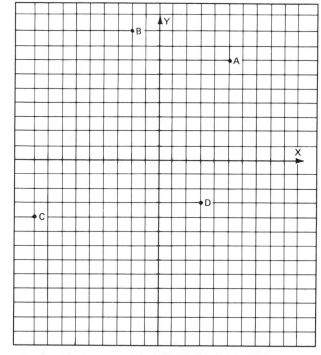

FIG. 11-3

SOLUTION

$$A (5,7), B (-2, 9), C (-9, -4), D (3, -3)$$

EXAMPLE 11-2 Use graph paper and plot the points $E(-2, 7)$, $F(0, 8)$, $G(8, 0)$, $H(-4, -4)$, $I(0, -11)$, $J(5, 7)$, and $K(6, -8)$.

SOLUTION First draw the axes and then plot the points as shown in Fig. 11-4.

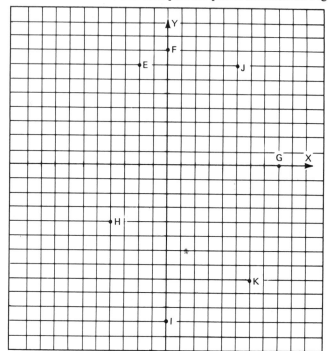

FIG. 11-4

EXAMPLE 11-3 Find and write the coordinates of points L, M, N, O, P, and Q, of Fig. 11-5. Estimate the values wherever necessary.

SOLUTION

$$L (2, 3), M (-3\tfrac{1}{2}, 3), N (4\tfrac{1}{2}, 4\tfrac{1}{2}), O (-3\tfrac{1}{3}, -1\tfrac{3}{4}), P (2, -4\tfrac{1}{2}), Q (4\tfrac{1}{2}, -5\tfrac{1}{3})$$

EXAMPLE 11-4 Use graph paper and plot each of the following points. Estimate wherever necessary.

$$R (4, -1\tfrac{1}{2}), S (-3, -7\tfrac{1}{2}), T (5, 6\tfrac{2}{3}), U (8\tfrac{3}{4}, 5\tfrac{1}{4}), V (-3\tfrac{1}{3}, -2\tfrac{1}{2}),$$
$$W (-6\tfrac{1}{2}, -7\tfrac{2}{3}), X (-4, 4\tfrac{1}{3}), Y (-7, 1\tfrac{1}{2}), Z (4\tfrac{1}{3}, 3\tfrac{1}{4})$$

SOLUTION First draw the axes and then plot the points as shown in Fig. 11-6.

EXERCISE 11-1

1. Given two points $A (5, 7)$ and $B (8, 3)$, what is the x-coordinate of point B? What is the y-coordinate of point A?

Write down the coordinates of points A, B, C, D, E, F, and G of the coordinate system of

Relations and Functions. Graphs of Equations

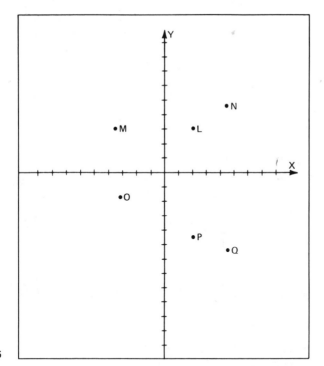

FIG. 11-5

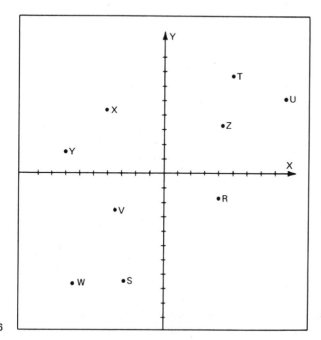

FIG. 11-6

exercise 2 (Fig. 11-7), and of points H, I, J, K, L, M, and N of the coordinate system of problem 3 (Fig. 11-8). Estimate values wherever necessary. (Notice that the two coordinate systems do not have the same scale.)

2.

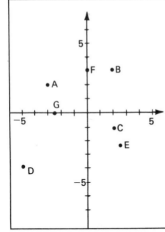

FIG. 11-7

3.

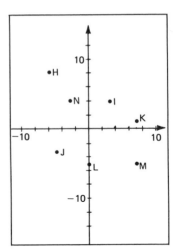

FIG. 11-8

Use graph paper and plot the points of exercises 4–8. Plot the ordered pairs of each exercise on a separate coordinate system.

4. A (3, −2), B (−1, 0), C (−8, −3), D (5, −2), E (0, −1), F (−6, 0), G (0, −6), H (4, 4), I (−5, 5)

5. J (4, 0), K (−1, 0), L (0, −1), M (1, 2), N (−1, −2)

6. O (−1, $2\frac{1}{2}$), P ($1\frac{1}{2}$, $5\frac{1}{3}$), Q ($5\frac{1}{4}$, −$2\frac{1}{2}$), R (−$7\frac{1}{2}$, −$5\frac{1}{3}$)

7. S (0, −$\frac{1}{2}$), T (8, −$7\frac{1}{3}$), U ($7\frac{1}{2}$, $2\frac{2}{3}$), V (−$4\frac{1}{3}$, $5\frac{1}{4}$)

8. W (−$1\frac{1}{4}$, −$1\frac{1}{4}$), X (0, $3\frac{1}{3}$), Y (7, $3\frac{1}{2}$), Z ($5\frac{2}{3}$, $2\frac{1}{3}$)

9. Plot the points (1, 1), (7, 7), (−3, −3), (0, 0), (4, 4), (−6, −6).

10. Plot ten points having both coordinates the same.

11. Plot the points (3, 7), (3, −2), (3, 1), (3, −8), (3, 5), (3, 0), (3, 9), (3, 8).

12. Plot ten points each having 6 as the first coordinate.

13. Plot the points (1, 4), (−1, 4), (5, 4), (8, 4), (−6, 4), (0, 4), (3, 4), (−3, 4).

14. Plot ten points each having −2 as the second coordinate.

11-3 Graphing Equations

To graph a given equation, a number of ordered pairs generated by the equation are plotted, and the points are connected with a smooth curve. The ordered pairs are generated by the equation when convenient values are substituted for x at random, and corresponding values are obtained for y. Suppose, for instance, that we want to graph the linear equation $y = 2x + 3$. The procedure for obtaining ordered pairs by substituting arbitrary values for x in the equation is shown in Fig. 11-9a. Fig. 11-9b shows the x values and the corresponding y values arranged in a table, and Fig. 11-9c shows the points plotted in a coordinate system.

Relations and Functions. Graphs of Equations

For $x = 0$, $y = 2(0) + 3 = 3$

For $x = 2$, $y = 2(2) + 3 = 7$

For $x = -3$, $y = 2(-3) + 3 = -3$

For $x = 1$, $y = 2(1) + 3 = 5$

For $x = -1$, $y = 2(-1) + 3 = 1$

(a)

x	y
0	3
2	7
-3	-3
1	5
-1	1

(b)

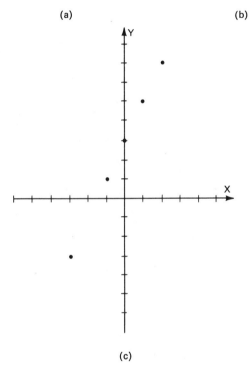

(c)

FIG. 11-9

It is evident from the five points of the graph of Fig. 11-9 that the graph of the linear equation $y = 2x + 3$ is a straight line (see Fig. 11-10). In fact, *the graph of any first degree equation in two variabes is a straight line*. This is why such equations are called linear equations.

The technique of graphing linear equations can be summarized as follows:

1. *If the given equation is not in the form $y = mx + b$, solve the equation for y.*
2. *Make a table like the one shown on the right to record corresponding values of x and y.*
3. *Assign convenient values to x and calculate the corresponding values of y.*
4. *Plot the ordered pairs of the table, and connect these points with a straight line to obtain the graph of the linear equation.*

x	y

Graphing Equations

179

NOTE: Since the graph of a linear equation is always a straight line, not too many points are needed. Because a straight line is determined by two points, three points are plotted, the third point serving as a check against the other two.

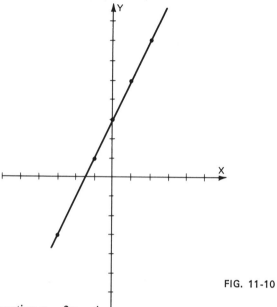

FIG. 11-10

EXAMPLE 11-5 Graph the equation $y = 2x - 4$.

SOLUTION The equation was given in the form $y = mx + b$. We proceed, then, with the other steps. The graph is shown in Fig. 11-11.

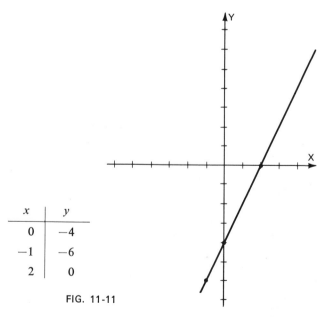

	x	y
$x = 0, y = 2(0) - 4 = -4$	0	-4
$x = -1, y = 2(-1) - 4 = -6$	-1	-6
$x = 2, y = 2(2) - 4 = 0$	2	0

FIG. 11-11

Relations and Functions. Graphs of Equations

EXAMPLE 11-6   Graph the equation $2x + y = 6$.

SOLUTION   First we solve for y to obtain $y = 6 - 2x$. Now we proceed as before (see Fig. 11-12).

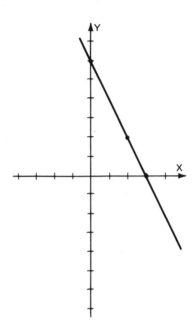

$x = 0, y = 6 - 2(0) = 6$	x	y
	0	6
$x = 3, y = 6 - 2(3) = 0$	3	0
$x = 2, y = 6 - 2(2) = 2$	2	2

FIG. 11-12

EXAMPLE 11-7   Graph the equation $\frac{1}{2}x - y = 4$.

SOLUTION   First we solve the equation for y to get $y = \frac{1}{2}x - 4$. When the numerical coefficient of x is a fraction, as in this case, we wish to avoid ordered pairs with fractional coordinates. Therefore, we give to the variable x values which are multiples of the denominator of the fractional coefficient. Thus,

	x	y
$x = 0, y = \frac{1}{2}(0) - 4 = -4$	0	-4
$x = -2, y = \frac{1}{2}(-2) - 4 = -5$	-2	-5
$x = 8, y = \frac{1}{2}(8) - 4 = 0$	8	0

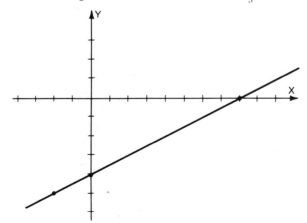

FIG. 11-13

The procedure to graph a quadratic equation (a quadratic function) in two variables is the same as to graph a linear equation. The graph in this case, however, is not a straight line, and it is called a **parabola**. Because of this, in plotting a quadratic equation, it is necessary to plot more points than in the case of a linear equation. The points plotted are connected with a smooth curve.

EXAMPLE 11-8 Graph the quadratic equation $y = x^2 - 3$.

SOLUTION

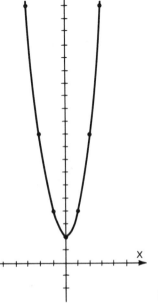

$x = 0, y = (0)^2 - 3 = -3$

$x = 1, y = (1)^2 - 3 = -2$

$x = 2, y = (2)^2 - 3 = 1$

$x = 3, y = (3)^2 - 3 = 6$

$x = -1, y = (-1)^2 - 3 = -2$

$x = -2, y = (-2)^2 - 3 = 1$

$x = -3, y = (-3)^2 - 3 = 6$

x	y
0	-3
1	-2
2	1
3	6
-1	-2
-2	1
-3	6

FIG. 11-14

EXAMPLE 11-9 Graph the quadratic equation $y = 2x^2 + 2$.

SOLUTION

$x = 0, y = 2(0)^2 + 2 = 2$

$x = 1, y = 2(1)^2 + 2 = 4$

$x = 2, y = 2(2)^2 + 2 = 10$

$x = 3, y = 2(3)^2 + 2 = 20$

$x = -1, y = 2(-1)^2 + 2 = 4$

$x = -2, y = 2(-2)^2 + 2 = 10$

$x = -3, y = 2(-3)^2 + 2 = 20$

x	y
0	2
1	4
2	10
3	20
-1	4
-2	10
-3	20

FIG. 11-15

EXAMPLE 11-10 Graph the quadratic equation $y = x^2 + 2x$ for x values greater than or equal to zero. Use a different scale on the vertical axis to control the size of the graph.

SOLUTION

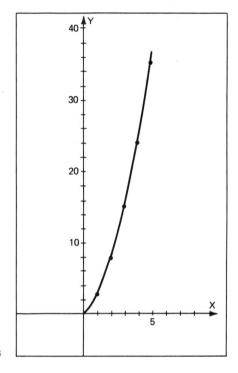

x	y
0	0
1	3
2	8
3	15
4	24
5	35

$x = 0, y = (0)^2 + 2(0) = 0$

$x = 1, y = (1)^2 + 2(1) = 3$

$x = 2, y = (2)^2 + 2(2) = 8$

$x = 3, y = (3)^2 + 2(3) = 15$

$x = 4, y = (4)^2 + 2(4) = 24$

$x = 5, y = (5)^2 + 2(5) = 35$

FIG. 11-16

EXERCISE 11-2

Graph linear equations 1–13.

1. $y = x$
2. $y = 3x$
3. $y = 3x - 1$
4. $y = 2x - 3$
5. $y - 2x = 4$
6. $3x - y = 5$
7. $y = \frac{1}{3}x - 2$
8. $y = \frac{2}{3}x + 2$
9. $y = \frac{x}{2} + 3$
10. $3y + 6x = 9$
11. $3x + 4y = 8$
12. $2x = 3y - 3$
13. $y = 4$

Graph quadratic equations 14 and 15.

14. $y = x^2 - 2x$
15. $y = 2x^2 + x - 4$

Graph quadratic equations 16–18 for x values greater than or equal to zero. Use a different scale on the vertical axis to control the size of the graph.

16. $y = 3x^2 + x$
17. $y = 2x^2$
18. $y = 3x^2 - 2x$

11-4 The Slope of a Line

Consider the two points $(2, 4)$ and $(6, 12)$ of the graph of Fig. 11-17. As the graph extends from point $(2, 4)$ to the point $(6, 12)$, the x coordinate increases from 2 to 6, that is, it increases by $6 - 2 = 4$ units. Also, between these two points of the graph,

The Slope of a Line

the y coordinate increases from 4 to 12, that is, it increases by $12 - 4 = 8$ units. Using the x increase and the y increase, we can find the **slope** (steepness) of this line, because *the slope of a line is defined as the ratio of the increase of the dependent variable to the increase of the independent variable.* According to this definition, the slope of the graph of Fig. 11-17 is

$$\text{slope} = \frac{12 - 4}{6 - 2} = \frac{8}{4} = \frac{2}{1} \quad \text{or} \quad 2$$

When we say that the slope of the line of Fig. 11-17 is $\frac{2}{1}$ or 2, we mean that the steepness of that line is 2 units of y per 1 unit of x. It should be said at this point that *the slope of a line is a constant ratio.*

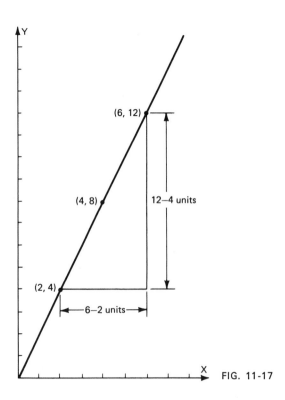

FIG. 11-17

To understand the formula for finding the slope of a line, consider first the two points of the line of Fig. 11-18a. Of the two points, we will call the lower point P_1 and the higher point P_2 (see Fig. 11-18b). Then the coordinates of point P_1 will be (x_1, y_1), and the coordinates of point P_2 will be (x_2, y_2) (see Fig. 11-18c). Distinguishing the coordinates of the two points of a given line with the symbols (x_1, y_1) and (x_2, y_2), and using the letter m to represent the slope, the formula for the slope

Relations and Functions. Graphs of Equations

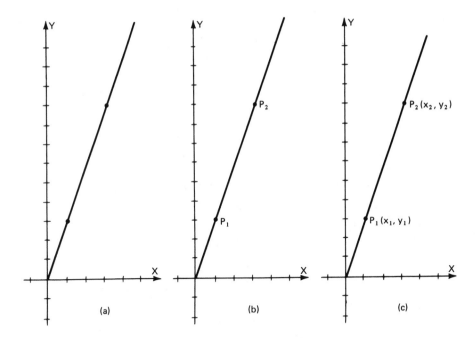

FIG. 11-18

of a line is
$$m = \frac{y_2 - y_1}{x_2 - x_1}$$

EXAMPLE 11-11   Find the slope of the graph in Fig. 11-19.

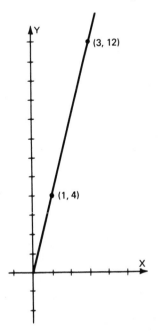

FIG. 11-19

The Slope of a Line

SOLUTION Point P_1 in this case is point $(1, 4)$, $x_1 = 1$ and $y_1 = 4$. P_2 in this case is the point $(3, 12)$, $x_2 = 3$ and $y_2 = 12$. Then, using the slope formula, we have

$$m = \frac{y_2 - y_1}{x_2 - x_1} = \frac{12 - 4}{3 - 1} = 4$$

It really makes no difference which of the two points of a line is called P_1 or P_2.

EXAMPLE 11-12 Find the slope of the graph in Fig. 11-20.

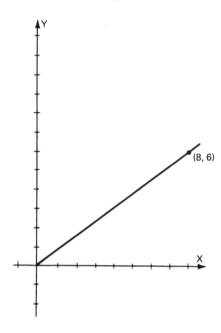

(8, 6)

FIG. 11-20

SOLUTION The points to be used are $(0, 0)$ and $(8, 6)$. Calling $(8, 6)$ point P_1 and $(0, 0)$ point P_2, we have

$$m = \frac{y_2 - y_1}{x_2 - x_1} = \frac{0 - 6}{0 - 8} = \frac{-6}{-8} = \frac{3}{4}$$

EXAMPLE 11-13 Find the slope of the line passing through the points $(2, 12)$ and $(3, 18)$.

SOLUTION Calling point $(3, 18)$ point P_1 and point $(2, 12)$ point P_2, we have

$$m = \frac{y_2 - y_1}{x_2 - x_1} = \frac{12 - 18}{2 - 3} = \frac{-6}{-1} = 6$$

EXERCISE 11-3

Draw the graph of each of linear equations 1–10, and find the slope of each graph.

1. $y = x$
2. $y = 5x$
3. $y = 3x$
4. $y = \frac{1}{3}x$
5. $y = \frac{1}{2}x$
6. $y = 3x + 2$

Relations and Functions. Graphs of Equations

7. $y = 2x + 3$

8. $y = \dfrac{2}{3}x + 1$

9. $y = 5x + 3$

10. $2y = 6 + 3x$

11. Find the slope of the line passing through $(3, 9)$ and $(5, 15)$.

12. Find the slope of the line passing through $(1, 9)$ and $(7, 63)$.

13. Find the slope of the graph in Fig. 11-21.

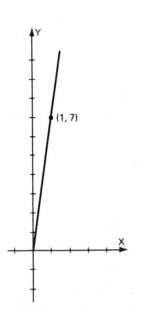

FIG. 11-21

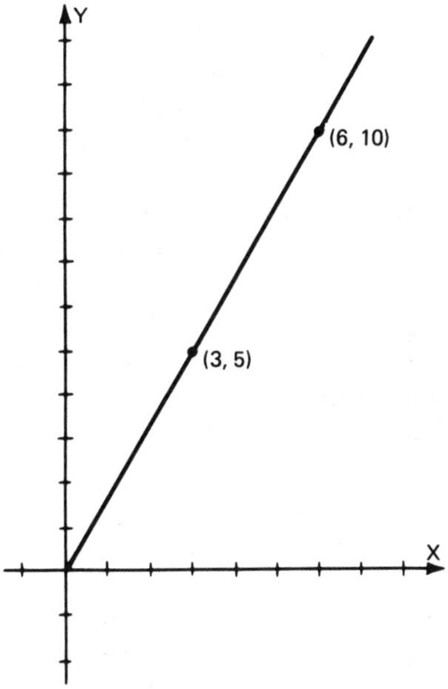

FIG. 11-22

14. Find the slope of the graph in Fig. 11-22.

15. Find the slope of the graph in Fig. 11-23.

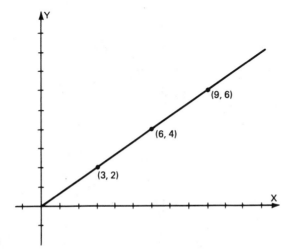

FIG. 11-23

The Slope of a Line

The graph of an equation can show the relation of two variables much better than the equation or a table of ordered pairs can. For this reason, graphs are extensively used in science and technology. Graphs can give ready information by making the reading of corresponding values very simple.

Graphs of technical relations are generally plotted in the first quadrant since negative values of the independent variable, such as negative weight, negative time, or negative voltage, are meaningless. Technical graphs must be plotted carefully, because *graph readings are estimated.*

In making a technical graph, the student should use the following four steps:

1. *Make a table of corresponding values near the graph.*
2. *Plot the quantity represented by the independent variable against the horizontal axis, and the quantity represented by the dependent variable against the vertical axis.*
3. *Choose a convenient scale for the unit of length of each axis to control the size of the graph.*
4. *Label the axes and show the units in which the variables are measured.*

Figure 11-24 shows the weight–elongation relation of the equation $y = 3x$. The student should notice how the axes are labeled, and how readily the units of the variables can be read.

We know that any ordered pair generated by the equation $y = 3x$ is a point of this graph, and that the coordinates of any point of this graph are an ordered pair which can be generated by this equation. Therefore, given a weight value, the corresponding elongation value can be read from the graph, and given an elongation value,

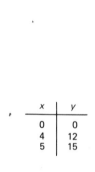

x	y
0	0
4	12
5	15

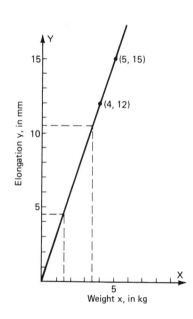

FIG. 11-24

Relations and Functions. Graphs of Equations

the corresponding weight value can be read from the graph. For instance, when the weight x is 1.5 kg, the elongation y is 4.5 mm, and when the elongation y is 10.5 mm, the corresponding weight x is 3.5 kg (see the dotted lines of Fig. 11-24).

Let us now find the slope of the graph in Fig. 11-24. Using the two points (0, 0) and (5, 15), we have

$$m = \frac{15 - 0}{5 - 0} = 3$$

that is 3 mm/kg, as is evident from the units of the variables of the graph.

We know from the previous chapter that the two variables x and y of the equation $y = 3x$ are proportional, and we have seen that the graph of their relation is a straight line. We also know that the constant 3 of the equation $y = 3x$ is the proportionality constant. (Dividing both sides of the equation $y = 3x$ by x, we get $\frac{y}{x} = 3$, which means that the ratio of y and x is 3.) We now find that 3 is also the slope of the graph of $y = 3x$. These facts suggest the following conclusions:

1. *The graph of two proportional quantities is a straight line.*
2. *The ratio of two proportional quantities and the slope of the graph of their equation have the same numerical value.*

EXAMPLE 11-14 (a) Are the values x and y of the equation $y = 6x$ proportional? (b) What is the ratio of y to x? (c) What do you expect that the slope of the graph of the equation $y = 6x$ will be? (d) Graph the equation and find the slope.

SOLUTION

(a) The variables x and y of the equation $y = 6x$ are proportional.

(b) Dividing both sides of the equation $y = 6x$ by x, we get $\frac{y}{x} = 6$.

(c) According to conclusion 2 above, the slope of the graph of $y = 6x$ must be the same as the ratio $\frac{y}{x}$; that is, 6.

(d) The graph of the equation $y = 6x$ is shown in Fig. 11-25. The slope of the graph is

$$m = \frac{18 - 6}{3 - 1} = \frac{12}{2} = 6$$

Thus, both the slope of the graph of the equation $y = 6x$, and the ratio $\frac{y}{x}$, are equal to the constant 6.

EXAMPLE 11-15 The distance covered by a car moving at a constant speed is described by the equation $d = 20t$, where d is the distance in kilometers, and t is the time in hours.

(a) Make a graph of these two related quantities, d and t. Label the axes and show the scale and the units of measurement used.
(b) How much is the ratio d/t?
(c) What do you expect the slope of the graph of the equation $d = 20t$ to be?
(d) Find the slope of the graph, and attach the proper units.
(e) Read from the graph the distance the car will cover in $3\frac{1}{2}$ hr.
(f) Read from the graph the time needed to cover 95 km.

Use of Graphs

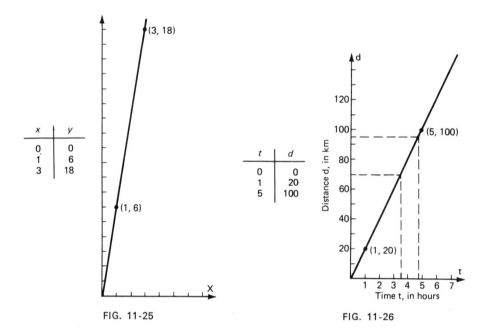

x	y
0	0
1	6
3	18

t	d
0	0
1	20
5	100

FIG. 11-25   FIG. 11-26

SOLUTION

(a) Since time is the independent variable, and distance the dependent variable, we plot time on the horizontal axis, and distance on the vertical axis. The graph of the equation $d = 20t$ is shown in Fig. 11-26. Notice that one unit of length of the horizontal axis stands for one hour, and one unit of length of the vertical axis stands for 10 km.

(b) The ratio d/t is obtained by dividing both sides of the equation $d = 20t$ by t. Thus, $d/t = 20$.

(c) Since the ratio of the two proportional quantities d and t is 20, the slope of the graph of the equation $d = 20t$ must also be 20.

(d)
$$m = \frac{y_2 - y_1}{x_2 - x_1} = \frac{100 - 20}{5 - 1} = 20$$

that is, 20 km/hr.

(e) The distance that will be covered in $3\frac{1}{2}$ hr is 70 km, as shown by one of the dotted lines of Fig. 11-26.

(f) The time needed to cover 95 km is $4\frac{3}{4}$ hr, as shown by one of the dotted lines of Fig. 11-26.

EXAMPLE 11-16 The weight of an object is given by the equation $w = dv$, where w is weight, d is density, and v is volume. Ten cm³ of aluminum are found to weigh 27 g, and 23 cm³ are found to weigh 62 g. (a) Make a graph of this information, and label the graph properly. (b) Find the slope of the graph and attach the proper units. (c) Find the density of aluminum from the graph. (Hint: Density is equal to the ratio of w to v, and is also equal to the slope of the graph.) (d) Read from the graph the volume of 44 g of aluminum.

Relations and Functions. Graphs of Equations

SOLUTION

(a) The two ordered pairs (10, 27) and (23, 62) are graphed. The graph is shown in Fig. 11-27 properly labeled.

(b)
$$m = \frac{w_2 - w_1}{v_2 - v_1} = \frac{62 - 27}{23 - 10} = \frac{35}{13} = 2.7 \text{ g/cm}^3$$

(c) Dividing both sides of the equation $w = dv$ by v, we obtain $\frac{w}{v} = d$, which means that density is the ratio of w to v. This ratio, that is the proportionality constant, d, and the slope of the graph are numerically equal. Therefore, the density of aluminum is 2.7 g/cm³.

(d) The volume of 44 g of aluminum is 16.5 cm³ (approx.), as shown by the dotted lines of Fig. 11-27.

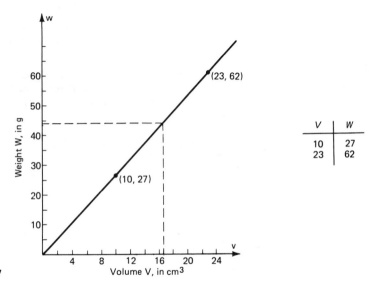

FIG. 11-27

V	W
10	27
23	62

EXAMPLE 11-17 A resistor is linear if the ratio $\frac{E}{I}$ is a constant. (a) Find the ratio $\frac{E}{I}$ in the equation $E = 5I$. (b) What do you expect the slope of the graph of this equation to be? (c) Graph the equation and find the slope.

SOLUTION

(a) Dividing both sides of the equation $E = 5I$ by I, the ratio $\frac{E}{I}$ is 5.

(b) The slope of the graph of the equation $E = 5I$, must also be 5.

(c) The graph of the equation $E = 5I$ is shown in Fig. 11-28, and the slope of the graph is

$$m = \frac{E_2 - E_1}{I_2 - I_1} = \frac{5 - 15}{1 - 3} = \frac{-10}{-2} = 5$$

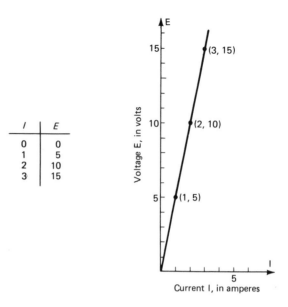

I	E
0	0
1	5
2	10
3	15

FIG. 11-28

EXAMPLE 11-18 The formula $F = \frac{9}{5}C + 32$ is used to convert Celsius temperatures to Fahrenheit temperatures. (a) Graph this equation. (b) Read from the graph the value of C when $F = 83°$. (c) Read from the graph the value of F when $C = 93°$.

SOLUTION

(a) The graph of the equation $F = \frac{9}{5}C + 32$ is shown in Fig. 11-29.

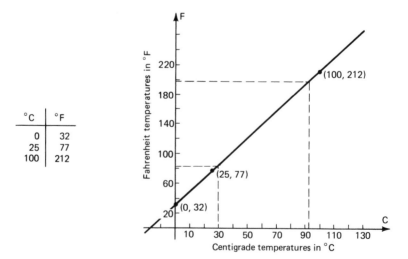

°C	°F
0	32
25	77
100	212

FIG. 11-29

Relations and Functions. Graphs of Equations

(b) $C = 30°$ (approx.) when $F = 83°$, as shown by one of the dotted lines of Fig. 11-29.

(c) $F = 195°$ (approx.) when $C = 93°$, as shown by one of the dotted lines of Fig. 11-29.

We have seen that graphs of nonlinear functions are curves. Curves do not have a constant slope, as graphs of linear functions do. The slope of a curve is continuously changing as the independent variable changes. But this subject is beyond the scope of this book. We can still use graphs of nonlinear functions, however, since corresponding values can be read from such graphs as well.

EXAMPLE 11-19   The distance a freely falling object covers is given by the equation $s = \frac{1}{2}gt^2$, where s is the distance in ft, g is the constant 32 ft/sec², and t is the time in seconds. (a) Graph this equation. (b) Read from the graph the distance covered by the falling object in 3.5 sec. (c) Read from the graph the time needed for the object to fall 80 ft.

SOLUTION

(a) The graph of the equation $s = \frac{1}{2}gt^2$ is shown in Fig. 11-30.

(b) The distance s is 190 ft (approx.) when time t is $3\frac{1}{2}$ sec.

(c) The time t is 2.25 sec (approx.) when distance s is 80 ft.

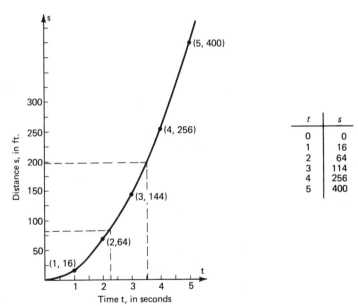

FIG. 11-30

EXAMPLE 11-20   The voltage, V, of a certain circuit changes with the time according to the equation $V = 4t^2 + t$, where V is the voltage in volts and t is the time in seconds. (a) Graph this equation. (b) Find the voltage from the graph when t is 4.3 sec. (c) Find from the graph the time needed for the voltage to reach 55 V.

SOLUTION

(a) The graph of the equation $V = 4t^2 + t$ is shown in Fig. 11-31.

Use of Graphs

193

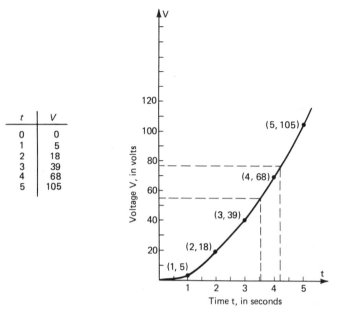

t	V
0	0
1	5
2	18
3	39
4	68
5	105

FIG. 11-31

(b) When $t = 4.3$ sec, $V = 76$ V (approx.)

(c) When $V = 55$ V, $t = 3.7$ sec (approx.)

EXERCISE 11-4

1. Given the equation $w = 15v$, (a) are w and v proportional? (b) Find the ratio of w to v. (c) What do you expect the slope of the graph of $w = 15v$ to be? (d) Graph the equation $w = 15v$, and find the slope.

2. The voltage in a certain linear resistor is described by the equation $E = 8I$. (a) Find the ratio of E to I. (b) What do you expect the slope of the graph of this equation to be? (c) Graph the equation and find the slope.

3. The weight–elongation relation of a certain spring is described by the equation $e = 1.5w$, where w is the weight in pounds, and e is the elongation in inches. (a) Graph the equation $e = 1.5w$, and label it properly. (b) Find the slope of the graph and attach the appropriate units. (c) Read from the graph the elongation of the spring when the weight is 4.5 lb. (d) Read from the graph the weight when the elongation is $7\frac{3}{4}$ in.

4. The motion of a train traveling at a constant speed is described by the equation $d = 60t$, in which d is distance in miles, and t is time in hours. (a) Draw the graph of this equation labeling the axes, and showing the scale and the units in which the variables are measured. (b) Predict the slope of the graph from the equation, and then find the slope to check your prediction. Referring to the graph, interpret the slope by attaching to it its appropriate units. (c) Read from the graph the approximate time needed for the train to cover 220 mi. (d) Read from the graph the approximate distance the train will travel in $4\frac{3}{4}$ hr.

5. Ohm's law in the form $V = RI$ says that the voltage V (in volts) is equal to the product of the resistance R (in ohms), and the current I (in amperes). (a) Complete the table below for $R = 20$. (b) Using the table, make a graph of this particular form of Ohm's law. (c) Read the current from the graph when the voltage is 115 V.

Relations and Functions. Graphs of Equations

$I(A)$	$V(V)$
0	
2	
5	

6. When the density and the volume of an object are known, its weight can be determined using the equation $w = dv$, where w is weight, d is density, and v is volume. Ten cm³ of iron weigh 78 g, and 5 cm³ weigh 39 g. (a) Make a graph of this information. (b) Find the slope of the graph and attach to it the appropriate units. (c) Find the density of iron from the graph. (d) Read from the graph the weight of 8.5 cm³ of iron.

7. The pressure at the bottom of a water tank is given by the equation $P = 62.4h$, where P is pressure in lb/ft², and h is height (or level) of the water in the tank. (a) Graph the equation $P = 62.4h$, label the axes, and show the units in which the variables are measured. (b) Find the ratio P/h. What do you expect the slope of the graph to be? (c) Find the slope of the graph, and attach the proper units to it.

8. The formula $C = \frac{5}{9}(F - 32)$ is used to convert Fahrenheit temperatures to Celsius temperatures. (a) Graph this equation. (b) Read from the graph the value of F when $C = 22°$. (c) Read from the graph the value of C when $F = 185°$, (Hint: For values of F, use multiples of 9 increased by 32).

9. Power in electrical circuits is described by the equation $P = I^2R$, where P is power (in watts), I is current (in amperes), and R is resistance (in ohms). (a) Use the equation $P = I^2R$ and $R = 20$ and complete the table below.

I	P
0	
0.1	
0.2	
0.3	
0.4	
0.5	
0.6	
0.7	
0.8	
0.9	
1.0	

(b) Use the completed table to make a graph of the equation $P = I^2R$. (c) Read from the graph the value of P when $I = 0.45$.

10. The distance a freely falling object falls is given by the equation $s = \frac{1}{2}gt^2$, where s is distance in meters, g is the constant 9.8 m/sec², and t is time in seconds. (a) Graph this equation. (b) Read from the graph the distance covered by the falling object in 3.75 sec. (c) Read from the graph the time needed for the object to fall 300 m.

11. The voltage in a certain circuit changes with the time according to the equation $V = 2t^2 + t$, where V is voltage in volts, and t is time in seconds. (a) Graph this equation. (b) Use the graph to find V when $t = 2.7$ sec. (c) Use the graph to find t when $V = 45$ V.

12. A steel ball is rolling down a grooved inclined board. The inclination of the board is such that the distance covered by the rolling ball is described by the equation $s = 3t^2$,

where s is the distance in feet, and t is the time in seconds. (a) Graph the equation $s = 3t^2$. (b) Read from the graph the value of s when $t = 3\frac{1}{3}$ sec. (c) Read from the graph the value of t when $s = 55$ ft.

REVIEW QUESTIONS

1. In the linear equation $E = 2I$, which of the two variables is the independent variable, and which is the dependent?
2. What is an ordered pair?
3. What are the points of the plane of a rectangular coordinate system associated with?
4. State the four steps for graphing a linear equation.
5. What is the only difference between the procedure for graphing a linear equation, and the procedure for graphing a quadratic equation? What are the graphs of quadratic equations called?
6. Write the formula for the slope of a line.
7. Are the readings from technical graphs exact or estimated?
8. State the four steps to follow in making a technical graph.
9. What can you say about the graph of two proportional quantities?
10. What can you say about the numerical value of the ratio of two proportional quantities, and the slope of the graph of these two quantities?

REVIEW EXERCISES

1. Given the two points $A(6, 9)$ and $B(3, 8)$, (a) what is the y-coordinate of point A? (b) What is the x-coordinate of point B?
2. Write down the coordinates of the points $A, B, C, D, E, F, G,$ and H, of the coordinate system of Fig. 11-32. Estimate wherever necessary.

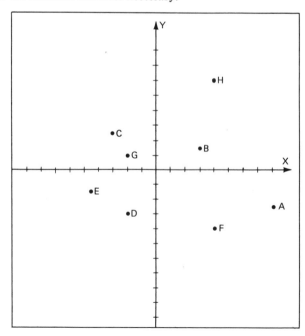

FIG. 11-32

3. Use graph paper and plot the following points. Estimate wherever necessary.

$K(-8, 3)$, $L(4, -1)$, $M(5, -7\frac{1}{3})$, $N(-7, -8\frac{1}{2})$, $O(-1, -9)$, $P(3\frac{1}{2}, 9)$, $Q(-7\frac{2}{3}, 0)$

Graph each of equations 4–7.

4. $y = 2x + 2$

5. $y = 3x - 2$

6. $6 = 2x - y$

7. $4x + 3y = 15$

Graph each of quadratic equations 8–10.

8. $y = x^2 + 1$ 9. $y = 2x^2 + 2$ 10. $y = x^2 - 2x - 6$

11. Graph the quadratic equation $k = 8m^2$, for m values greater than or equal to zero. Use a different scale for the vertical axis to control the size of the graph.

12. Graph the quadratic equation $s = 4t^2 - 2t$ for t values greater than or equal to zero. Use a different scale for the s axis to control the size of the graph.

Draw the graph and find the slope for linear equations 13 and 14.

13. $y = 7x + 2$ 14. $6 - 3x = -2y$

In problems 15 and 16, find the slope of the line passing through the two given points.

15. $(3, -1)$, $(7, 2)$ 16. $(9, 5)$, $(4, 2)$

17. Find the slope of the graph of Fig. 11-33.

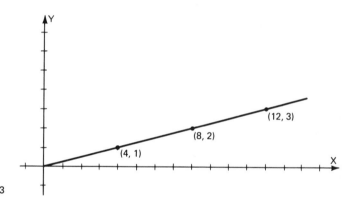

FIG. 11-33

18. (a) Are the two variables s and r of the linear equation $s = 6r$ proportional? (b) What is the ratio $\frac{s}{r}$? (c) What do you expect the slope of the graph of the equation $s = 6r$, to be? (d) Graph the equation and find the slope.

19. The equation $E = RI$ is a form of Ohm's law in which E is voltage (in volts), R is resistance (in ohms), and I is current (in amperes). (a) Let $R = 3$, and plot the graph of the equation labeling it properly. (b) What is the ratio E/I? (c) What do you expect the slope of the graph of the equation $E = RI$ to be? (d) Find the slope of the graph.

20. The motion of a car traveling at a constant speed is described by the equation $d = 90t$, where d is distance in kilometers and t is time in hours. (a) Draw a graph of this equation. (b) Predict the slope of the graph by finding the ratio d/t. (c) Find the slope of the graph. Referring to the graph, interpret the slope by attaching to it the appropriate units. (d) Read from the graph the approximate time needed for the car to cover 240 km. (e) Read from the graph the approximate distance that the car will travel in 3.6 hr.

21. When the density and the volume of an object are known, its weight can be determined using the formula $w = dv$, where w is weight, d is density, and v is volume. Seven cubic

Review Exercises

feet of copper weigh 3885 lb, and 3 cubic feet of copper weigh 1665 lb. (a) Make a graph of this information. (b) Find the slope of the graph and, referring to the units of the graph, interpret the slope by attaching to it its proper units. (c) Read from the graph the weight of 6.3 ft³ of copper.

22. A heavy steel cylinder is rolling down an inclined plane. The inclination of the plane is such that the motion of the cylinder is described by the equation $d = 5t^2$, where d is distance in meters and t is time in seconds. (a) Graph this relation. (b) Read from the graph the value of d when $t = 2.75$ sec. (c) Read from the graph the value of t when $d = 115$ m.

Exponents and Radicals

We have seen in Sect. 4-5 the laws of exponents which, for easy reference, are given below:

LAWS OF EXPONENTS

1. $a^m a^n = a^{m+n}$
2. $(a^m)^n = a^{mn}$
3. $(ab)^m = a^m b^m$
4. $\left(\dfrac{a}{b}\right)^m = \dfrac{a^m}{b^m} \qquad (b \neq 0)$
5. $\dfrac{a^m}{a^n} = \begin{cases} a^{m-n} & \text{when} \quad m > n \\ 1 & \text{when} \quad m = n \qquad (a \neq 0) \\ \dfrac{1}{a^{n-m}} & \text{when} \quad m < n \end{cases}$

We have seen that Law 5 of the exponents can involve a negative exponent, and we have agreed on the definition

$$a^{-x} = \frac{1}{a^x} \qquad (a \neq 0) \qquad\qquad (a)$$

In operations involving negative exponents we can use this definition to change negative exponents to positive, and then proceed to perform the operations using the five laws of exponents. For instance, to carry out the indicated operation in

$$a^3 \cdot a^{-2}$$

we proceed as follows:

$$a^3 \cdot a^{-2} = a^3 \cdot \frac{1}{a^2} \quad \textit{[by definition (a)]}$$

$$= \frac{a^3}{a^2}$$

$$= a \qquad (by \; Law \; 5)$$

Notice at this point, however, that we could arrive at the same result by using Law 1 directly:

$$a^3 \cdot a^{-2} = a^{3+(-2)}$$

$$= a$$

Also

$$(a^{-3})^2 = \left(\frac{1}{a^3}\right)^2 \qquad [by \; definition \; (a)]$$

$$= \frac{1}{a^6} \qquad (by \; Law \; 4)$$

But we could arrive at the same result by using Law 2 directly:

$$(a^{-3})^2 = a^{-6}$$

$$= \frac{1}{a^6} \qquad [by \; definition \; (a)]$$

These two examples suggest that *the laws of exponents are valid when the exponents are integers, that is, either positive, zero, or negative.*

EXAMPLE 12-1 Carry out the following operation:

$$(A^2 \cdot B)^{-3}$$

SOLUTION

$$(A^2 \cdot B)^{-3} = A^{-6} \cdot B^{-3} \qquad (by \; Law \; 3)$$

$$= \frac{1}{A^6} \cdot \frac{1}{B^3} \qquad (by \; Law \; 5)$$

$$= \frac{1}{A^6 B^3}$$

Consider now the following two cases of negative exponents

$$(a) \; a^{-3} \qquad (b) \; \frac{1}{a^{-3}}$$

Using the definition of negative exponents, we have

$$(a) \; a^{-3} = \frac{1}{a^3} \qquad (b) \; \frac{1}{a^{-3}} = \frac{1}{\frac{1}{a^3}}$$

$$= a^3$$

Thus, by (a)

$$a^{-3} = \frac{1}{a^3} \qquad or \qquad \frac{1}{a^3} = a^{-3}$$

and by (b)

$$\frac{1}{a^{-3}} = a^3 \qquad \text{or} \qquad a^3 = \frac{1}{a^{-3}}$$

It should be clear at this point that *we can take a power from the numerator to the denominator or from the denominator to the numerator, simply by changing the sign of the exponent.* Thus,

$$k^2m^{-3} = \frac{k^2}{m^3} \qquad \frac{t^3}{s^{-2}} = t^3s^2 \qquad \frac{3}{a^{-2}} = 3a^2 \qquad \text{and} \qquad \frac{ax^2}{b} = ab^{-1}x^2$$

EXAMPLE 12-2 Express the following term with positive exponents only and simplify.

$$\frac{2wx^{-3}}{4^{-2}y^{-2}}$$

SOLUTION

$$\frac{2wx^{-3}}{4^{-2}y^{-2}} = \frac{2 \cdot 4^2wy^2}{x^3} \qquad \text{(by Law 5)}$$

$$= \frac{32wy^2}{x^3}$$

EXAMPLE 12-3 Express the following term without negative or zero exponents and simplify.

$$\frac{3^{-1}A^2B^{-3}C^4}{6^{-2}A^2B^{-2}C^{-2}}$$

SOLUTION

$$\frac{3^{-1}A^2B^{-3}C^4}{6^{-2}A^2B^{-2}C^{-2}} = 3^{-1}6^2A^2A^{-2}B^{-3}B^2C^4C^2$$

$$= 3^{-1}6^2A^0B^{-1}C^6$$

$$= \frac{36C^6}{3B}$$

$$= \frac{12C^6}{B}$$

EXERCISE 12-1

In each of exercises 1–18 use the laws of exponents to evaluate the expression.

1. 2^{-3}

2. 5^{-1}

3. $\left(\dfrac{1}{3}\right)^{-3}$

4. 7^{-2}

5. $\left(\dfrac{3}{4}\right)^{-2}$

6. $\left(\dfrac{3}{10}\right)^{2}$

7. $(2^{-3} \cdot 3^{-2})^2$

8. $(5^{-2} \cdot 3^{-1})^{-2}$

9. $(3^{-3} \cdot 6^{-1})^{-2}$

10. $3^{-2} \cdot 3^4$

11. $6^{-2} \cdot 6$

12. $10^2 \cdot 10^{-5}$

13. $\dfrac{2^{-2}}{2^{-3}}$

14. $\dfrac{5^0}{5^{-1}}$

15. $\dfrac{9^{-1}}{9^0}$

16. $(3^2)^{-1}$

17. $(2^2)^{-2}$

18. $(5^{-2})^{-3}$

In each of exercises 19–24, express the term without negative or zero exponents and simplify.

19. $\dfrac{3^{-1}a^2}{4b^{-3}}$

20. $\dfrac{A^0 B^{-2}}{AC^{-3}}$

21. $\dfrac{10^{-2}x^3 y^{-2}}{50x^3 y^{-3}}$

22. $\dfrac{k^{-3}m^0 r^{-2}}{k^{-1}m^2 r}$

23. $\dfrac{9x^{-4}y^{-2}z}{3x^{-3}y^{-3}z^2}$

24. $\dfrac{(vw)^{-2}}{v^{-3}w^{-3}}$

12-2 Scientific Notation

Integral exponents are used in writing very large or very small numbers in **scientific notation**. Scientific notation makes computations with such numbers much simpler. For instance, the number 484,000 written in scientific notation is 4.84×10^5, and the number 0.00217, written in scientific notation is 2.17×10^{-3}.

Notice that each of the two numbers in scientific notation is written as a product of two factors, the first of which is a number between 1 and 10, and the second, a power of 10. The examples of these two numbers suggest the following two-step rule for writing a number in scientific notation.

To write a given number in scientific notation

1. *Write the first factor of the number in scientific notation with the decimal point of the given number moved to the right of its first significant digit.*
2. *Write as a second factor of the number in scientific notation a power of 10 having an exponent equal to the number of places the decimal point was moved in step 1. This exponent will be positive if the decimal point was moved to the left, or it will be negative if it was moved to the right.*

EXAMPLE 12-4   Write (a) 1274 and (b) 0.0000023 in scientific notation.

SOLUTION

(a) The first factor of the number 1274 in scientific notation will be 1.274, with the decimal point to the right of 1. The second factor will be 10^3 since the decimal point of the given number was moved three places to the left. Thus,

$$1274 = 1.274 \times 10^3$$

(b) The first factor of the number 0.0000023 in scientific notation will be 2.3 with the decimal point to the right of 2. The second factor will be 10^{-6} since the decimal point of the given number was moved six places to the right. Thus,

$$0.0000023 = 2.3 \times 10^{-6}$$

In changing numbers from the standard notation to scientific notation, the student should *make sure to include in the first factor all the significant digits of the given number.* Thus,

(a) 362.60   and   (b) 38$\bar{0}$,000

written in scientific notation are

(a) 3.6260×10^2 (b) 3.80×10^5

Also, *the significant digits of a number in scientific notation are all the digits of the first factor of the number.* Thus, given the two numbers

(c) 1.20×10^{-2} and   (d) 1.406×10^{-6}

the number in (c) has three significant digits, and the number in (d) has four significant digits.

It should be easy at this point to see that all one has to do *to change a number in scientific notation to standard form is to move the decimal point in the first factor a number of places equal to the exponent of 10. If the exponent is positive the decimal point should be moved to the right, if negative to the left.*

EXAMPLE 12-5   Express (a) 2.11×10^{-5} and (b) 3.2×10^7 in standard notation.

SOLUTION

(a) The decimal point of the first factor 2.11, is moved five places to the left because the exponent of 10 is -5. Thus, $2.11 \times 10^{-5} = 0.0000211$.
(b) The decimal point of the first factor 3.2 is moved seven places to the right because the exponent of 10 is 7. Thus, $3.2 \times 10^7 = 32,000,000$.

EXERCISE 12-2

In each of exercises 1–12 write the number in scientific notation.

1. 56.7	**2.** 217	**3.** 706
4. 61,000	**5.** 0.091	**6.** 0.00063
7. 490,000,000	**8.** 0.0000204	**9.** 278.5
10. 8,140,000	**11.** 0.00017	**12.** 130,000

In each of exercises 12–18 write the number in standard form.

13. 5.1×10^6	**14.** 1.23×10^3	**15.** 1.09×10^{-3}
16. 3×10^{-1}	**17.** 6.7×10^{-6}	**18.** 3.07×10^4

In each of exercises 19–23 write the number in scientific notation. Be careful of the significant digits in each number.

19. 0.00090	**20.** 21$\bar{0}$,000	**21.** 0.000970
22. 7,230,$\bar{0}$00	**23.** 0.0170	

In each of exercises 24–30 give the number of significant digits in the number.

24. 7.50×10^{-4}	**25.** 3.10×10^5	**26.** 3.02×10^{-6}
27. 8.0×10^7	**28.** 1.011×10^7	**29.** 7.00×10^5
30. 6.0×10^{-3}		

12-3 Computations in Scientific Notation

We will see now how numbers in scientific notation are used to simplify computations with very large or very small numbers.

The rule for the multiplication or division of numbers in scientific notation follows from the laws of exponents. Thus,

To multiply, or divide numbers in scientific notation,

1. *Multiply, or divide their first factors.*
2. *Multiply, or divide their second factors by adding or subtracting the exponents of 10.*
3. *Put as the first factor of the product or quotient in scientific notation the result of step 1, as the second factor, the result of step 2.*

EXAMPLE 12-6   Multiply 3,000,000 and 0.027.

SOLUTION   The multiplication expressed in scientific notation is

$$(3 \times 10^6)(2.7 \times 10^{-2})$$

Then,

1. $3 \times 2.7 = 8.1$
2. $10^6 \times 10^{-2} = 10^{6+(-2)} = 10^4$
3. 8.1×10^4

Thus,

$$(3,000,000)(0.027) = 8.1 \times 10^4$$

EXAMPLE 12-7   Divide 0.0000096 by 6000.

SOLUTION   The division expressed in scientific notation is

$$\frac{9.6 \times 10^{-6}}{6 \times 10^3}$$

Then,

1. $9.6 \div 6 = 1.6$
2. $\dfrac{10^{-6}}{10^3} = 10^{-6-3} = 10^{-9}$
3. 1.6×10^{-9}

Thus,

$$0.0000096 \div 6000 = 1.6 \times 10^{-9}$$

A little experience will soon make it possible for the student to perform these operations taking all three steps at the same time. The multiplication and the division of the last two examples for instance can be performed directly as follows:

$$(3 \times 10^6)(2.7 \times 10^{-2}) = 3 \times 2.7 \times 10^6 \times 10^{-2}$$
$$= 8.1 \times 10^4$$

and

$$\frac{9.6 \times 10^{-6}}{6 \times 10^3} = \frac{9.6}{6} \times 10^{-6} \times 10^{-3}$$
$$= 1.6 \times 10^{-9}$$

If the numbers in these computations are approximate numbers, the student should remember to round off the product, or the quotient, of the first step, to the correct number of significant digits.

EXAMPLE 12-8   Perform the indicated operations.

$$\frac{(3030)(0.000214)}{(0.0317)}$$

Exponents and Radicals

SOLUTION

$$\frac{(3030)(0.000214)}{(0.0317)} = \frac{(3.03 \times 10^3)(2.14 \times 10^{-4})}{(3.17 \times 10^{-2})}$$

$$= \frac{(3.03)(2.14)}{(3.17)} \times 10^3 \times 10^{-4} \times 10^2$$

$$= 2.05 \times 10$$

EXERCISE 12-3

In each of exercises 1–10, perform the indicated operations using scientific notation. Assume all the numbers to be approximate numbers.

1. $(0.00038)(12,000)$

2. $(0.000070)(0.0013)$

3. $\dfrac{0.0058}{2700}$

4. $\dfrac{330}{0.014}$

5. $\dfrac{(0.00032)(12,700)}{(0.019)}$

6. $\dfrac{(732,000)(0.0070)}{(0.000030)}$

7. $\dfrac{(410,000,000)(21,000)}{512,000}$

8. $\dfrac{(0.003112)(0.001234)}{(0.000270)}$

9. $\dfrac{(0.0078)}{(0.0000140)(0.000036)}$

10. $\dfrac{(9000)(0.214)}{(0.000310)(0.00081)}$

11. The difference in length of the main cable of a suspension bridge at 38°C, from its 6200 ft at 0°C, is given by the expression $(0.000011)(6200)(38)$. Evaluate this expression using scientific notation.

12. Young's modulus of elasticity for a wire 140-in. long, with a cross-sectional area of 0.12 in.^{2}, loaded with a weight of 460 lb, and stretching 0.018 in. under the load, is given in. lb/in.^{2} by the expression

$$\frac{460 \times 140}{0.12 \times 0.018}$$

Evaluate this expression using scientific notation.

12-4 Radicals

The symbol $\sqrt[n]{a}$ denotes the nth root of a. A number of the form $\sqrt[n]{a}$ (read "nth root of a") is called a **radical**. Thus, the numbers $\sqrt{3}$ and $\sqrt[3]{17}$ are radicals. We have called the little number at the notch of the radical sign the index, and we will call the number under the radical sign, the **radicand**.

We say that the square root of 49 is 7 because $7^2 = 49$. In general, a is the square root of b if $a^2 = b$. But 7 is not the only square root of 49, since $(-7)^2 = 49$ also. This example suggests that *every positive real number has two square roots, one positive and the other negative*. The positive square root of a number is called its **principal square root**. In technical problems we are mainly interested in the positive square roots, but we will consider negative square roots in the last chapter.

EXAMPLE 12-9 Give the principal square root of each of the following:

      (a) $\sqrt{25}$ (b) $\sqrt{64}$ (c) $\sqrt{121}$

SOLUTION

      (a) $\sqrt{25} = 5$ (b) $\sqrt{64} = 8$ (c) $\sqrt{121} = 11$

Notice now that $\sqrt[3]{-8} = -2$, since $(-2)^3 = -8$, but $\sqrt{-1}$ does not exist since no real number squared will give -1. These two examples suggest that *any radical of the form $\sqrt[n]{-a}$ has no real number as its nth root if the index, n, is even.*

EXAMPLE 12-10 Find the indicated root in each of the following:

(a) $\sqrt[3]{-27}$ (b) $\sqrt{-16}$

SOLUTION

(a) $\sqrt[3]{-27} = -3$ (b) $\sqrt{-16}$ is not a real number

So far we have seen integral exponents, and we have seen that the laws of exponents are valid for both positive and negative integral exponents. Let us see now if we can give meaning to fractional, or **rational exponents**, using the laws of exponents. Consider $9^{1/2}$. If we use Law 2 to square $9^{1/2}$, we have

$$(9^{1/2})^2 = 9^1 = 9.$$

We conclude, then, that since $(9^{1/2})^2 = 9$, $9^{1/2} = \sqrt[2]{9} = 3$.

Consider also $8^{1/3}$. If we use Law 2 and cube $8^{1/3}$, we have

$$(8^{1/3})^3 = 8^1 = 8$$

We conclude, then, that since $(8^{1/3})^3 = 8$, $8^{1/3} = \sqrt[3]{8} = 2$ We can now give meaning to $9^{3/2}$. Since $9^{3/2} = 9^{(1/2 \cdot 3)}$, (or $9^{3/2} = (9^{1/2})^3$), and $9^{1/2} = \sqrt[2]{9}$, we conclude that

$$9^{3/2} = \sqrt[2]{9^3} = \sqrt[2]{729} = 27 \qquad \text{or} \qquad 9^{3/2} = (\sqrt[2]{9})^3 = 3^3 = 27$$

The above examples then suggest the following two definitions:

$$\sqrt[n]{a} = a^{1/n} \qquad \text{and} \qquad \sqrt[n]{a^m} = a^{m/n}. \qquad (b)$$

Notice that the second definition is more general and includes the first. With the second definition, and assuming, without formal justification, that $a^{m/n} = \sqrt[n]{a^m}$, we can change the radical form of a number to its exponential form, and the exponential form to its radical form. The student should notice in the above definitions that the denominator n, of the fractional exponent is the index of the radical form, and the numerator m, of the fractional exponent is the exponent of the radicand.

EXAMPLE 12-11 Use the definition $\sqrt[n]{a^m} = a^{m/n}$ to change each of the following radicals to its exponential form.

(a) $\sqrt[4]{a^3}$ (b) $\sqrt[5]{v^2}$ (c) $\sqrt[3]{w^7}$

SOLUTION

(a) $\sqrt[4]{a^3} = a^{3/4}$ (b) $\sqrt[5]{v^2} = v^{2/5}$ (c) $\sqrt[3]{w^7} = w^{7/3}$

EXAMPLE 12-12 Use the definition $\sqrt[n]{a^m} = a^{m/n}$ to change each of the following to its radical form.

(a) $a^{2/3}$ (b) $x^{5/4}$ (c) $y^{1/5}$

SOLUTION

(a) $a^{2/3} = \sqrt[3]{a^2}$ (b) $x^{5/4} = \sqrt[4]{x^5}$ (c) $y^{1/5} = \sqrt[5]{y}$

EXAMPLE 12-13   Change $16^{3/4}$ to its radical form and evaluate.

SOLUTION

$$16^{3/4} = (16^{1/4})^3$$
$$= (\sqrt[4]{16})^3$$
$$= 2^3$$
$$= 8$$

EXERCISE 12-4

In each of exercises 1–6 give the principal square root.
1. $\sqrt{9}$ 2. $\sqrt{49}$ 3. $\sqrt{144}$
4. $\sqrt{169}$ 5. $\sqrt{289}$ 6. $\sqrt{729}$

In each of exercises 7–12 find the indicated root.
7. $\sqrt[5]{-32}$ 8. $\sqrt[2]{-4}$ 9. $\sqrt[3]{(-a)^6}$
10. $\sqrt[3]{(-x)^9}$ 11. $\sqrt[4]{-16}$ 12. $\sqrt[2]{(-a)^2}$

In each of exercises 13–18 use the definition $\sqrt[n]{a^m} = a^{m/n}$ to change the radical to its exponential form.
13. $\sqrt{7}$ 14. $\sqrt[3]{5}$ 15. $\sqrt[5]{c^3}$
16. $\sqrt[4]{E^3}$ 17. $\sqrt[5]{t}$ 18. $\sqrt{P^5}$

In each of exercises 19–24 use the definition $\sqrt[n]{a^m} = a^{m/n}$ to change the exponential to the radical form.
19. $x^{1/2}$ 20. $d^{1/3}$ 21. $k^{2/3}$
22. $g^{3/5}$ 23. $h^{6/7}$ 24. $r^{4/3}$

In each of exercises 25–28 change the exponential to its radical form and evaluate.
25. $8^{5/3}$ 26. $16^{3/2}$ 27. $25^{3/2}$ 28. $4^{5/2}$

12-5 Simplifying Radicals

Some radicals which do not have an exact root can be reduced to a simpler form. Consider first the radical $\sqrt{36}$. Notice that $\sqrt{36} = \sqrt{4 \cdot 9} = \sqrt{4} \cdot \sqrt{9} = 2 \cdot 3 = 6$. We say then that $\sqrt{4 \cdot 9} = \sqrt{4} \cdot \sqrt{9}$.

The above example suggests the following definition.

$$\sqrt[n]{a \cdot b} = \sqrt[n]{a} \cdot \sqrt[n]{b} \tag{c}$$

Now suppose we are given the radical $\sqrt{54}$ to simplify. First we factor the radicand 54 into two factors, one of which is a perfect square. Then using definition (c) we have

$$\sqrt{54} = \sqrt{9 \cdot 6}$$
$$= \sqrt{9} \cdot \sqrt{6}$$
$$= 3 \cdot \sqrt{6}$$

Simplifying Radicals

EXAMPLE 12-14 Simplify each of the following radicals:

(a) $\sqrt{50}$ (b) $\sqrt{g^3}$ (c) $3\sqrt[3]{24h^5}$

SOLUTION

$$(a)\ \sqrt{50} = \sqrt{25 \cdot 2} = \sqrt{25} \cdot \sqrt{2} = 5 \cdot \sqrt{2}$$

$$(b)\ \sqrt{g^3} = \sqrt{g^2 \cdot g} = \sqrt{g^2} \cdot \sqrt{g} = g\sqrt{g}$$

$$(c)\ 3\sqrt[3]{24h^5} = 3\sqrt[3]{8 \cdot h^3 \cdot 3 \cdot h^2}$$
$$= 3\sqrt[3]{8h^3} \cdot \sqrt[3]{3h^2}$$
$$= 3 \cdot 2h\sqrt[3]{3h^2}$$
$$= 6h\sqrt[3]{3h^2}$$

EXERCISE 12-5

In each of the following simplify the radical.

1. $\sqrt{18}$
2. $\sqrt{20}$
3. $\sqrt{45}$
4. $\sqrt{98}$
5. $\sqrt[3]{40}$
6. $\sqrt[3]{108}$
7. $\sqrt{18d^3}$
8. $\sqrt{27k^4}$
9. $3\sqrt{75g^6}$
10. $\sqrt[3]{72A^{10}}$
11. $\sqrt[3]{ab^5}$
12. $\sqrt{63t^7}$
13. $5\sqrt[3]{270M^5N}$
14. $3\sqrt{147DE^4}$
15. $\sqrt{8eg^3}$
16. $4\sqrt[3]{16k^2m^3}$
17. $7\sqrt[3]{56cd^6}$
18. $3\sqrt{48hm^6}$
19. $\sqrt{ab^2z^4}$
20. $4\sqrt[3]{375T^3S}$
21. $3\sqrt[3]{16G^3R^5V^5}$
22. $\sqrt{12H^5K^3}$
23. $6\sqrt{245VY^3Z}$
24. $7\sqrt[3]{72P^5QR^3}$
25. $2\sqrt[3]{B^2CX^6}$

12-6 Addition and Subtraction of Radicals

Only like quantities may be added or subtracted. Two radical quantities are like terms if their indexes and their radicands are the same. Thus, $\sqrt{10}$, $2\sqrt{10}$ and $5\sqrt{10}$ are like terms, and so are $2\sqrt[3]{a}$, $7\sqrt[3]{a}$, and $\sqrt[3]{a}$.

The student is reminded that the coefficients of the radicals are signed numbers, that their signs indicate whether they are positive or negative, and that the operation of addition is understood in all cases.

EXAMPLE 12-15 Combine each of the following.

(a) $4\sqrt{5} + 2\sqrt{5}$
(b) $5\sqrt[3]{9} - 3\sqrt[3]{9} + 7\sqrt[3]{9}$
(c) $8\sqrt{A} - 4\sqrt[3]{B} - 6\sqrt{A} + \sqrt[3]{B}$

SOLUTION

(a) $4\sqrt{5} + 2\sqrt{5} = 6\sqrt{5}$
(b) $5\sqrt[3]{9} - 3\sqrt[3]{9} + 7\sqrt[3]{9} = 9\sqrt[3]{9}$
(c) $8\sqrt{A} - 4\sqrt[3]{B} - 6\sqrt{A} + \sqrt[3]{B} = 2\sqrt{A} - 3\sqrt[3]{B}$

EXERCISE 12-6

Combine terms in each of the following exercises.

1. $2\sqrt{3} + 5\sqrt{3}$
2. $3\sqrt{11} + 8\sqrt{11} - 2\sqrt{11}$

3. $7\sqrt{13} - 8\sqrt{13} + 2\sqrt{13} - 5\sqrt{13}$

4. $3\sqrt{2} - \sqrt{2} + 5\sqrt{2} - 4\sqrt{2}$

5. $5\sqrt[3]{7} - 2\sqrt[3]{7} + 6\sqrt[3]{7}$

6. $12\sqrt[3]{5} - 8\sqrt[3]{5} + 2\sqrt[3]{5}$

7. $4\sqrt[3]{6} + 3\sqrt[3]{6} - \sqrt[3]{6} - \sqrt[3]{6}$

8. $3\sqrt{3} + 5\sqrt{2} + 8\sqrt{3} - \sqrt{2}$

9. $7\sqrt[3]{12} + 3\sqrt{5} - 5\sqrt{5} - 4\sqrt[3]{12}$

10. $6\sqrt[3]{4} - 3\sqrt[3]{7} + 9\sqrt[3]{4} - 4\sqrt[3]{7} + 2\sqrt[3]{4}$

11. $3\sqrt{15} - 5\sqrt{11} + 7\sqrt{11} - 2\sqrt{15} + 5\sqrt{15}$

12. $4\sqrt{\frac{1}{3}} - \sqrt{\frac{1}{3}} + 3\sqrt{\frac{1}{3}} + 9\sqrt{\frac{1}{3}}$

13. $5\sqrt{\frac{7}{8}} + 3\sqrt{\frac{7}{8}} - 11\sqrt{\frac{7}{8}}$

14. $0.07\sqrt{3} + 0.04\sqrt{3} - 0.02\sqrt{3}$

15. $2.3\sqrt{8} + 1.5\sqrt{8} - 6.2\sqrt{8}$

12-7 Multiplication and Division of Radicals

Since $\sqrt{4} \cdot \sqrt{9} = 2 \cdot 3 = 6$, and $\sqrt{4 \cdot 9} = \sqrt{36} = 6$, we say that $\sqrt{4} \cdot \sqrt{9} = \sqrt{4 \cdot 9}$. Also, since

$$\frac{\sqrt{36}}{\sqrt{9}} = \frac{6}{3} = 2$$

and

$$\sqrt{\frac{36}{9}} = \sqrt{4} = 2$$

we say that

$$\frac{\sqrt{36}}{\sqrt{9}} = \sqrt{\frac{36}{9}}$$

The above two examples suggest the following two definitions for the multiplication and the division of radicals.

$$\sqrt[n]{a} \cdot \sqrt[n]{b} = \sqrt[n]{ab} \qquad (d)$$

$$\frac{\sqrt[n]{a}}{\sqrt[n]{b}} = \sqrt[n]{\frac{a}{b}} \qquad (e)$$

Notice that *two radicals may be multiplied only when their indexes are the same. Their product is a radical with the same index as the two radicals multiplied, and a radicand equal to the product of their radicands.*

Notice also that *two radicals may be divided only when their indexes are the same. Their quotient is a radical with the same index as the two radicals divided, and a radicand equal to the quotient of their radicands.*

EXAMPLE 12-16 Perform each indicated multiplication and simplify wherever possible.

(a) $\sqrt{2}\sqrt{17}$ (b) $-\sqrt{23}\sqrt{5}$

(c) $4\sqrt[3]{4}\sqrt[3]{10}$ (d) $5(\sqrt{2} + 3)$

SOLUTION

(a) $\sqrt{2} \cdot \sqrt{17} = \sqrt{2 \cdot 17}$
$= \sqrt{34}$

(b) $-\sqrt{23} \cdot \sqrt{5} = -\sqrt{23 \cdot 5}$
$= -\sqrt{115}$

(c) $4\sqrt[3]{4} \cdot \sqrt[3]{10} = 4\sqrt[3]{4 \cdot 10}$
$= 4\sqrt[3]{40}$
$= 8\sqrt[3]{5}$

(d) $5(\sqrt{2} + 3) = 5\sqrt{2} + 15$

EXAMPLE 12-17 Multiply $(3\sqrt{5} + \sqrt{2})(\sqrt{5} - \sqrt{2})$.

SOLUTION This is multiplication of two binomials involving radicals. The multiplication of such binomials is similar to the multiplication of regular binomials. Thus,

$(3\sqrt{5} + \sqrt{2})(\sqrt{5} - \sqrt{2})$
$= 3\sqrt{5} \cdot \sqrt{5} + 3\sqrt{5}(-\sqrt{2}) + \sqrt{2} \cdot \sqrt{5} + \sqrt{2}(-\sqrt{2})$
$= 3\sqrt{25} - 3\sqrt{10} + \sqrt{10} - 2$
$= 15 - 2\sqrt{10} - 2$
$= 13 - 2\sqrt{10}$

When two binomials involving radicals are exactly the same except for the sign of their second terms, each binomial is called the **conjugate** of the other. Thus, the binomials $(2\sqrt{7} + 3)$ and $(2\sqrt{7} - 3)$ are conjugates of each other.

EXAMPLE 12-18 Multiply $(2\sqrt{7} + 3)(2\sqrt{7} - 3)$.

SOLUTION Because the binomials are the sum and difference of the same two terms, the product will be the square of the first minus the square of the second term. Thus,

$$(2\sqrt{7} + 3)(2\sqrt{7} - 3) = 4 \cdot 7 - 9$$
$$= 19$$

Notice that *when a binomial involving radicals is multiplied by its conjugate, the product is always a rational number.*

EXAMPLE 12-19 Perform each indicated division and simplify wherever possible.

(a) $\dfrac{5\sqrt{64}}{\sqrt{16}}$ (b) $\dfrac{12\sqrt{96a^2}}{3\sqrt{3}}$ (c) $\dfrac{\sqrt[3]{54x^3}}{\sqrt[3]{2}}$

SOLUTION

(a) $\dfrac{5\sqrt{64}}{\sqrt{16}} = 5\sqrt{\dfrac{64}{16}} = 5\sqrt{4} = 10$

(b) $\dfrac{12\sqrt{96a^2}}{3\sqrt{3}} = 4\sqrt{\dfrac{96a^2}{3}} = 4\sqrt{32a^2} = 16a\sqrt{2}$

(c) $\dfrac{\sqrt[3]{54x^3}}{\sqrt[3]{2}} = \sqrt[3]{\dfrac{54x^3}{2}} = \sqrt[3]{27x^3} = 3x$

EXAMPLE 12-20 Simplify $\dfrac{\sqrt{8}}{\sqrt{75}}$.

Exponents and Radicals

SOLUTION

$$\frac{\sqrt{8}}{\sqrt{75}} = \frac{\sqrt{4} \cdot \sqrt{2}}{\sqrt{25} \cdot \sqrt{3}} = \frac{2\sqrt{2}}{5\sqrt{3}}$$

The result $\frac{2\sqrt{2}}{5\sqrt{3}}$ is correct, but it should not be called the final answer. Since a fractional answer without the radical in the denominator is considered a simpler form, it is common practice to remove radicals from denominators. *Radicals are eliminated from a denominator by multiplying both terms of the fraction by a quantity which will make the denominator a rational number.* Such a quantity is called a **rationalizing factor**, and the process of removing radicals from the denominator is called **rationalizing the denominator**. Thus, to remove the radical from the denominator of $\frac{2\sqrt{2}}{5\sqrt{3}}$, we multiply both terms of the fraction by $\sqrt{3}$, as shown below:

$$\frac{2\sqrt{2}}{5\sqrt{3}} = \frac{2\sqrt{2}}{5\sqrt{3}} \cdot \frac{\sqrt{3}}{\sqrt{3}}$$

$$= \frac{2\sqrt{6}}{5 \cdot 3}$$

$$= \frac{2\sqrt{6}}{15}$$

EXAMPLE 12-21   Rationalize the denominator of the fraction $\frac{\sqrt{2}}{\sqrt{12}}$.

SOLUTION

$$\frac{\sqrt{2}}{\sqrt{12}} = \frac{\sqrt{2}}{\sqrt{12}} \cdot \frac{\sqrt{3}}{\sqrt{3}}$$

$$= \frac{\sqrt{6}}{\sqrt{36}}$$

$$= \frac{\sqrt{6}}{6}$$

The student should notice that $\sqrt{3}$ instead of $\sqrt{12}$ was used as a rationalizing factor to avoid simplification of the numerator after rationalizing. The student should try to see that with $\sqrt{12}$, even though the result will be the same, there is more work involved in rationalizing the denominator.

When the denominator of a fraction is a binomial involving radicals, it is rationalized by multiplying both terms of the fraction by the conjugate of the denominator.

EXAMPLE 12-22   Rationalize the denominator of the fraction $\frac{5}{\sqrt{3} - 2}$.

SOLUTION

$$\frac{5}{\sqrt{3} - 2} = \frac{5}{\sqrt{3} - 2} \cdot \frac{\sqrt{3} + 2}{\sqrt{3} + 2}$$

$$= \frac{5\sqrt{3} + 10}{3 - 4}$$

$$= \frac{5\sqrt{3} + 10}{-1}$$

$$= -5\sqrt{3} - 10$$

Multiplication and Division of Radicals

In each of exercises 1–12 perform the indicated multiplication and simplify wherever possible.

1. $\sqrt{8} \cdot \sqrt{3}$ 2. $\sqrt{5} \cdot \sqrt{7}$

3. $\sqrt{12} \cdot \sqrt{7}$ 4. $\sqrt[3]{24} \cdot \sqrt[3]{3}$

5. $\sqrt[4]{4} \cdot \sqrt[4]{8}$ 6. $\sqrt{7A} \cdot \sqrt{3A}$

7. $\sqrt{3B} \cdot \sqrt{12B}$ 8. $\sqrt[3]{gh^2} \cdot \sqrt[3]{g^2} \cdot \sqrt[3]{h^2}$

9. $7(5 - \sqrt{5})$ 10. $8(\sqrt{3} + 2)$

11. $5\sqrt[3]{2V^2W} \cdot 2\sqrt[3]{9V} \cdot \sqrt[3]{3W^5}$ 12. $4\sqrt[3]{3e^2g^4h} \cdot 2\sqrt[3]{2e^5gh^2} \cdot 3\sqrt[3]{4gh}$

In each of exercises 13–25 multiply the binomials.

13. $(\sqrt{7} + 3)(\sqrt{7} - 5)$ 14. $(2 - \sqrt{3})(4 - \sqrt{3})$

15. $(8 + 2\sqrt{3})(3 - \sqrt{3})$ 16. $(4\sqrt{3} + \sqrt{2})(2\sqrt{3} - \sqrt{2})$

17. $(\sqrt{6} - 2)^2$ 18. $(3 + \sqrt{8})^2$

19. $(1 + \sqrt{C})(2 + \sqrt{C})$ 20. $(A + \sqrt{B})(A - \sqrt{B})$

21. $(\sqrt{2} + 5)(\sqrt{2} - 5)$ 22. $(\sqrt{5} - 3)(\sqrt{5} + 3)$

23. $(3 + 2\sqrt{8})(3 - 2\sqrt{8})$ 24. $(\sqrt{11} + \sqrt{7})(\sqrt{11} - \sqrt{7})$

25. $(3\sqrt{6} + 4)(3\sqrt{6} - 4)$

In each of exercises 26–42 simplify the indicated division.

26. $\dfrac{\sqrt{3}}{\sqrt{8}}$ 27. $\dfrac{\sqrt[3]{80}}{\sqrt[3]{2}}$ 28. $\dfrac{\sqrt[3]{72}}{\sqrt[3]{3}}$ 29. $\dfrac{\sqrt{18}}{\sqrt{8}}$

30. $\dfrac{\sqrt{t^3}}{\sqrt{t}}$ 31. $\dfrac{5\sqrt{32t}}{\sqrt{2}}$ 32. $\dfrac{4\sqrt{27w^5}}{\sqrt{3w^3}}$ 33. $\dfrac{\sqrt{5}}{\sqrt{3}}$

34. $\dfrac{\sqrt{13}}{\sqrt{7}}$ 35. $\dfrac{5}{\sqrt{a^3}}$ 36. $\dfrac{\sqrt{R}}{\sqrt{T^3}}$ 37. $\dfrac{3}{\sqrt{2} - 1}$

38. $\dfrac{8}{2 - \sqrt{6}}$ 39. $\dfrac{2\sqrt{2} + 4}{5 - \sqrt{2}}$ 40. $\dfrac{20 - \sqrt{7}}{3 + \sqrt{7}}$ 41. $\dfrac{\sqrt{N} + 5}{\sqrt{N} - 3}$

42. $\dfrac{\sqrt{5} + \sqrt{2}}{\sqrt{5} - \sqrt{2}}$

REVIEW QUESTIONS

1. Are the laws of exponents valid for integers?

2. What are the two parts of a number written in scientific notation?

3. Give, in your own words, the two-step rule for writing a number in scientific notation.

4. How many of the significant digits of a number written in scientific notation are included in the first part of the number?

5. Describe in your own words the procedure for changing a number in scientific notation to standard form.

6. Give, in your own words, the rule for multiplying numbers in scientific notation.

7. Give, in your own words, the rule for dividing numbers in scientific notation.

8. What is the principal square root of a number?

9. Describe in your own words the procedure of simplifying radicals.

10. How are radicals added or subtracted?

11. Give the formula defining multiplication of radicals.

12. Give the formula defining division of radicals.

13. What is the conjugate of a binomial involving radicals?

14. What is a rationalizing factor?

15. Describe in your own words the process of rationalizing the denominator of a fraction.

REVIEW EXERCISES

In each of exercises 1–4 use the laws of exponents to evaluate the expression.

1. 2^{-5} **2.** 3^{-1} **3.** $\dfrac{3^{-2}}{3^{-3}}$ **4.** $\dfrac{7^0}{7^{-1}}$

In each of exercises 5–7 express the term without negative or zero exponents and simplify.

5. $\dfrac{5^{-1}e^2}{5^{-2}y^{-1}}$ **6.** $\dfrac{14x^5y^{-3}z^0}{21x^{-3}y^{-5}z^2}$ **7.** $\dfrac{(st)^{-3}}{s^{-5}t^{-5}}$

In each of exercises 8–11, write the number in scientific notation. Be careful to include the correct number of significant digits in your answer.

8. 43.1   **9.** 32,0̃00   **10.** 0.00230   **11.** 210,000

In each of exercises 12–14 give the number of significant digits in the number.

12. 4.8×10^3 **13.** 7.3×10^6 **14.** 6.0×10^3

In each of exercises 15–18 perform the indicated operations using scientific notation. Assume all the numbers to be approximate numbers.

15. $(1{,}700{,}000)(0.00160)$ **16.** $\dfrac{3100}{0.002}$

17. $\dfrac{320{,}000}{(0.0031)(0.000040)}$ **18.** $\dfrac{(0.0071)(370)}{0.0048}$

In each of exercises 19–20 give the principal square root.

19. $\sqrt{81}$ **20.** $\sqrt{225}$

In each of exercises 21–22 change the radical to its exponential form.

21. $\sqrt[4]{3^3}$ **22.** $\sqrt[3]{a^2}$

In each of exercises 23–27 change the exponential to the radical form. Evaluate wherever possible.

23. $B^{3/4}$ **24.** $16^{3/2}$ **25.** $T^{1/3}$ **26.** $8^{2/3}$ **27.** $h^{1/2}$

In each of exercises 28–31 simplify the radical.

28. $\sqrt[3]{500}$ **29.** $\sqrt{363}$ **30.** $\sqrt{bh^2x^3}$ **31.** $\sqrt[3]{aR^5T^7}$

In each of exercises 32–33 combine terms.

32. $5\sqrt{8} + 3\sqrt{8} - \sqrt{8}$ **33.** $6\sqrt[3]{7} - 3\sqrt{7} + 5\sqrt[3]{7} + 4\sqrt{7}$

In each of exercises 34–37 perform the indicated multiplication and simplify wherever possible.

34. $\sqrt[3]{5} \cdot \sqrt[3]{16}$ **35.** $5\sqrt{14a} \cdot 2\sqrt{2a}$

36. $4\sqrt[3]{2H^2K} \cdot 3\sqrt{4H^2K^2}$ **37.** $\sqrt{5tu} \cdot \sqrt{15t^3uv}$

In each of exercises 38–40 multiply the binomials.

38. $(\sqrt{3} + 9)(\sqrt{3} - 2)$ **39.** $(\sqrt{5} - \sqrt{3})(\sqrt{5} + \sqrt{3})$

40. $(\sqrt{2v} + \sqrt{3w})(\sqrt{2v} - \sqrt{3w})$

In each of exercises 41–45 simplify the indicated division.

41. $\dfrac{\sqrt{7}}{\sqrt{8}}$ **42.** $\dfrac{\sqrt[3]{216}}{\sqrt[3]{27}}$ **43.** $\dfrac{\sqrt{R}}{\sqrt{S^2}}$ **44.** $\dfrac{\sqrt{h} + 3}{\sqrt{h} - 3}$

45. $\dfrac{\sqrt{3} - \sqrt{8}}{\sqrt{3} + \sqrt{8}}$

13

Systems
of Linear Equations

13-1 Basic Ideas About Linear Equations
in Two Variables

So far we have solved linear equations in one variable, and problems involving linear equations in one variable. In this chapter we will see how to solve equations and problems involving two variables.

Suppose we are told that the sum of two numbers, x and y, equals 11. It is not difficult to express this relation as the equation

$$x + y = 11$$

but how do we solve such an equation; that is, how do we find the two numbers meant? We know from chapter 11 that there is a great number of ordered pairs which will satisfy the given equation. The ordered pairs (5, 6), (6, 5), (10, 1), (4, 7) are a few examples.

Suppose now that we are given the additional information that the difference between these two numbers is 1. Accordingly, we write the second equation $x - y = 1$. For this equation also, we can find as many ordered pairs as we please that will satisfy the equation. For instance, (11, 10), (6, 5), (7, 6) are a few such ordered pairs. In order, however, to solve the problem, or find the two numbers which when added together will give 11, and when subtracted will give 1, we must find one ordered pair which will satisfy both the equations,

$$x + y = 11$$
$$x - y = 1$$

This ordered pair is (6, 5) because

$$6 + 5 = 11$$
$$6 - 5 = 1$$

Linear equations in x and y, which are both satisfied by the same x value and the same y value are called **linear simultaneous equations in two variables** or **systems of equations in two variables**.

Systems of equations may be solved in a number of ways. In this chapter, we will see three such methods.

13-2 Graphic Solution of Systems of Equations

We learned in chapter 11 that the graph of a linear equation is a straight line. If, as we have seen, the two equations of a system have a common ordered pair as their solution, the graphs of these two equations will have a common point; the two lines will intersect. Consider, for instance, the system of equations

$$3x - y = 5$$
$$x + y = 7$$

The graphs of these two equations, plotted in the same coordinate system are shown in Fig. 13-1. Notice that the two graphs intersect at the point where $x = 3$ and $y = 4$. Then $x = 3$ and $y = 4$ is the solution of the system $3x - y = 5$ and $x + y = 7$.

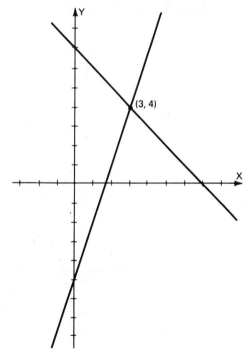

(3, 4)

FIG. 13-1

If the two equations of a system are plotted on the same coordinate system, and their graphs are parallel, they will not intersect; the two lines will not have a common point, and the system will not have a solution. In such a case, the equations of the system are called **inconsistent**. For instance, the two equations

$$y = 2x + 3$$
$$y - 2x = 7$$

are inconsistent because their graphs do not intersect (see Fig. 13-2), and, therefore, the system has no solution.

EXAMPLE 13-1 Solve the system of equations

$$4x - y = 5$$
$$2x + y = 7$$

by graphing and check.

SOLUTION

$$x = 2 \qquad y = 3$$

The graphs of the two equations of the system intersect at (2, 3) (see Fig. 13-3). Check:

$$4(2) - 3 \overset{?}{=} 5$$
$$8 - 3 = 5$$
$$2(2) + 3 \overset{?}{=} 7$$
$$4 + 3 = 7 \checkmark$$

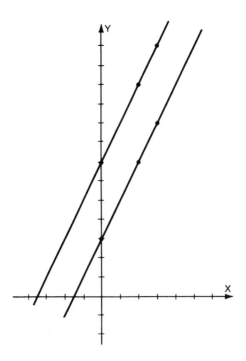

FIG. 13-2

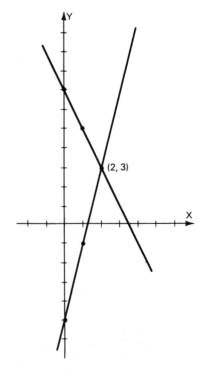

(2, 3)

FIG. 13-3

Systems of Linear Equations

The method of solving systems of linear equations by graphing gives solutions which often must be estimated, while the two algebraic methods we will see give exact solutions.

13-3 Solving Systems of Equations Using Elimination by Addition or Subtraction

Suppose we are asked to solve the system of equations

$$\text{(a) } 2x + 3y = 18$$
$$\text{(b) } 9x - 3y = 15$$

We can add together the left sides of the two equations and then add together the right sides of the two equations, since this is nothing more than adding equals to equals. Thus,

$$
\begin{array}{r}
2x + 3y = 18 \\
9x - 3y = 15 \\
\hline
11x \qquad = 33
\end{array}
$$

Notice that the y's were eliminated by this addition because their coefficients have the same absolute value, and opposite signs. From the resulting simple equation, it is evident that $x = 3$. Substituting 3 for x in either of the two original equations, we get the value of y. Thus,

$$\text{(a) } 2(3) + 3y = 18 \qquad \text{(b) } 9(3) - 3y = 15$$
$$3y = 12 \qquad\qquad -3y = -12$$
$$y = 4 \qquad\qquad\qquad y = 4$$

Therefore, the solution for the given system is $x = 3$ and $y = 4$.

The method of solving systems of equations by elimination by addition or subtraction can be summarized as follows:

1. *Multiply one or, if necessary, both of the equations, so that the numerical coefficients of any of the variables will be of equal absolute value.*
2. *If those equal coefficients are of unlike signs, add the two equations. If the two equal coefficients are of like signs, subtract one equation from the other.*
3. *Solve the resulting equation for the remaining variable.*
4. *Substitute the value of this variable in either of the original equations of the system and solve the equation for the other variable.*
5. *Check the result in both of the original equations.*

EXAMPLE 13-2   Solve the system of equations

$$\text{(a) } 3x - 6y = 9$$
$$\text{(b) } 4x - 2y = 24$$

using the method of elimination by addition or subtraction and check.

SOLUTION

$$(a) \quad 3x - 6y = 9$$

$$(b) \quad 4x - 2y = 24 \quad \text{[multiplying equation (b) by 3]}$$

$$(a) \quad 3x - 6y = 9 \quad \text{[subtracting equation (b) from equation (a)]}$$

$$(b) \; 12x - 6y = 72$$

$$\overline{-9x = -63} \quad \text{(solving for x)}$$

$$x = 7$$

$$4(7) - 2y = 24 \quad \text{[substituting 7 for x in equation (b)]}$$

$$28 - 2y = 24 \quad \text{(solving for y)}$$

$$y = 2$$

Thus the solution is $x = 7$, $y = 2$. Check:

(a) $3(7) - 6(2) \stackrel{?}{=} 9$ (b) $4(7) - 2(2) \stackrel{?}{=} 24$

$\quad 21 - 12 \stackrel{?}{=} 9 \quad\quad\quad 28 - 4 \stackrel{?}{=} 24$

$\quad\quad\quad 9 = 9\checkmark \quad\quad\quad\quad 24 = 24\checkmark$

EXAMPLE 13-3 Solve the system of equations

$$(a) \; 6x + 8y = -12$$

$$(b) \; 5x - 6y = 28$$

using the method of elimination by addition or subtraction and check.

SOLUTION

$$(a) \quad 6x + 8y = -12 \quad \text{[multiplying equation (a) by 3 and equation (b) by 4]}$$

$$(b) \quad 5x - 6y = 28$$

$$(a) \; 18x + 24y = -36 \quad \text{[adding equations (a) and (b)]}$$

$$(b) \; 20x - 24y = 112$$

$$\overline{38x = 76} \quad \text{(solving for x)}$$

$$x = 2$$

$$6(2) + 8y = -12 \quad \text{[substituting 2 for x in equation (a)]}$$

$$8y = -24 \quad \text{(solving for y)}$$

$$y = -3$$

Thus the solution is $x = 2$, and $y = -3$. Check:

(a) $6(2) + 8(-3) \stackrel{?}{=} -12$ (b) $5(2) - 6(-3) \stackrel{?}{=} 28$

$\quad\quad 12 - 24 \stackrel{?}{=} -12 \quad\quad\quad 10 + 18 \stackrel{?}{=} 28$

$\quad\quad\quad\quad -12 = -12\checkmark \quad\quad\quad\quad 28 = 28\checkmark$

EXAMPLE 13-4 Solve the system of equations

$$\text{(a) } 3x - 9 = 2y$$
$$\text{(b) } 6x + y = -2$$

using the method of elimination by addition or subtraction and check.

SOLUTION

$$\text{(a) } 3x - 9 = 2y$$
$$\text{(b) } 6x + y = -2$$

$$\begin{array}{ll} \text{(a) } 3x - 2y = 9 & \textit{[equation (a) is rearranged so} \\ \text{(b) } 6x + y = -2 & \textit{that similar terms are in col-} \\ & \textit{umns]} \end{array}$$

$$\begin{array}{ll} \text{(a) } \quad 3x - 2y = 9 & \\ \underline{\text{(b) } 12x + 2y = -4} & \textit{[multiplying equation (b) by 2]} \\ \quad 15x \qquad\quad = 5 & \textit{[adding equations (a) and (b)]} \\ \qquad\qquad x = \tfrac{1}{3} & \textit{(solving for x)} \\ 3(\tfrac{1}{3}) - 2y = 9 & \textit{[substituting } \tfrac{1}{3} \textit{ for x in equa-} \\ & \textit{tion (a)]} \\ \qquad -2y = 8 & \textit{(solving for y)} \\ \qquad\quad y = -4 & \end{array}$$

Thus the solution is $x = \tfrac{1}{3}$ and $y = -4$. Check:

$$\text{(a) } 3(\tfrac{1}{3}) - 9 \overset{?}{=} 2(-4) \qquad \text{(b) } 6(\tfrac{1}{3}) - 4 \overset{?}{=} -2$$
$$1 - 9 \overset{?}{=} -8 \qquad\qquad 2 - 4 \overset{?}{=} -2$$
$$-8 = -8\checkmark \qquad\qquad\quad -2 = -2\checkmark$$

EXAMPLE 13-5 Solve the system of equations

$$\text{(a) } 3x + 2y = 18.5$$
$$\text{(b) } 7x - 6y = 13.3$$

using the method of elimination by addition or subtraction and check.

SOLUTION

$$\begin{array}{ll} \text{(a) } 3x + 2y = 18.5 & \textit{[multiplying equation (a) by 3]} \\ \text{(b) } 7x - 6y = 13.3 & \\ \text{(a) } 9x + 6y = 55.5 & \\ \underline{\text{(b) } 7x - 6y = 13.3} & \\ \quad 16x \qquad\quad = 68.8 & \textit{(solving for x)} \\ \qquad\quad x = 4.3 & \\ 3(4.3) + 2y = 18.5 & \textit{[substituting 4.3 for x in equa-} \\ & \textit{tion (a)]} \\ \qquad 2y = 5.6 & \textit{(solving for y)} \\ \qquad\; y = 2.8 & \end{array}$$

Thus the solution is $x = 4.3$ and $y = 2.8$. Check:

(a) $3(4.3) + 2(2.8) \overset{?}{=} 18.5$ (b) $7(4.3) - 6(2.8) \overset{?}{=} 13.3$

$12.9 + 5.6 \overset{?}{=} 18.5$ $30.1 - 16.8 \overset{?}{=} 13.3$

$18.5 = 18.5\checkmark$ $13.3 = 13.3\checkmark$

13-4 Solving Systems of Equations by the Substitution Method

Since the value of x should be the same for both equations of a system, we can solve for x in terms of y in one equation of the system, and substitute for x in the other equation. This changes the last equation into an equation in y with no x. The same is true for y; we can solve for y in terms of x in one equation, and substitute for y in the other equation. This method is known as the method of **substitution**.

EXAMPLE 13-6 Solve the system of equations

(a) $3x - y = 3$

(b) $\quad 2y = 3x$

using the substitution method and check.

SOLUTION

(a) $3x - y = 3$

(b) $\quad 2y = 3x$

Solving equation (a) for y, we have

$$y = 3x - 3$$

Substituting $3x - 3$ for y in equation (b), we have

$$2(3x - 3) = 3x$$

Solving the last equation for x, we have

$$6x - 6 = 3x$$

$$3x = 6$$

$$x = 2$$

Substituting 2 for x in equation (a), we have

$$3(2) - y = 3$$

$$y = 3$$

Thus the solution is $x = 2$ and $y = 3$. Check:

(a) $3(2) - 3 \overset{?}{=} 3$ (b) $2(3) \overset{?}{=} 3(2)$

$6 - 3 \overset{?}{=} 3$ $6 = 6\checkmark$

$3 = 3\checkmark$

Solving systems of equations by the method of substitution can be summarized as follows:

1. *Solve one of the equations of the system for one of the two variables in terms of the other.*
2. *Substitute the expression obtained in step 1 into the other equation.*
3. *Solve the resulting equation for its only variable.*
4. *Substitute the value of the variable of step 3 into the equation of step 1 and solve for the second variable.*
5. *Check the result in the two original equations.*

EXAMPLE 13-7   Solve the system of equations

$$\text{(a) } x + 2y = 9$$
$$\text{(b) } 2x - 5y = -18$$

using the substitution method and check.

SOLUTION

$$x = 9 - 2y \qquad \textit{[solving equation (a) for x]}$$
$$2(9 - 2y) - 5y = -18 \qquad \textit{[substituting } 9 - 2y \textit{ for x in equation (b)]}$$
$$18 - 4y - 5y = -18 \qquad \textit{(solving for y)}$$
$$-9y = -36$$
$$y = 4$$
$$x + 2(4) = 9 \qquad \textit{[substituting 4 for y in equation (a)]}$$
$$x + 8 = 9 \qquad \textit{(solving for x)}$$
$$x = 1$$

Thus the solution is $x = 1$ and $y = 4$. Check:

$$\text{(a) } 1 + 2(4) \overset{?}{=} 9 \qquad \text{(b) } 2(1) - 5(4) \overset{?}{=} -18$$
$$1 + 8 \overset{?}{=} 9 \qquad 2 - 20 \overset{?}{=} -18$$
$$9 = 9 \checkmark \qquad -18 = -18 \checkmark$$

EXAMPLE 13-8   Solve the system of equations

$$\text{(a) } 3x + 2y = 18$$
$$\text{(b) } 7x - 3y = 19$$

using the substitution method and check.

SOLUTION

$$x = \frac{19 + 3y}{7} \qquad \textit{[solving equation (b) for x]}$$
$$3\left(\frac{19 + 3y}{7}\right) + 2y = 18 \qquad \textit{[substituting } (19 + 3y)/7 \textit{ for x in equation (a)]}$$
$$\frac{57 + 9y}{7} + 2y = 18 \qquad \textit{(solving for y)}$$
$$57 + 9y + 14y = 126$$
$$y = 3$$

Solving Systems of Equations by the Substitution Method

$$7x - 3(3) = 19 \qquad \textit{[substituting 3 for y in equation (b)]}$$

$$7x - 9 = 19 \qquad \textit{(solving for x)}$$

$$x = 4$$

Thus the solution is $x = 4$ and $y = 3$. Check:

(a) $3(4) + 2(3) \overset{?}{=} 18$ \qquad (b) $7(4) - 3(3) \overset{?}{=} 19$

$12 + 6 \overset{?}{=} 18$ \qquad\qquad $28 - 9 \overset{?}{=} 19$

$18 = 18 \checkmark$ \qquad\qquad\quad $19 = 19 \checkmark$

EXERCISE 13-1

Solve systems 1–6 by the method of graphing:

1. $2x + y = 8$
   $3x - y = 2$
2. $3x - y = 7$
   $2x - 4y = -12$
3. $4x + 4y = 28$
   $6x - 3y = 6$
4. $6x - 3y = -3$
   $4x + y = 7$
5. $3x + 3y = 15$
   $x + 2y = 3$
6. $2x - 14 = -2y$
   $x - 2y = 4$

Solve systems 7–20 by the method of elimination by addition or subtraction and check your answers.

7. $x + y = 12$
   $x - y = 2$
8. $k + 3m = 9$
   $2k - 5m = 7$
9. $4t - 6s = 8$
   $2t + 3s = 28$
10. $6x - 8y = 38$
    $3x + 2y = 31$
11. $4r + 5s = 21$
    $7r - s = 27$
12. $10a - 3b = 6$
    $6a + 3b = 42$
13. $2v + 2w = 26$
    $3v + 4w = 46$
14. $3c - 2d = 17$
    $5c - 6d = 23$
15. $4x + 7y = 49$
    $6x + 3y = 51$
16. $7u + 6v = 27$
    $8u + 2v = -8$
17. $\frac{2}{3}x + 3y = 16$
    $3x + \frac{3}{2}y = 36$
18. $\frac{3}{4}x - 4y = -2$
    $\frac{5}{8}x + 3y = 11$
19. $0.4x + 0.6y = 220$
    $0.7x + 0.3y = 160$
20. $0.5x + 0.6y = 320$
    $0.8x + 0.3y = 380$

Solve systems of equations 21–30 by the method of substitution and check your results.

21. $5y = 15x$
    $x + y = 8$
22. $4y - 3x = 9$
    $x + 4y = 13$
23. $2x + y = 15$
    $8x + 2y = 42$
24. $x + 2y = 11$
    $2x + y = 16$
25. $3x + 2y = 41$
    $x - y = 7$
26. $2x - y = 25$
    $3x - 2y = 33$
27. $4x - 5y = 1$
    $6x - 7y = 3$
28. $5v - 8w = 8$
    $4v + 3w = 138$
29. $\frac{x}{2} + 2y = 22$

    $x - \frac{y}{2} = 8$
30. $\frac{2x}{3} + y = 18$

    $x - \frac{y}{4} = 6$

Systems of Linear Equations

13-5 Solving Problems Using Linear Equations in Two Variables

Often a word problem is expressed more conveniently, and is solved more easily, by using two variables rather than one.

EXAMPLE 13-9 Find two numbers, such that if the first is added to four times the second, the sum is 29, and if the second is added to six times the first, the sum is 36.

SOLUTION Let x be the first number and y the second number. Then,

(a) $4y + x = 29$

(b) $6x + y = 36$

We now proceed to solve the system of equations.

(a) $\quad x + 4y = 29 \qquad$ [rearranging the terms of equation (a), multiplying equation (b) by 4, and subtracting]

(b) $\quad \dfrac{24x + 4y = 144}{-23x \qquad\quad = -115} \qquad$ (solving for x)

$\qquad\qquad x = 5$

$\qquad 5 + 4y = 29 \qquad$ [substituting 5 for x in equation (a)]

$\qquad\qquad 4y = 24 \qquad$ (solving for y)

$\qquad\qquad y = 6$

Thus the two numbers are 5 and 6. Check:

(a) $4(6) + 5 \overset{?}{=} 29$ $\qquad$ (b) $6(5) + 6 \overset{?}{=} 36$

$\quad 24 + 5 \overset{?}{=} 29$ $\qquad\qquad 30 + 6 \overset{?}{=} 36$

$\qquad 29 = 29\checkmark$ $\qquad\qquad\quad 36 = 36\checkmark$

EXAMPLE 13-10 A certain fraction becomes 1 when 3 is added to its numerator. It becomes $\frac{1}{2}$ when 2 is added to its denominator. Find the fraction.

SOLUTION Let x represent the numerator and y the denominator of the fraction. Then,

(a) $\dfrac{x+3}{y} = 1$

(b) $\dfrac{x}{y+2} = \dfrac{1}{2}$

We now proceed to solve the system of equations.

(a) $x + 3 = y \qquad$ (clearing the equations of fractions by cross-multiplication)

(b) $2x = y + 2$

$\quad 2x = (x+3) + 2 \qquad$ [substituting the value $x + 3$ for y in equation (b) and solving for x]

$$x = 5$$

$$\frac{5+3}{y} = 1 \qquad \text{[substituting 5 for } x \text{ in equation (a)]}$$

$$y = 8 \qquad \text{(solving for } y)$$

Thus the fraction is $\frac{5}{8}$. Check:

(a) $\dfrac{5+3}{8} \overset{?}{=} 1$ (b) $\dfrac{5}{8+2} \overset{?}{=} \dfrac{1}{2}$

$\dfrac{8}{8} = 1 \checkmark$ $\dfrac{5}{10} = \dfrac{1}{2} \checkmark$

EXAMPLE 13-11 An electrician and his assistant, in wiring a new house, worked 22 and 18 hr, respectively, and received $557.00. In wiring another house, the electrician and his assistant worked 26 and 21 hr, respectively, and received $655.00. What are the hourly wages of each?

SOLUTION Let x be the amount the electrician charges per hour and let y be the amount that his assistant charges per hour. Then,

(a) $22x + 18y = 557$

(b) $26x + 21y = 655$

$154x + 126y = 3899 \qquad$ [*multiplying equation (a) by 7, equation (b) by 6, subtracting, and solving for x*]

$$\frac{156x + 126y = 3930}{-2x \qquad\qquad = -31}$$

$$x = 15.5$$

$$22(15.5) + 18y = 557 \qquad \text{[substituting 15.5 for } x \text{ in equation (a)]}$$

$$341 + 18y = 557 \qquad \text{(solving for } y)$$

$$y = 12$$

Thus the electrician's wage is $15.50 an hour and the assistant's wage is $12.00 an hour. Check:

(a) $22(15.5) + 18(12) \overset{?}{=} 557$ (b) $26(15.5) + 21(12) \overset{?}{=} 655$

$341 + 216 \overset{?}{=} 557$ $403 + 252 \overset{?}{=} 655$

$557 = 557 \checkmark$ $655 = 655 \checkmark$

EXERCISE 13-2

1. Find two numbers such that twice the first plus the second is equal to 17, and twice the second plus the first is equal to 19.
2. If the first of two numbers is multiplied by 3, and the second number is subtracted from the product, the remainder is 7. If the sum of the two numbers is multiplied by 2, and 11 is added to the product, the result is 37. Find the two numbers.

Systems of Linear Equations

3. If the numerator of a certain fraction is increased by one, its value becomes one-third. If the denominator is increased by one, the value of the fraction becomes one-fourth. What is the fraction?

4. The value of 18 coins consisting of nickels and quarters is $3.10. How many nickels and how many quarters are there in the pile?

5. A freight train leaves a station traveling at the speed of 20 mph. Three hours later, a passenger train leaves the same station traveling at the speed of 50 mph in the direction of the freight train. How long will it take the second train to catch up with the first? What will their distance be from the station?

6. A plane covered 1600 km with a tailwind in $3\frac{1}{3}$ hr, and came back against a headwind in 4 hr. Find the speed of the plane in still air, and the speed of the wind.

7. Two people have $570. If the first person's money was three times what it really is, and the second person's money was five times what it really is, the sum would be $2350. Find the amount of money that each person has.

8. Five years ago, Tom was six times as old as Helen. In four years Tom will be three times as old as Helen is now. Find their present age.

9. Three dozen radial ball bearings and two dozen cylindrical roller bearings were purchased for $86.00. Three years later, the price of the same kind of ball bearings advanced $3.00 per dozen, and the price of the same kind of roller bearings increased by $4.00 per dozen, so that $185.00 was paid for five dozen ball bearings and four dozen roller bearings. Find the original price of each.

10. Find the speed of a motor boat in still water, and the speed of the current of the river, if it takes $4\frac{1}{4}$ hr for the boat to travel 102 mi upstream and 4 hr to travel 144 mi downstream.

11. A health-food store dealer sells unbleached flour for 42 cents a pound and whole wheat flour for 32 cents a pound. How many pounds of each kind of flour should he mix in order to make 300 pounds of flour selling at 38 cents a pound?

12. There are two parts in an electrical circuit, and the current in one part is 8 A more than the current in the other part. At the point in the circuit where the two parts join, the current is 27 A. What is the current in each part of the circuit?

13. An oil tank has two pumps. When the first pump is used 3 hr, and the second pump 2 hr, 32,000 ℓ of oil are discharged. When the first pump is used $1\frac{1}{2}$ hr, and the second pump is used $2\frac{1}{2}$ hr, 26,500 ℓ of oil are discharged. How many liters of oil can each pump discharge in an hour?

14. Twenty-five dozen resistors and 18 dozen printed circuits were purchased for $121.80. Then 40 dozen more of the same resistors and 28 dozen more of the same printed circuits were purchased from the same supplier for $190.80. Find the price per dozen of the resistors and the price per dozen of the printed circuits.

13-6 Solving Systems of Three Linear Equations in Three Variables

The algebraic methods used in the previous sections for the solution of systems of equations in two unknowns can be used also for the solution of systems of equations in three unknowns. There is, of course, more work involved in the solution of the systems in this section, but there is nothing new in the method.

EXAMPLE 13-12 Solve, by the method of elimination by addition or subtraction, the following system of equations:

$$\text{(a) } 2x + 3y + 4z = 16$$
$$\text{(b) } 3x + 2y - 5z = 8$$
$$\text{(c) } 5x - 6y + 3z = 6$$

SOLUTION We first multiply equation (a) by 3, and equation (b) by 2 and then we subtract the resulting equations to eliminate x.

$$6x + 9y + 12z = 48$$
$$\underline{6x + 4y - 10z = 16}$$
$$5y + 22z = 32 \qquad \qquad \text{(d)}$$

Then we multiply equation (a) by 5, and equation (c) by 2 and subtract the resulting equations to eliminate x.

$$10x + 15y + 20z = 80$$
$$\underline{10x - 12y + 6z = 12}$$
$$27y + 14z = 68 \qquad \qquad \text{(e)}$$

We now multiply equation (d) by 27 and equation (e) by 5, and subtract the resulting equations to get an equation only in z.

$$135y + 594z = 864$$
$$\underline{135y + 70z = 340}$$
$$524z = 524 \qquad \qquad \text{(f)}$$

Solving equation (f) for z, $z = 1$. Substituting 1 for z in equation (d), we have

$$5y + 22(1) = 32$$
$$5y = 10$$
$$y = 2$$

Substituting 1 for z and 2 for y in equation (a) we have

$$2x + 3(2) + 4(1) = 16$$
$$2x = 6$$
$$x = 3$$

EXAMPLE 13-13 Solve, by the method of substitution, the following system of equations:

$$\text{(a) } 3x - y + z = 10$$
$$\text{(b) } 5x + 2y - z = 5$$
$$\text{(c) } x - 2y - 2z = -2$$

SOLUTION We first solve equation (c) for x to get

$$x = 2y + 2z - 2$$

and substitute $2y + 2z - 2$ for x in the other two equations of the system. Thus,

$$3(2y + 2z - 2) - y + z = 10$$
$$5y + 7z = 16 \qquad \qquad \text{(d)}$$

226

and
$$5(2y + 2z - 2) + 2y - z = 5$$
$$12y + 9z = 15 \qquad \text{(e)}$$

Then we solve the system of equations (d) and (e) by substitution again. We solve equation (d) for y to get

$$y = \frac{16 - 7z}{5}$$

and substitute $(16 - 7z)/5$ for y in (e). Thus,

$$\frac{12(16 - 7z)}{5} + 9z = 15$$
$$12(16 - 7z) + 45z = 75$$
$$-39z = -117$$
$$z = 3$$

Substituting 3 for z in equation (d) we have

$$5y + 7(3) = 16$$
$$y = -1$$

Also, substituting -1 for y and 3 for z in equation (a) we have

$$3x - (-1) + (3) = 10$$
$$x = 2$$

Thus, the solution of the system is $x = 2$, $y = -1$, and $z = 3$. It is left to the student to check this result.

EXERCISE 13-3

Solve systems of equations 1–6 by any method.

1. $2x - 4y + 9z = 28$
   $7x + 3y - 5z = 3$
   $9x + 10y - 11z = 4$

2. $x - 2y + 3z = 6$
   $2x + 3y - 4z = 20$
   $3x - 2y + 5z = 26$

3. $4x - 3y + 2z = 40$
   $5x + 9y - 7z = 47$
   $9x + 8y - 3z = 97$

4. $3x + 2y + z = 23$
   $10x + 5y + 4z = 75$
   $5x + 2y + 4z = 46$

5. $5x - 6y + 4z = 15$
   $7x + 4y - 3z = 19$
   $2x + y + 6z = 46$

6. $2A + B - 2C = 1$
   $3A - C = 7$
   $3A + 4B = 24$

7. The sum of three numbers is 9. The sum of the first, twice the second, and three times the third is 22. The sum of the first, four times the second, and nine times the third is 58. Find the three numbers.

8. The value of 30 coins consisting of nickels, dimes, and quarters is $5.25. How many of each kind of coin are there in the pile if there are three times as many nickels as there are dimes?

9. Three people, A, B, and C, have $480. If A's money was twice what it really is, and C's money was four times what it really is, the sum would be $1380. But if A's money was three times what it really is, and B's money was five times what it really is, the sum would be $1080. Find the amount of money that each person has.

Solving Systems of Three Linear Equations in Three Variables

10. A foreign car with its driver and a passenger weighs 3100 lb. Three times the driver's weight and two times the passenger's weight total 740 lb. The weight of the car and five times the weight of its driver is 3500 lb. Find the weight of the driver, the passenger, and the car.

REVIEW QUESTIONS

1. When are two linear equations in x and y called linear simultaneous equations in two variables, or a system of linear equations in two variables?
2. Two linear equations are plotted in the same coordinate system and their graphs intersect. What is the significance of the point of intersection?
3. When the graphs of two linear equations do not intersect, what does this mean about the two linear equations?
4. Of the methods of solving systems of linear equations, which is less accurate, the graphic or the algebraic method?
5. State in your own words the method of solving systems of equations in two variables by elimination by addition or subtraction.
6. State in your own words the method of solving systems of equations in two variables by substitution.

REVIEW EXERCISES

Solve systems of equations 1–7 by any method:

1. $x + 2y = 35$
   $2x - y = 10$

2. $5x + 7y = 43$
   $11x + 9y = 69$

3. $4\frac{1}{3}(x + y) = 65$
   $7x + 3y = 77$

4. $2\frac{1}{4}(x + y) = 45$
   $3(y - 2x) = 24$

5. $5x + 3y = 47.8$
   $7x + 2y = 51.3$

6. $11x + 8y = 27.6$
   $9x - 3y = 5.4$

7. $2x - 3y + 2z = 13$
   $4y + 2z = 14$
   $3x - 2y = 7$

8. A certain fraction becomes 1 when 1 is added to its numerator. It becomes $\frac{1}{2}$ when 6 is added to its denominator. Find the fraction.

9. In a right triangle, the sum of two angles is 90°, and their difference is 16°. Find the angles.

10. One girl is seated 6 ft from the pivot point of a see-saw, and a boy is seated on the same side of the see-saw 8 ft from the pivot point. The two together balance a man weighing 160 lb seated 6 ft from the pivot point. The see-saw will also balance if the man does not change position, but the girl and the boy sit 8.5 ft and 6 ft from the pivot point, respectively. Find the weight of the girl and the boy.

11. A homeowner paid $402.00 to a landscape contractor to plant rhododendron bushes in the flower garden of his new home. The contractor worked for $3\frac{1}{2}$ hr and planted 20 bushes. The same contractor planted 15 rhododendron bushes for another customer and, having worked for 3 hr, charged $306.00. If the contractor charges for labor and plants only, what is the price of each rhododendron, and how much does the contractor charge per hour?

12. The current in one of the two parts of an electric circuit is 5 A less than the current in the other part. At the point of the circuit where the two parts join, the current is 19 A. Find the current in each part of the circuit.

13. A water tank has two pumps. If the first pump is used for 1 hr and the second for 4 hr, 17,000 ft³ of water are removed from the tank. If the first pump is used for 2 hr and the second for 2½ hr, 1475 ft³ of water are removed. How many cubic feet of water can each pump remove in one hour?

14. Find three numbers such that their sum is 15/8. The sum of four times the first and the third is 7/2, and the sum of two times the second and third is 7/4.

15. Twenty ring stands, fifteen large flasks, and twelve bunsen burners were purchased for the chemistry laboratory of a technical institute for $153.25. If fifteen more ring stands were purchased, and three fewer flasks, the number of the bunsen burners purchased remaining the same, the total cost would have been $185.05. But if ten more ring stands were purchased, and three fewer bunsen burners, the number of flasks purchased remaining the same, the total cost would have been $168.00. Find the price of each item purchased.

14

Quadratic Equations in One Variable

14-1 Basic Ideas About Quadratic Equations

Here we will consider the solution of **quadratic equations in one variable** or **second degree equations in one variable**. The equations

$$8x^2 + 7x + 1 = 0, \qquad 6x^2 + 2x = 0, \qquad 5x^2 = 12,$$

and

$$(x - 3)(x + 4) = 0$$

are quadratic equations in one variable.

Quadratic equations in one variable are those equations which can be given the form

$$ax^2 + bx + c = 0$$

where a is the numerical coefficient of the variable in the second power, b is the numerical coefficient of the variable in the first power, and c is a constant.

The $ax^2 + bx + c = 0$ form of the quadratic equation is called the **standard form**. Thus, the quadratic equation $5x^2 + 3x + 2 = 0$ is in standard form.

In the following three quadratic equations

(a) $3x^2 + 2x + 5 = 0$

(b) $6x^2 + 4 = 0$

(c) $5x^2 + 8x = 0$

a, b, and c are as follows:

(a) $a = 3, b = 2, c = 5$

(b) $a = 6, b = 0, c = 4$

(c) $a = 5, b = 8, c = 0$

EXAMPLE 14-1 In each case below, state whether the equation is a quadratic equation.

$$\text{(a) } 2x^2 - 3x + 2 = 0$$
$$\text{(b) } 5x^3 + 2x^2 - x = 0$$
$$\text{(c) } 7x + 3 = 0$$
$$\text{(d) } 8x^2 - 3 = 0$$

SOLUTION

(a) yes   (b) no   (c) no   (d) yes

EXAMPLE 14-2 Give the standard form to each of the following quadratic equations:

$$\text{(a) } 9x^2 = 3x - 5$$
$$\text{(b) } 2x^2 = 4(x + 5)$$

SOLUTION

$$\text{(a) } 9x^2 = 3x - 5 \qquad \textit{(given)}$$
$$9x^2 - 3x + 5 = 0 \qquad \textit{(by transposition)}$$
$$\text{(b) } 2x^2 = 4(x + 5) \qquad \textit{(given)}$$
$$2x^2 = 4x + 20 \qquad \textit{(by multiplication)}$$
$$2x^2 - 4x - 20 = 0 \qquad \textit{(by transposition)}$$

EXERCISE 14-1

State whether the equations in exercises 1–10 are quadratic equations.

1. $12x + 2x^2 = -3$
2. $x - 3x = 4x^2$
3. $7x + 3 = 0$
4. $7x^2 - 9x = 5$
5. $16x^2 = -2$
6. $\frac{1}{3}x^2 + 2x = \frac{3}{4}$
7. $18x - 5 = 0$
8. $23x^2 - 11x = 6$
9. $\frac{3x^2}{4} + 2x - 3 = 0$
10. $7x^2 = x$

Give the standard form to each of quadratic equations 11–20.

11. $5x^2 = x + 2$
12. $8x - 3 = -3x^2$
13. $9x + 7 = 8x^2$
14. $x(2x - 3) = 12$
15. $(2x + 5)(x - 3) = 0$
16. $x - 18 = 12x^2$
17. $3x + x^2 = 2$
18. $\frac{x}{6} + x^2 = 7$
19. $x^2 + \frac{x}{5} + 12 = 0$
20. $\frac{3}{2} + x = 11x^2$

14-2 Solution by Factoring

The method of solution by factoring follows from a very simple principle:

If $ab = 0$, then either $a = 0$, or $b = 0$, or both a and b are zero.

Suppose we are asked to solve the quadratic equation $x^2 + x - 12 = 0$. Factoring the trinomial, we have $(x - 3)(x + 4) = 0$. Then, using the above principle, we assume that

$$\text{(a)} \ x - 3 = 0 \quad \text{and} \quad \text{(b)} \ x + 4 = 0$$

It is evident from (a) that $x = 3$, and from (b) that $x = -4$. Thus, the roots for the quadratic equation $x^2 + x - 12 = 0$ are $x = 3$ and $x = -4$.
Check:

$$x = 3: \quad (3)^2 + (3) - 12 \overset{?}{=} 0 \qquad x = -4: \quad (-4)^2 + (-4) - 12 \overset{?}{=} 0$$

$$0 = 0 \checkmark \qquad\qquad\qquad\qquad 16 - 4 - 12 \overset{?}{=} 0$$

$$0 = 0 \checkmark$$

The method of solving quadratic equations in one variable by factoring can be summarized as follows:

1. *Write the equation in the form $ax^2 + bx + c = 0$; that is, bring all terms to the left side of the equation and have zero on the right side.*
2. *Factor the polynomial of the equation.*
3. *Set each factor of the equation involving the variable equal to zero, using the above principle.*
4. *Solve each resulting linear equation.*
5. *Check both roots in the original equation.*

EXAMPLE 14-3   Solve $x^2 - 5x = 14$ by factoring.

SOLUTION

$$x^2 - 5x = 14 \qquad\qquad (given)$$

$$x^2 - 5x - 14 = 0 \qquad\qquad (by \ writing \ the \ equation \ in \\ standard \ form)$$

$$(x - 7)(x + 2) = 0 \qquad\qquad (by \ factoring)$$

$$x - 7 = 0, \ x = 7 \qquad (by \ setting \ each \ factor$$
$$x + 2 = 0, \ x = -2 \qquad equal \ to \ zero \ and \ solving \\ for \ x)$$

Thus, the roots are $x = 7$ and $x = -2$. Check:

$$x = 7: \quad (7)^2 - 5(7) \overset{?}{=} 14 \qquad x = -2: \quad (-2)^2 - 5(-2) \overset{?}{=} 14$$

$$49 - 35 \overset{?}{=} 14 \qquad\qquad\qquad 4 + 10 \overset{?}{=} 14$$

$$14 = 14 \checkmark \qquad\qquad\qquad\qquad 14 = 14 \checkmark$$

EXAMPLE 14-4   Solve $2x^2 - x - 6 = 0$ by factoring.

SOLUTION

$$2x^2 - x - 6 = 0$$

$$(x - 2)(2x + 3) = 0$$

$$x - 2 = 0, \ x = 2$$

$$2x + 3 = 0, \ x = -\tfrac{3}{2}$$

Thus, the roots are $x = 2$ and $x = -\frac{3}{2}$. Check:

$$x = 2: \quad 2(2)^2 - (2) - 6 \overset{?}{=} 0 \qquad\qquad x = -\frac{3}{2}: \quad 2(-\frac{3}{2})^2 - (-\frac{3}{2}) - 6 = 0$$

$$8 - 2 - 6 \overset{?}{=} 0 \qquad\qquad\qquad\qquad \frac{9}{2} + \frac{3}{2} - 6 \overset{?}{=} 0$$

$$0 = 0\checkmark \qquad\qquad\qquad\qquad\qquad 0 = 0\checkmark$$

EXAMPLE 14-5 Solve $16x^2 = 36$ by factoring.

SOLUTION

$$16x^2 = 36$$

$$16x^2 - 36 = 0$$

Factoring out the common factor, 4, we have

$$4(4x^2 - 9) = 0$$

and dividing both sides by that common factor

$$4x^2 - 9 = 0$$

$$(2x + 3)(2x - 3) = 0$$

$$(2x + 3) = 0, \ x = -\frac{3}{2}$$

$$(2x - 3) = 0, \ x = \frac{3}{2}$$

Thus, the roots are $x = -\frac{3}{2}$, and $x = \frac{3}{2}$. Check:

$$x = -\frac{3}{2}: \quad 16(-\frac{3}{2})^2 \overset{?}{=} 36 \qquad\qquad x = \frac{3}{2}: \quad 16(\frac{3}{2})^2 \overset{?}{=} 36$$

$$36 = 36\checkmark \qquad\qquad\qquad\qquad 36 = 36\checkmark$$

NOTE: In solving quadratic equations which are in the standard form, first factor out the largest common numerical factor, and then divide both sides of the equation by that common factor. The resulting equation is equivalent to the original, and simpler to solve.

EXERCISE 14-2

Solve quadratic equations 1–20 by factoring.

1. $x^2 + 6x - 7 = 0$
2. $x^2 - 6x + 8 = 0$
3. $x^2 - 4x = 5$
4. $x^2 = x + 6$
5. $30 + 11x + x^2 = 0$
6. $x^2 - 5x - 66 = 0$
7. $2x^2 + 5x = 3$
8. $6x^2 = 30 + 8x$
9. $3x^2 - 8x = -4$
10. $8x^2 + 14x + 3 = 0$
11. $18x^2 + 3x = 3$
12. $10x^2 + 13x = 3$
13. $4x^2 = 8x$
14. $x^2 = 9x$
15. $x^2 = -2x$
16. $x^2 = -9x$
17. $4x^2 = 25$
18. $49x^2 = 16$
19. $25x^2 = 81$
20. $9x^2 = 64$

Solution by Factoring

14-3 Solution by the Square Root Method

The solution of the quadratic equation in which $b = 0$, that is, of the form $ax^2 = c$, is very simple. For instance, suppose we are asked to solve the equation

$$2x^2 = 8$$
$$x^2 = 4 \qquad \text{(dividing both sides by 2)}$$
$$x = \pm\sqrt{4} \qquad \text{(taking the square root of both sides)}$$
$$= \pm 2$$

NOTE: Because the square root of positive numbers can be either positive or negative, both the plus and the minus signs are written in front of the root. Thus, $\sqrt{16} = \pm 4$, $\sqrt{3} = \pm 1.732$ (approx.), and $\sqrt{2/3} = \pm 0.816$ (approx.).

The method of solving quadratic equations of the form $ax^2 = c$ can be summarized as follows.

1. *Divide both sides of the equation by a.*
2. *Take the square root of both sides.*
3. *Simplify (if applicable).*

EXAMPLE 14-6 Solve $2x^2 = 6$.

SOLUTION

$$2x^2 = 6$$
$$x^2 = 3$$
$$x = \pm\sqrt{3}$$
$$= \pm 1.732 \text{ (approx.)}$$

EXAMPLE 14-7 Solve $3x^2 = 14$.

SOLUTION

$$3x^2 = 14$$
$$x^2 = \frac{14}{3}$$
$$x = \pm\sqrt{\frac{14}{3}}$$
$$= \pm 2.16 \text{ (approx.)}$$

EXERCISE 14-3

Solve the following quadratic equations by the square root method.

1. $3x^2 = 27$	**2.** $7x^2 - 63 = 0$	**3.** $6x^2 = 24$
4. $5x^2 - 3 = 77$	**5.** $3x^2 - 5 = 70$	**6.** $3x^2 = 23$
7. $27x^2 = 7$	**8.** $29x^2 = 11$	**9.** $3x^2 = 21$
10. $18x^2 = 33$	**11.** $4x^2 = 19$	**12.** $17x^2 = 31$

Quadratic Equations in One Variable

Another method which will solve both kinds of the quadratic equation we have seen so far, as well as quadratic equations that the above methods cannot solve, is the **quadratic formula.**

The quadratic formula is obtained by solving the general form of the quadratic equation, $ax^2 + bx + c = 0$, for x in terms of a, b, and c. The quadratic formula is

$$x = \frac{-b \pm \sqrt{b^2 - 4ac}}{2a}$$

The two roots are

$$x = \frac{-b + \sqrt{b^2 - 4ac}}{2a} \quad \text{and} \quad x = \frac{-b - \sqrt{b^2 - 4ac}}{2a}$$

In technical problems, usually only one of the two solutions fits the requirements of a given problem. *The case in which the expression $b^2 - 4ac$ under the radical sign of the formula is negative is of no interest to us. In such cases, we will say that the given equation has no real solution.*

EXAMPLE 14-8 Solve $3x^2 = 21x - 30$ and check.

SOLUTION We first convert the equation to its standard form, factor out the common factor 3, and identify a, b, and c.

$$3x^2 = 21x - 30$$
$$3x^2 - 21x + 30 = 0$$
$$x^2 - 7x + 10 = 0$$
$$a = 1$$
$$b = -7$$
$$c = 10$$

Substituting in the formula, we have

$$x = \frac{-(-7) \pm \sqrt{(-7)^2 - 4(1)(10)}}{2(1)}$$
$$= \frac{7 \pm \sqrt{49 - 40}}{2}$$
$$= \frac{7 \pm \sqrt{9}}{2}$$
$$= \frac{7 \pm 3}{2}$$

Thus, the roots are $x = (7 + 3)/2 = 5$ and $x = (7 - 3)/2 = 2$. Check:

$$x = 5: \quad 3(5)^2 \stackrel{?}{=} 21(5) - 30 \qquad x = 2: \quad 3(2)^2 \stackrel{?}{=} 21(2) - 30$$
$$75 = 75 \checkmark \qquad\qquad\qquad\qquad 12 = 12 \checkmark$$

EXAMPLE 14-9 Solve $2x^2 + x - 6 = 0$.

SOLUTION

$$a = 2$$
$$b = 1$$
$$c = -6$$

Substituting in the formula, we have

$$x = \frac{-1 \pm \sqrt{(1)^2 - 4(2)(-6)}}{2(2)}$$

$$= \frac{-1 \pm \sqrt{49}}{4}$$

$$= \frac{-1 \pm 7}{4}$$

Thus, $x = \frac{6}{4} = \frac{3}{2}$ and $x = -\frac{8}{4} = -2$.

In actual technical problems, solutions of quadratic equations are rarely perfect square roots or simple fractions as in examples 14-8 and 14-9.

EXAMPLE 14-10 Solve $2x^2 - 7x + 4 = 0$.

SOLUTION

$$x = \frac{7 \pm \sqrt{(-7)^2 - 4(2)(4)}}{2(2)}$$

$$= \frac{7 \pm \sqrt{49 - 32}}{4}$$

$$= \frac{7 \pm \sqrt{17}}{4}$$

$$= \frac{7 \pm 4.12}{4}$$

Thus, $x = (7 + 4.12)/4 = 2.78$ (approx.) and $x = (7 - 4.12)/4 = 0.72$ (approx.).

EXAMPLE 14-11 Solve $2x^2 + 3x + 2 = 0$.

SOLUTION

$$x = \frac{-3 \pm \sqrt{(3)^2 - 4(2)(2)}}{2(2)}$$

$$= \frac{-3 \pm \sqrt{9 - 16}}{4}$$

$$= \frac{-3 \pm \sqrt{-7}}{4}$$

Since the number under the radical is negative, we say that the quadratic equation $2x^2 + 3x + 2 = 0$ has no real solution.

Quadratic Equations in One Variable

Solve quadratic equations 1–20 by using the quadratic formula.

1. $x^2 - 6x = 7$
2. $2x^2 = 7x - 3$
3. $6x^2 + 6 = 13x$
4. $x^2 - 2x = 3$
5. $x^2 - 8x + 15 = 0$
6. $3x^2 = 7x + 20$
7. $3x^2 - 53x + 34 = 0$
8. $2x^2 + 5x = 3$
9. $2x^2 + 3x - 1 = 0$
10. $x^2 + 7x = 11$
11. $3x^2 - x = 1$
12. $x^2 - 2x - 2 = 0$
13. $8x^2 = 15 - 14x$
14. $3x^2 - 2x + 7 = 0$
15. $x^2 - 3x + \frac{1}{3} = 0$
16. $3x^2 - 2x + \frac{3}{4} = 0$
17. $3x^2 + 14x - 24 = 0$
18. $2x^2 = 6x - 2$
19. $4x^2 - 3x = 2$
20. $x^2 - 3x = 5$

14-5 Solving Problems Using Quadratic Equations

In this section, we will consider word problems whose solution involves quadratic equations.

EXAMPLE 14-12 An open top trough is to be made from a sheet of metal by cutting off squares from each corner of the metal sheet, and bending up the sides, as shown in Fig. 14-1. If the length of the sheet is 26 in., and the width is 16 in., what should the side, x, of each square be, if the bottom of the trough is to be 200 in.^{2}?

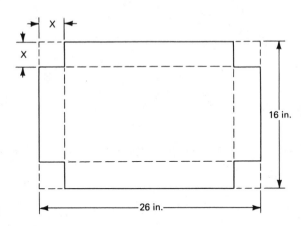

FIG. 14-1

SOLUTION

$$A = LW$$

Then,

$$200 = (26 - 2x)(16 - 2x)$$
$$200 = 416 - 52x - 32x + 4x^2$$
$$4x^2 - 84x + 216 = 0$$

Solving Problems Using Quadratic Equations

Removing the common factor 4, we have

$$x^2 - 21x + 54 = 0$$

$$x = \frac{21 \pm \sqrt{(-21)^2 - 4(1)(54)}}{2(1)}$$

$$x = \frac{21 \pm 15}{2}$$

$$x = 18, \ x = 3$$

The root $x = 18$ is rejected as impossible. Therefore, the answer is 3 in.

EXAMPLE 14-13 A motorboat can go 20 mph in still water. If it takes the motorboat 3 hr, 12 min to go 30 mi upstream and return to its starting point, find the speed of the current.

SOLUTION Let x be the speed of the current. Solving the formula $d = rt$ for time, we have $t = \dfrac{d}{r}$. Then,

$$\text{time upstream} = \frac{30}{20 - x}$$

$$\text{time downstream} = \frac{30}{20 + x}$$

$$3 \text{ hr } 12 \text{ min} = 3\frac{12}{60} \quad \text{or} \quad 3.2 \text{ hr}$$

Since the sum of the time to go and the time to come back equals the total time for the trip, we have

$$\frac{30}{20 - x} + \frac{30}{20 + x} = 3.2$$

$$30(20 + x) + 30(20 - x) = 3.2(20 - x)(20 + x)$$

$$1200 = 1280 - 3.2x^2$$

$$3.2x^2 = 80$$

$$x = \pm\sqrt{\frac{80}{3.2}}$$

$$x = 5, \ x = -5$$

Since the speed of the current cannot be negative, -5 is rejected, and the answer is 5 mph.

EXERCISE 14-5

1. The sum of a number and three times its reciprocal is 19/4. Find the number.
2. A positive number added to its square is 132. Find the number.
3. The area of a circle is 351 in.². Find its radius. (Hint: The formula for the area of the circle is $A = \pi r^2$.)
4. A metal washer has an outside diameter of 8.6 cm. The area of the inside circle is 30.2 cm². Find the width, x, of the washer (see Fig. 14-2).
5. The length of a rectangular piece of sheet metal is 30 cm longer than its width. An open tank is to be made by cutting 10-cm squares from each corner and folding up the sides. If the volume of the container is to be 18,000 cm³, what are the dimensions of the sheet metal?

Quadratic Equations in One Variable

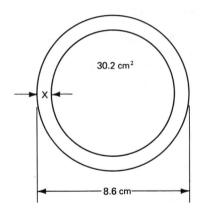

FIG. 14-2

6. Find two consecutive odd numbers whose product is 195.

7. An object is shot vertically upward. Its height h, in feet, is given by the formula $h = 102t - 16t^2$, where t is time in seconds. Find the time needed for the object to reach a height of 152 ft.

8. A rectangular sheet of metal has an area of 2232 cm². If its length is 10 cm more than twice its width, find the dimensions of the sheet of metal.

9. A rectangular sheet of metal 38 in. × 30 in. is to be made into an open top trough by cutting off squares from each corner and folding up the sides. If the area of the bottom of the trough is to be 768 in.², find the depth of the trough.

10. A variable electric current is given by the formula $I = t^2 - 8t + 18$, where t is time in seconds. Find the time when the current is equal to 3 A.

11. A 16 in. × 22 in. frame of uniform thickness encloses an area of 247 in.² (see Fig. 14-3). Find the thickness, x, of the frame.

12. A rectangular swimming pool has outside dimensions of 16.80 m and 10.70 m, as shown in Fig. 14-4. Find the thickness, x, of the concrete wall, if the area of the bottom of the swimming pool is 166.8 m².

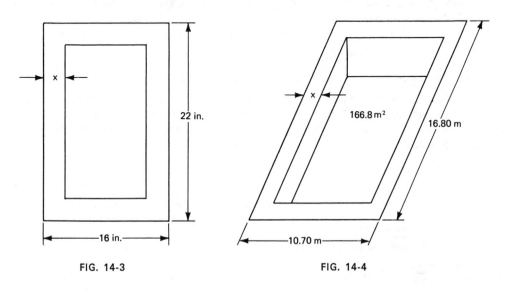

FIG. 14-3   FIG. 14-4

Solving Problems Using Quadratic Equations

13. The length of a rectangle is three times its width. Find the dimensions of the rectangle if its area is 202 ft².
14. The length of a rectangular piece of land is 12.8 m more than its width. Find the dimensions of that tract of land, if its area is 1510 m².
15. The angle beam shown in Fig. 14-5 has an area of 8.6 cm². Find the thickness, t. (Hint: divide the area into two rectangles as shown by the dotted line in the figure.)

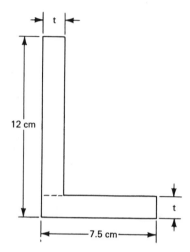

FIG. 14-5

16. In a river with a current of 3 mph, it takes a motorboat 6 hr 24 min to go 36 mi upstream and come back to its starting point. Find the speed of the boat in still water.

14-6 Equations with Radicals

Equations containing a radical that has a variable in the radicand are called **radical equations**. For instance, the equation $\sqrt{x-4}+3=4$ is a radical equation.

A radical equation is solved by first converting the equation to one without radicals. To eliminate square root radicals from a radical equation, we will use the following principle: *Any root of an equation $A = B$ is also a root of the equation $A^2 = B^2$.*

EXAMPLE 14-14 Solve the equation $\sqrt{x+13}+2=x+3$.

SOLUTION Squaring both sides of our equation in its present form will not eliminate the radical $\sqrt{x+13}$. If, however, the term 2 is transposed to the right side, and the *radical is left alone on the left side*, it will be eliminated when, by using the above principle, the two sides are squared. Thus,

$$\sqrt{x+13}+2=x+3$$
$$\sqrt{x+13}=x+1 \qquad \text{(by transposing 2)}$$
$$x+13=x^2+2x+1 \qquad \text{(by squaring both sides)}$$
$$x^2+x-12=0$$

Solving for x at this point, we have

$$x=3,\ x=-4$$

Quadratic Equations in One Variable

Let us check the two solutions of this example. Substituting 3 for x in $\sqrt{x+13} + 2 = x + 3$, and taking the principal square root of $\sqrt{x+13}$, we have

$$\sqrt{x+13} + 2 = x + 3$$

$$\sqrt{3+13} + 2 \overset{?}{=} 3 + 3$$

$$\sqrt{16} + 2 \overset{?}{=} 6$$

$$4 + 2 \overset{\checkmark}{=} 6$$

However, substituting -4 for x in $\sqrt{x+13} + 2 = x + 3$, and taking the principal square root of $\sqrt{x+13}$, we have

$$\sqrt{x+13} + 2 = x + 3$$

$$\sqrt{-4+13} + 2 \overset{?}{=} -4 + 3$$

$$\sqrt{9} + 2 \overset{?}{=} -4 + 3$$

$$3 + 2 \neq -1$$

We say, therefore, that $x = -4$ is not a real solution.

We have seen in Sect. 8-1 that often extraneous solutions are introduced when the original equation is multiplied by the LCM. In this case, the extraneous solutions "slip in" when we square both sides of the original equation. The student should see once more the value of checking in order that solutions which are not real may be detected and rejected.

EXAMPLE 14-15 Solve the equation $\sqrt{5x+9} = 2 + \sqrt{3x+3}$.

SOLUTION Squaring both sides of the given equation, we have

$$5x + 9 = 4 + 4\sqrt{3x+3} + 3x + 3$$

$$2x + 2 = 4\sqrt{3x+3} \qquad \text{(by combining like terms and isolating the remaining radical)}$$

$$4x^2 + 8x + 4 = 16(3x+3) \qquad \text{(by squaring both sides once more)}$$

$$4x^2 - 40x - 44 = 0 \qquad \text{(converting the quadratic equation to the standard form)}$$

$$x = 11, x = -1 \qquad \text{(solving for } x\text{)}$$

Check:

$$x = 11: \quad \sqrt{5(11)+9} = 2 + \sqrt{3(11)+3}$$

$$\sqrt{64} = 2 + \sqrt{36}$$

$$8 = 2 + 6 \checkmark$$

Equations with Radicals

$$x = -1: \quad \sqrt{5(-1) + 9} = 2 + \sqrt{3(-1) + 3}$$
$$\sqrt{4} = 2 + \sqrt{0}$$
$$2 = 2 \checkmark$$

EXERCISE 14-6

Solve each of the following radical equations. Check and reject extraneous solutions.

1. $\sqrt{3x + 7} - 1 = 3$ **2.** $3x + 2\sqrt{x} - 1 = 0$

3. $2x + \sqrt{x - 4} = 3x - 6$ **4.** $2x - \sqrt{x - 3} + 1 = 2x - 1$

5. $\sqrt{x + 10} + 2x - 1 = 3x - 3$ **6.** $\sqrt{x - 2} = \sqrt{3}$

7. $\sqrt{5} - \sqrt{3x - 7} = 0$ **8.** $\sqrt{2x - 6} + 3 = \sqrt{x + 14} + 2$

9. $\sqrt{2x - 2} + 7 = \sqrt{3x + 7} + 5$ **10.** $\sqrt{3x - 2} - \sqrt{2x - 2} - 1 = 0$

11. $2x = \sqrt{3x - 3} + 5$ **12.** $x = \sqrt{5x + 1} + 1$

REVIEW QUESTIONS

1. What are the equations that can be given the form $ax^2 + bx + c = 0$ called?

2. What is the standard form of the quadratic equation in one variable?

3. State the five steps for solving quadratic equations in one variable by factoring.

4. State the name of the method and the three steps of solving quadratic equations in one variable of the form $ax^2 = c$.

5. Write the quadratic formula.

6. Describe in your own words the procedure for solving radical equations involving square roots.

7. Why is checking important in radical equations?

REVIEW EXERCISES

Solve quadratic equations 1–6 by factoring.

1. $3x^2 + 5x = 0$ **2.** $2x = -x^2$

3. $3x = x^2$ **4.** $15x^2 + 7x = 2$

5. $18x^2 - 4 = x$ **6.** $18x^2 - 9x = 20$

Solve quadratic equations 7–10 by the square root method.

7. $6x^2 = 24$ **8.** $17x^2 - 2 = 34$

9. $16x^2 - 2 = 46$ **10.** $14x^2 - 6 = 41$

Solve quadratic equations 11–16 by the quadratic formula.

11. $2x^2 - 15x + 7 = 0$ **12.** $14x - x^2 = 33$

13. $15x^2 + 8x = 16$ **14.** $x^2 - 2x - 17 = 0$

15. $x^2 - 2x + \frac{1}{4} = 0$ **16.** $1.6x^2 - 9.6x - 25.6 = 0$

17. The sum of the squares of two positive odd consecutive numbers is 290. Find the numbers.

18. The sum of two numbers is 19, and their product is 48. Find the numbers.

19. The sum of a number and its reciprocal equals 4.25. Find the number.

                           Quadratic Equations in One Variable

20. The voltage of an electric circuit is given by the formula $V = t^2 - 10t + 30$. Find the value of t at which the voltage is equal to 6.

21. A picture frame is 14 in. × 17 in., and is of uniform thickness (see Fig. 14-6). If an area of 180 in.² of the picture in the frame can be seen, what is the thickness of the frame?

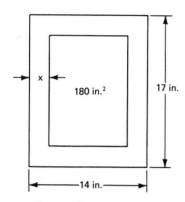

FIG. 14-6

22. A motorboat in a river with a 6 mph current goes 54 mi upstream and back in 4 hr 48 min. Find the speed of the boat in still water.

Solve each of the following radical equations. Check and reject extraneous solutions.

23. $\sqrt{5x - 4} - \sqrt{x} - 2 = 0$

24. $\sqrt{3x + 1} - 2 = \sqrt{x - 4} + 1$

25. $2\sqrt{3 + 3x} + 1 = \sqrt{2x + 5}$

BASIC GEOMETRY

Geometry is of utmost importance to the scientist, the engineer, and the technician. Geometry deals with the relationships of lines to each other, and with properties and relationships of a variety of geometrical figures. Although geometry is a theoretical and highly organized branch of mathematics, the geometry which we will study in this book is not theoretical, but applied. We will study a limited number of topics, stressing relationships of lines, and angles, and some important properties of certain common geometrical figures. In solving length, area, and volume problems, we will use many formulas, without giving any formal proof for them.

Introduction to Geometry

We start our study of geometry by describing, rather than defining, some basic geometrical terms. **Point, line, straight line,** and **surface** are basic terms in geometry.

A **point** has position only, it has no size or dimension. A point is represented by a dot, but the dot is not a point. In geometric discussions, a point is designated by a capital letter next to the dot. Thus, $C^{\bullet}$ and $M^{\bullet}$ (read "point C" and "point M") are the usual ways of designating the two points C and M.

A **line** has only one dimension, length; it has no width or thickness. A line can either be **curved** or **straight**. Both kinds can be thought of as the path of a moving point. In the case of the curved line, the moving point is constantly changing direction, while in the case of the straight line, the point is moving indefinitely in the same direction. Any part of a circle is a curved line, and a stretched thread or a ray of light are examples of straight lines. Usually a line is designated by two points on the line or by a lower case letter close to the line. The symbol for the line containing the two points E and F is $\overleftrightarrow{EF}$. The curved lines AB and c and the straight lines $\overleftrightarrow{EF}$ and h are shown in Fig. 15-1.

A **surface** has two dimensions, length and width, but no thickness. The top of a table and the curved side of a cylinder are examples of a surface. When two points of a surface are connected by a straight line and all points of that line are on the surface, it is a **plane surface**.

A **straight line segment** is the part of a straight line between two given points of the line. A straight line segment between the two points A and B is shown in Fig. 15-2. The symbol for that straight line segment is $\overline{AB}$ (read "the straight line segment AB").

A **ray** is the part of a straight line extending indefinitely in one direction only from a point on the line. A ray is shown in Fig. 15-3. The symbol for that ray is $\overrightarrow{AB}$ (read "ray AB").

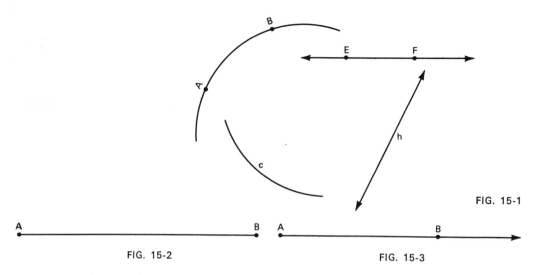

FIG. 15-1

A ●————————————————● B

FIG. 15-2

A ●————————————————● B →

FIG. 15-3

An **angle** is the union of two rays with a common end point. The end point is the **vertex** of the angle. An angle is shown in Fig. 15-4. Notice that the angle shown is identified by three capital letters. The first and third letters are used for any two points of the **sides**, and the middle letter is used for the vertex. Thus, if we use the angle sign ($\angle$), the angle of Fig. 15-4 will be designated by $\angle ABC$, (read "angle ABC"). An angle can be named also by placing a lower case letter, or a number inside the vertex. Thus, the two angles of Fig. 15-5 are $\angle a$ and $\angle 3$.

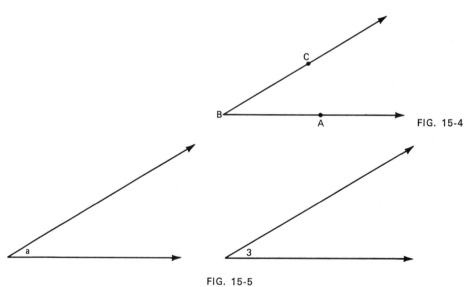

FIG. 15-4

FIG. 15-5

15-3 Measuring the Size of an Angle

The size of an angle depends on how far apart its sides are from each other and not at all on their length. Remember that the sides are rays of indefinite length.

Angles are measured by a standard unit called the **degree**. *An angle of one degree*

Introduction to Geometry

results when two consecutive points of a circle which has been divided into **360** equal parts are connected with the center with two straight line segments. Thus the *degree is an angle which intercepts* a 1/360 *part of a circle* (see Fig. 15-6).

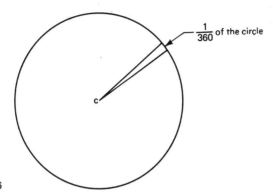

$\frac{1}{360}$ of the circle

FIG. 15-6

The degree is subdivided into 60 **minutes** and the minute into 60 **seconds**. The symbols for the degree, minute, and second are (°), ('), and ("), respectively. Thus, we write the compound denominate number 25 degrees, 11 minutes, 45 seconds as 25° 11' 45".

EXAMPLE 15-1 Add the three angles $\angle a$, $\angle b$, and $\angle c$, if $\angle a = 17° 12' 48''$, $\angle b = 25° 39' 37''$, and $\angle c = 19° 27' 28''$.

SOLUTION

$$
\begin{array}{r}
17° 12' \ 48'' \\
25° 39' \ 37'' \\
19° 27' \ 28'' \\
\hline
61° 78' 113''
\end{array}
$$

Since 113" is 1' 53", we have 61° (78' + 1') 53" or 61° 79' 53". Also, since 79' is 1° 19', we have (61° + 1°) 19' 53" or 62° 19' 53".

EXAMPLE 15-2 Subtract $\angle m$ from $\angle n$, if $\angle m = 13° 28' 43''$ and $\angle n = 38° 17' 24''$.

SOLUTION

$$
\begin{array}{r}
38° 17' 24'' \\
13° 28' 43''
\end{array}
$$

Since 43" is greater than 24", the subtraction in this form is impossible. We borrow 1°, which is equal to 60', from 38°. Now 38° becomes 37° and 17' becomes 77'. We also borrow in the same manner from the minutes and add to the seconds. Thus, 77' becomes 76' and 24" becomes 84". We rewrite and complete the subtraction as follows:

$$
\begin{array}{r}
37° 76' 84'' \\
13° 28' 43'' \\
\hline
24° 48' 41''
\end{array}
$$

Measuring the Size of an Angle

EXERCISE 15-1

1. Name the basic terms of geometry.
2. Distinguish between a straight line and a straight line segment.
3. Distinguish between a straight line and a ray.

Read each of the following symbols:

4. $\overleftrightarrow{CD}$ 5. $\overrightarrow{EF}$ 6. $\angle KLM$ 7. $\angle 5$ 8. $\angle d$

9. Name the geometric figure shown in Fig. 15-7.

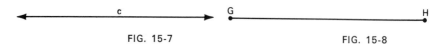

FIG. 15-7   FIG. 15-8

10. Name the geometric figure shown in Fig. 15-8.
11. Which of two angles in Fig. 15-9. is larger, $\angle a$ or $\angle b$?

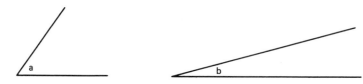

FIG. 15-9

12. Is the degree an angle?
13. Name each line segment of the geometric figure in Fig. 15-10.

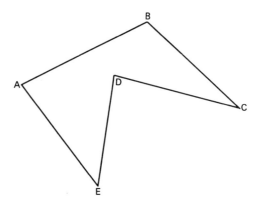

FIG. 15-10

14. Name the angle formed by $\overline{DC}$ and $\overline{DE}$ in Fig. 15-10.
15. What is point B for angle $\angle ABC$ in Fig. 15-10?

16. Add 8° 17′ 21″
        19° 24′ 18″
        23° 8′ 12″

17. Add   7° 15′ 5″
        51° 24′ 16″
         3° 38′ 14″

18. Add 28° 11′ 32″
         5° 13′
        41° 6′ 33″

19. Add 43°
        24° 6′ 33″
         9° 17′

20. Subtract   52° 37′ 49″
               17° 26′ 31″

21. Subtract   64° 8′ 27″
               18° 23′ 47″

22. Subtract   48°
               34° 8′ 11″

15-4 Kinds of Angles

There are several different kinds of angles, and they are listed below.

1. **The right angle.** A right angle always equals 90°. $\angle$ *DEF* of Fig. 15-11(a) is a right angle. The symbol (⌐) close to the vertex in the figure is the symbol for a right angle.
2. **Acute angle.** An acute angle is less than 90°. Thus, $\angle$ *GHI* of Fig. 15-11(b) is an acute angle. Also, a 40° angle is an acute angle.
3. **Obtuse angle.** An obtuse angle is more than 90°. Thus, $\angle$ *KLM* of Fig. 15-11(c) is an obtuse angle. Also, a 135° angle is an obtuse angle.
4. **The straight line angle.** The straight line is a 180° angle. Thus, $\angle$ *NOP* of Fig. 15-11(d) is a straight line angle.

Certain pairs of angles are important in geometric problems. Below are three such pairs of angles.

5. **Adjacent angles.** Two angles are adjacent when they have the same vertex and a common side between them. Thus, $\angle$ *AOB* and $\angle$ *BOC* of Fig. 15-12(a) are adjacent angles.
6. **Complementary angles.** Two angles are complementary angles when their sum is 90°. Thus, $\angle$ *KOL*, and $\angle$ *LOM* of Fig. 15-12(b) are complementary angles. Also, two angles which are 35° and 55° are complementary angles.
7. **Supplementary angles.** Two angles are supplementary angles when their sum equals 180°. Thus, $\angle$ *NOP* and $\angle$ *POR* of Fig. 15-12(c) are supplementary angles. Also, two angles which are 135° and 45°, respectively, are supplementary angles.

EXAMPLE 15-3 Two angles, $\angle$ *a* and $\angle$ *b*, are complementary angles. If $\angle$ *b* = 72° 13′ 38″, what is the size of $\angle$ *a*?

SOLUTION Since the two angles are complementary angles, $\angle$ *a* = 90° − $\angle$ *b*. First we express 90° as 89° 59′ 60″ and then we complete the subtraction.

$$89° \ 59′ \ 60″$$
$$72° \ 13′ \ 38″$$
$$\overline{17° \ 46′ \ 22″}$$

Thus, $\angle$ *a* = 17° 46′ 22″.

EXAMPLE 15-4 Two angles, .$\angle$ *a* and $\angle$ *b*, are supplementary angles. If $\angle$ *a* = 17° 48′ 31″, what is the size of $\angle$ *b*?

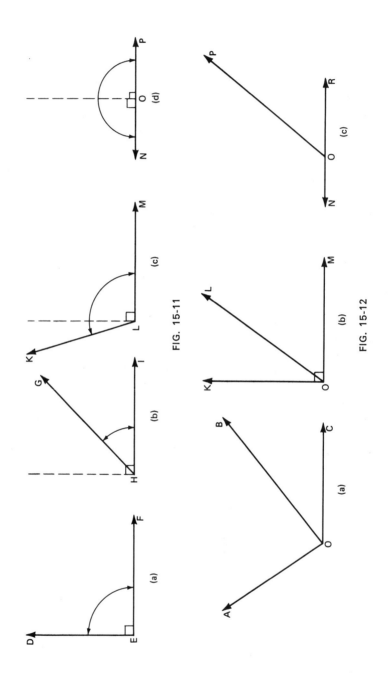

FIG. 15-11

FIG. 15-12

SOLUTION Since the two angles are supplementary angles, $\angle b = 180° - \angle a$. First we express $180°$ as $179° \, 59' \, 60''$ and then we complete the subtraction.

$$\begin{array}{r} 179° \, 59' \, 60'' \\ 17° \, 48' \, 31'' \\ \hline 162° \, 11' \, 29'' \end{array}$$

Thus, $\angle b = 162° \, 11' \, 29''$.

EXAMPLE 15-5 If $\angle a = 18° \, 23' \, 31''$ (a) what is its complement? (b) What is its supplement?

SOLUTION

$$\begin{array}{rr} 89° \, 59' \, 60'' & 179° \, 59' \, 60'' \\ 18° \, 23' \, 31'' & 18° \, 23' \, 31'' \\ \hline 71° \, 36' \, 29'' & 161° \, 36' \, 29'' \end{array}$$

Thus, the complement of $\angle a = 71° \, 36' \, 29''$ and the supplement of $\angle a = 161° \, 36' \, 29''$.

EXERCISE 15-2

1. When are two angles called adjacent?
2. What is the straight line angle?
3. What kind of angle is $\angle AOD$ of Fig. 15-13?
4. If line BD of Fig. 15-13 is a straight line segment, (a) what kind of angles are $\angle AOD$ and $\angle AOB$? (b) What is the size of $\angle AOB$? (c) What kind of angle is $\angle AOB$?
5. What kind of angles are $\angle AOB$ and $\angle BOC$ of Fig. 15-13?
6. If line BD of Fig. 15-13 is a straight line segment, and $\angle BOC = 38° \, 12' \, 33''$, what is the size of $\angle COD$?
7. $\angle a = 34° \, 17' \, 5''$. What is its complement?
8. $\angle b = 139° \, 13' \, 19''$. What is its supplement?
9. In the truss bridge shown in Fig. 15-14 beam A makes an angle of $90°$ with the roadway of the bridge. If the slanted beam B makes an angle of $47°$ with beam A, what is the size of the angle beam B makes with the roadway of the bridge?

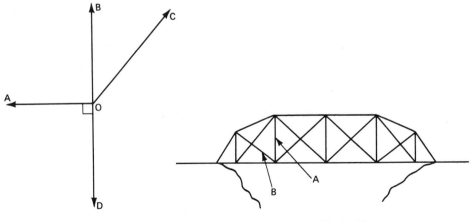

FIG. 15-13   FIG. 15-14

Kinds of Angles

10. In the roof truss shown in Fig. 15-15 $\angle$ 1 and $\angle$ 2 are 65° each. What is the size of $\angle$ 3, beam a makes with beam b?

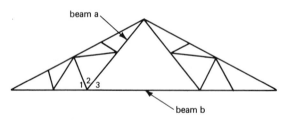

FIG. 15-15

15-5 Intersecting Lines and More Kinds of Angles

When two straight lines, $\overleftrightarrow{AB}$ and $\overleftrightarrow{CD}$, are on the same plane as shown in Fig. 15-16(a), one of two things can happen: They will either intersect when extended, or they will not.

If the two straight lines, $\overleftrightarrow{AB}$ and $\overleftrightarrow{CD}$, do not intersect, no matter how much they are extended, see Fig. 15-16(b), they are called **parallel lines.**

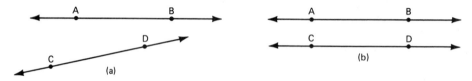

FIG. 15-16

If the two straight lines, $\overleftrightarrow{AB}$ and $\overleftrightarrow{CD}$, intersect, they do so at only one point, called the **point of intersection** as shown in Fig. 15-17(a) and (b).

When the two straight lines, $\overleftrightarrow{AB}$ and $\overleftrightarrow{CD}$, intersect, only one of two things can happen: they either form four equal angles [Fig. 15-17(a)], or they form two pairs of opposite angles called **vertical angles** [Fig. 15-17(b)].

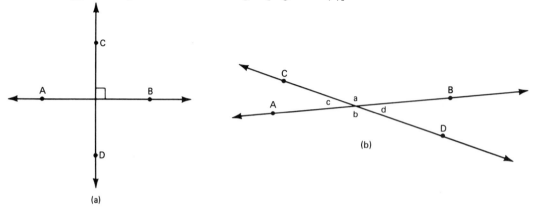

FIG. 15-17

When all four angles formed by intersecting lines are equal, such lines are said to be **perpendicular** to each other. Thus, $\overleftrightarrow{AB}$ and $\overleftrightarrow{CD}$ of Fig. 15-17(a) are perpendicular to each other. It is also true that when two intersecting lines are perpendicular, the four angles formed are equal.

When two lines intersect to form vertical angles, *the angles in each pair of vertical angles are equal.* Thus, in Fig. 15-17(b), $\angle a$ and $\angle b$ are a pair of vertical angles, and $\angle c$ and $\angle d$ are a pair of vertical angles. Therefore,

$$\angle a = \angle b \quad \text{and} \quad \angle c = \angle d$$

Notice that the pairs of vertical angles are nonadjacent angles. In Fig. 15-17(b), any two angles that have a common side are supplementary angles. Therefore, if the size of any one of the four angles formed by two intersecting lines is known, the size of the other three angles can be found.

EXAMPLE 15-6 Find the size of the other three angles, if $\angle c$ of Fig. 15-17(b) is 42° 11′ 43″.

SOLUTION Since $\angle d$ and $\angle c$ are vertical angles, $\angle d = \angle c = 42° 11′ 43″$. Since $\angle a$ and $\angle c$ are supplementary angles,

$$\angle a = 180° - \angle c$$
$$= 180° - 42° 11′ 43″$$
$$= 137° 48′ 17″$$

Since $\angle b$ and $\angle a$ are vertical angles, $\angle b = \angle a = 137° 48′ 17″$.

When two parallel straight lines, $\overleftrightarrow{AB}$ and $\overleftrightarrow{CD}$ are intersected by a third straight line, $\overleftrightarrow{EF}$, they form equal **alternate interior angles**, as shown in Fig. 15-18. Thus, $\angle a$ and $\angle b$ are alternate interior angles, and so are $\angle c$ and $\angle d$. Therefore, $\angle a = \angle b$ and $\angle c = \angle d$.

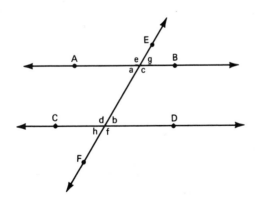

FIG. 15-18

When the size of any of the eight angles of Fig. 15-18 is known, the size of the other seven angles can be determined.

EXAMPLE 15-7 If $\overleftrightarrow{AB}$ and $\overleftrightarrow{CD}$ are parallel and are intersected by $\overleftrightarrow{EF}$, find the size of the other angles, if $\angle 7 = 152° 13′ 51″$ (see Fig. 15-19).

Intersecting Lines and More Kinds of Angles

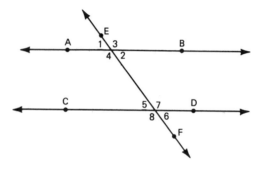

FIG. 15-19

SOLUTION Since ∠ 7 and ∠ 4 are alternate interior angles, and since ∠ 7 and ∠ 8, and ∠ 4 and ∠ 3 are vertical angles, it follows that ∠ 7 = ∠ 8 = ∠ 4 = ∠ 3 = 152° 13′ 51″. Since ∠ 7 and ∠ 5 are supplementary angles, it follows that

$$∠ 5 = 180° - ∠ 7$$
$$= 180° - 152° 13′ 51″$$
$$= 27° 46′ 9″$$

Since ∠ 5 and ∠ 2 are alternate interior angles, and ∠ 5 and ∠ 6, and ∠ 2 and ∠ 1 are vertical angles, it follows that ∠ 5 = ∠ 6 = ∠ 2 = ∠ 1 = 27° 46′ 9″.

Alternate interior angles have practical applications in certain problems of trigonometry, as we will see later.

EXAMPLE 15-8 A helicopter is flying horizontally above the sea. Angle *a* is the angle formed by the horizontal line of flight and the pilot's line of sight as he looks at a ship. If ∠ *a* = 38°, what is the size of ∠ *b*, the angle formed by the pilot's line of sight and the surface of the sea (see Fig. 15-20)?

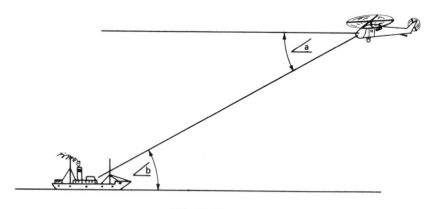

FIG. 15-20

SOLUTION The line of flight and the surface of the sea are parallel. Therefore, the line of sight makes ∠ *a* and ∠ *b* equal alternate interior angles. Then, ∠ *a* = ∠ *b* = 38°.

1. Draw two parallel lines and intersect them with a third straight line not perpendicular to them. Label the eight angles formed to separate them into two groups of four equal angles.

2. In the truss bridge shown in Fig. 15-21, beam *a* is parallel to beam *b*. Beam *c* is perpendicular to beam *a*. Is beam *c* also perpendicular to beam *b*? Justify your answer.

3. If ∠ *a* of Fig. 15-21 is 80°, what are the sizes of the other three angles in that intersection of beams?

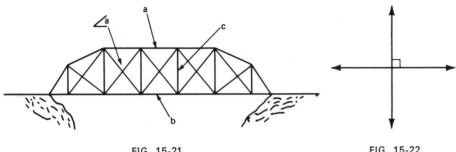

FIG. 15-21   FIG. 15-22

4. Are the four angles of Fig. 15-22 a special case of vertical angles? Why?

5. Two straight lines intersect and form four angles, one of which is 18° 23′ 48″. Find the size of each of the other three angles.

6. If ∠ *d* of Fig. 15-23 is 45° 12′, what is the size of each of the other four angles?

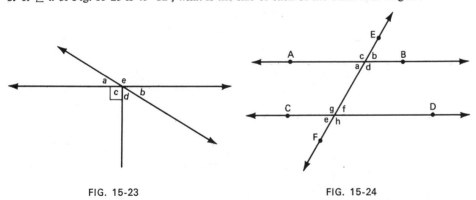

FIG. 15-23   FIG. 15-24

7. If $\overleftrightarrow{AB}$ and $\overleftrightarrow{CD}$ are parallel lines and are intersected by $\overleftrightarrow{EF}$, find the size of the other angles if ∠ *d* = 103° (see Fig. 15-24).

8. A plane is flying horizontally above the ocean. The pilot is looking at a lighthouse. Angle *a* is the angle formed by the horizontal line of flight and the pilot's line of sight. If ∠ *a* = 29°, what is the size of ∠ *b*, the angle formed by the surface of the ocean and the pilot's line of sight?

Intersecting Lines and More Kinds of Angles   257

9. Line d is parallel to the base of the triangle of Fig. 15-25, and through the vertex of ∠ b. The three sides of the triangle are extended as shown. Show that ∠ a + ∠ b + ∠ c = 180°. (Hint: What is the relation of ∠ 1, ∠ 2, and ∠ 3 to the three angles of the triangle? What is their sum?)

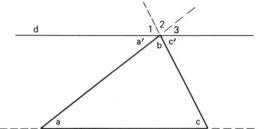

FIG. 15-25

10. Light rays from a distant source are assumed, for all practical purposes, to be parallel. How do the angles formed by the rays of the sun and the surface of the earth, at any given time of the day compare with each other (see Fig. 15-26)? (Hint: Think of the surface of the earth as a straight line segment intersecting the parallel rays of light.)

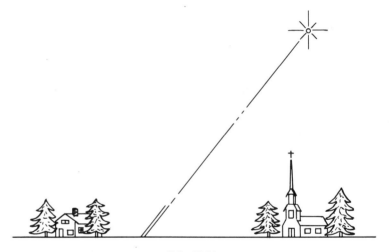

FIG. 15-26

REVIEW QUESTIONS

1. What is the difference between a straight line and a straight line segment?

2. What is the difference between a straight line and a ray?

3. What is an angle?

4. What is a vertex?

5. Name the unit for the measurement of angles.

6. When is an angle called acute?

7. When is an angle called obtuse?

8. When are two angles called complementary angles?

Introduction to Geometry

9. When are two angles called supplementary angles?
10. What is the size of the straight line angle?
11. When are two angles called adjacent angles?
12. Which straight lines are called parallel?
13. Which straight lines are said to be perpendicular?
14. What angles are called vertical angles?
15. What angles are called alternate interior angles?

REVIEW EXERCISES

1. Add $\angle a$, $\angle b$, and $\angle c$, if $\angle a = 22°\ 11'\ 28''$, $\angle b = 34°\ 51'\ 17''$, and $\angle c = 8°\ 43'\ 42''$.
2. Subtract $\angle d$ from $\angle c$, if $\angle c = 21°\ 18'$ and $\angle d = 17°\ 42'\ 27''$.
3. Name the two acute and the two obtuse angles of Fig. 15-27.

FIG. 15-27

4. Two straight line segments intersect, and one of the four angles formed is a right angle. (a) What is the size of the other three angles? (b) What are these two lines in relation to each other?
5. Two straight line segments intersect and one of the four angles formed is $24°\ 28'\ 53''$. What is the size of the other three angles?
6. Two parallel straight lines are intersected by a third straight line. The smaller of the two angles formed by the lower of the parallel lines is $28°\ 43'$. How many more angles are of this size? What is the size of the remaining angles?

16

Common Plane Figures and Common Solids

16-1 Triangles

All closed geometric figures whose sides are straight line segments are called **polygons**. A **triangle** is a polygon with three sides. Triangles are named by letters placed at each vertex. Thus, the triangle in Fig. 16-1(a) is triangle *ABC*, and the triangle of Fig. 16-1(c) is triangle *HIJ*. Triangles are classified according to the kind of angles they have or according to their sides.

A triangle having a right angle is called a **right triangle** [Fig. 16-1(a)]. A triangle having an obtuse angle is called an **obtuse triangle** [Fig. 16-1(d)]. A triangle having three acute angles is called an **acute triangle** [Fig. 16-1(b) and (c)]. Obtuse and acute triangles are called **oblique triangles**.

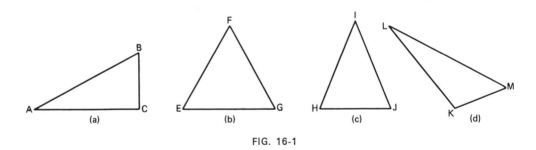

(a) (b) (c) (d)

FIG. 16-1

A triangle having all three sides equal in length is called an **equilateral triangle** [Fig. 16-1(b)]. A triangle having two sides equal is called an **isosceles triangle** [Fig. 16-1(c)]. A triangle having no two sides equal is called a **scalene triangle** [Fig. 16-1(d)].

In an isosceles triangle, the unequal side is called the **base**, and the angle opposite the base is called the **vertex** angle. Thus, $\overline{OP}$ of the isosceles triangle *OPQ* of Fig. 16-2(a) is the base, and $\angle Q$ is the vertex. In addition, *the two base angles of an isosceles triangle are equal*. Thus, in Fig. 16-2(a), $\angle O = \angle P$.

When a perpendicular is dropped from the vertex angle of an isosceles triangle to its base, the perpendicular divides both the vertex angle and the base into two equal

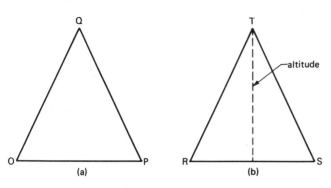

FIG. 16-2   (a)   (b)

halves, and the isosceles triangle into two *exactly equal* right triangles. These two right triangles are called **congruent**. In general *two geometric figures are congruent when they have exactly the same shape and the same size.* The perpendicular is called the **perpendicular bisector** and is the **altitude** of the isosceles triangle [Fig. 16-2(b)].

A very useful fact about the angles of a triangle is that the *sum of the three angles of any kind of triangle is* 180°. Thus, in the four triangles of Fig. 16-1

$$\angle A + \angle B + \angle C = 180°$$
$$\angle E + \angle F + \angle G = 180°$$
$$\angle H + \angle I + \angle J = 180°$$

and
$$\angle K + \angle L + \angle M = 180°$$

Because of this fact, the size of the third angle of any triangle can be found if the size of the other two angles is known.

EXAMPLE 16-1 If the vertex angle of an isosceles triangle is 70°, what is the size of each of the base angles?

SOLUTION Since the sum of the three angles of a triangle is 180°, and the vertex angle here is 70°, then the other two angles together are $180° - 70° = 110°$. Also, since the base angles of an isosceles triangle are equal, each angle is $110° \div 2 = 55°$.

EXAMPLE 16-2 The law of reflection is one of the simplest laws of physics. It states: The angle of incidence, $\angle i$, is equal to the angle of reflection, $\angle r$ [Fig. 16-3(a)]. Notice that $\angle i$ is formed by the incident ray and line N (N for normal), which is perpendicular to the reflecting surface, and $\angle r$ is formed by the reflected ray and line N.

A mirror is making an angle of 43° with the horizontal. A ray of light falling vertically strikes the mirror and is reflected, as shown in Fig. 16-3(b). Find the size of $\angle i$ and $\angle r$.

SOLUTION The triangle formed by the mirror, the horizontal, and the extension of the incident ray, is a right triangle. Since $\angle a = 43°$, $\angle b = 47°$. If $\angle b = 47°$, then $\angle b' = 47°$ also. ($\angle b$ and $\angle b'$ are vertical angles.) Since line N is perpendicular to the mirror and $\angle b' = 47°$, $\angle i = 90° - 47° = 43°$. By the law of reflection, $\angle i = \angle r$. Therefore, $\angle i = \angle r = 43°$.

Triangles

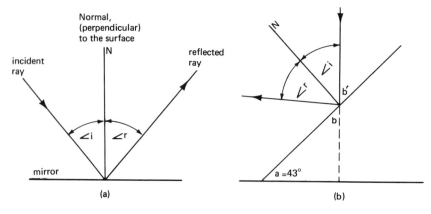

FIG. 16-3

16-2 The Right Triangle and the Pythagorean Theorem

Many problems in science and technology can be reduced to a right triangle problem. In this section, we will study only geometric properties and solutions of the right triangle. In the next chapter, using the powerful tool of trigonometry, we will be able to solve right triangle problems which cannot be solved geometrically.

In the right triangle of Fig. 16-4 side a is the altitude and side b is the base. Side c, the longest side, is called the **hypotenuse**. Notice that the altitude is perpendicular to the base.

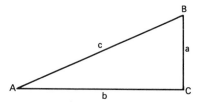

FIG. 16-4

In solving a triangle problem, it is always helpful to draw a sketch of the triangle to be solved. If the triangle is a right triangle, it is very important to label the triangle with the letters ABC for the angles with a for the altitude, b for the base, and c for the hypotenuse, as shown in Fig. 16-4. This method of labeling a right triangle is common practice, and will be used in this book.

For the right triangle, a very important property known as the **Pythagorean theorem** is true. The theorem says that *the square of the length of the hypotenuse is equal to the sum of the squares of the lengths of the other two sides.*

In terms of the sides, a, b, and c, of the right triangle of Fig. 16-5(a), the theorem expressed mathematically is

$$c^2 = a^2 + b^2$$

Figure 16-5(b) is a representation of the theorem. In Fig. 16-5(c), it can be seen clearly that the area of the square constructed on the hypotenuse which is 5 units long,

Common Plane Figures and Common Solids

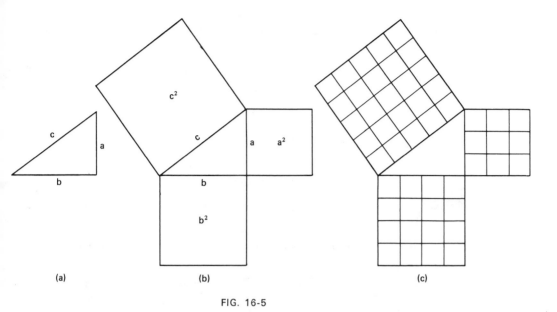

(a)   (b)   (c)

FIG. 16-5

is equal to the sum of the areas of the squares constructed on the other sides, which are 3 and 4 units long, respectively. Thus, $5^2 = 3^2 + 4^2$, or $25 = 9 + 16$.

From the original mathematical form of the theorem, $c^2 = a^2 + b^2$, one can arrive algebraically at the following three forms of the formula:

$$c = \sqrt{a^2 + b^2}$$

$$a = \sqrt{c^2 - b^2}$$

$$b = \sqrt{c^2 - a^2}$$

Using the appropriate form, the length of any missing side can be found if the other two sides are known.

EXAMPLE 16-3 Find the length of the hypotenuse of the right triangle for which $a = 23.3$ cm and $b = 17.8$ cm.

SOLUTION A sketch of the triangle is shown in Fig. 16-6. Notice that the triangle is

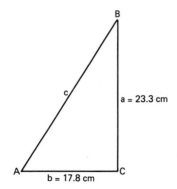

FIG. 16-6

The Right Triangle and the Pythagorean Theorem

labeled according to the right triangle of Fig. 16-4. Substituting into the formula $c = \sqrt{a^2 + b^2}$, we have

$$c = \sqrt{(23.3)^2 + (17.8)^2}$$
$$= \sqrt{859.73}$$
$$= 29.3$$

Therefore, the hypotenuse is 29.3 cm.

NOTE: It must be evident here that only principal square roots can be used, since lengths of sides can not be negative.

EXAMPLE 16-4 Find the length of the missing side of the right triangle in which $b = 11.3$ in. and $c = 13.3$ in.

SOLUTION The missing side is a. Substituting in the formula $a = \sqrt{c^2 - b^2}$, we have

$$a = \sqrt{(13.3)^2 - (11.3)^2}$$
$$= \sqrt{49.2}$$
$$= 7.01$$

Therefore, side a is 7.01 in.

EXAMPLE 16-5 Find the length of the altitude of the isosceles triangle of Fig. 16-7.

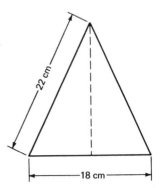

—18 cm—   FIG. 16-7

SOLUTION The base of each right triangle formed by the perpendicular bisector is 9 cm and its hypotenuse is 22 cm. Then

$$a = \sqrt{(22)^2 - (9)^2}$$
$$= \sqrt{403}$$
$$= 20$$

Therefore, the altitude is 20 cm.

Common Plane Figures and Common Solids

1. Two angles of a triangle are 38° and 64°, respectively. What is the size of the third angle?

2. One acute angle of a right triangle is 26° 17'. What is the size of the other acute angle?

3. One of the base angles of an isosceles triangle is 63°. What is the size of the vertex angle?

4. Two mirrors m and m' make an angle of 120° with each other, and mirror m is horizontal (see Fig. 16-8). A ray of incidence strikes mirror m at point A and at an angle of 55° with the perpendicular line N. The ray is reflected and strikes mirror m', as shown. Find the size of $\angle r$, $\angle n$, and $\angle r'$.

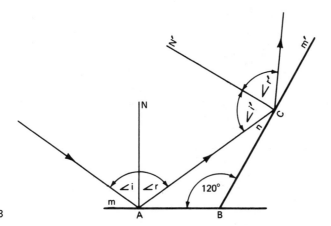

FIG. 16-8

Find the indicated side for each of the right triangles of exercises 5–10. These right triangles are labeled with the letters a, b, and c according to the right triangle of Fig. 16-4.

5. $a = 5$ in., $b = 12$ in. Find c.

6. $a = 9$ ft, $c = 15$ ft. Find b.

7. $a = 6.8$ cm, $c = 10.2$ cm. Find b.

8. $a = 3.25$ m, $c = 11.5$ m. Find b.

9. $b = 22.5$ cm, $c = 25.5$ cm. Find a.

10. $b = 2.8$ km, $c = 3.4$ km. Find a.

11. Find the length of rafter c of the roof of Fig. 16-9. Add 25 cm for overhang.

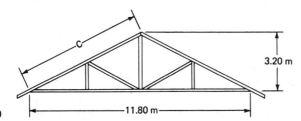

FIG. 16-9

12. Find the distance, center to center of the pulleys P and P_1 of Fig. 16-10. The distance BC is a perpendicular distance.

The Right Triangle and the Pythagorean Theorem

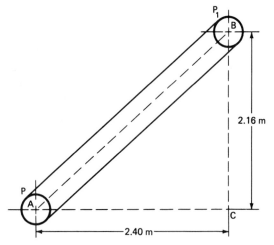

2.16 m

2.40 m

FIG. 16-10

13. Find the length of line *d* which joins the two nonadjacent angles of the rectangle of Fig. 16-11.

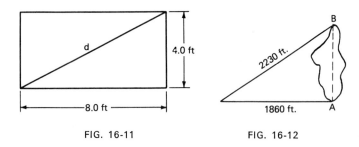

4.0 ft

8.0 ft

FIG. 16-11

2230 ft.

1860 ft.

FIG. 16-12

14. Find the distance *d* across the two points *A* and *B* of the lake of Fig. 16-12.

16-3 Similar Right Triangles

Geometric figures are **similar** if they have the same shape but are of different size. Two circles of different size, for instance, are similar.

The following three basic rules about the similarity of two triangles are very useful.

1. *Two triangles are similar if corresponding angles of the two triangles are equal.*
2. *If two triangles are similar, then the ratios of corresponding sides of the two triangles are equal (proportional).*
3. *If two angles of one triangle are equal to the two corresponding angles of a second triangle, then the remaining pair of angles of the two triangles is a pair of equal angles, and the two triangles are similar.*

Thus, given the two triangles *ABC* and *A′B′C′* (see Fig. 16-13), according to rule 1, the two triangles are similar if

Common Plane Figures and Common Solids

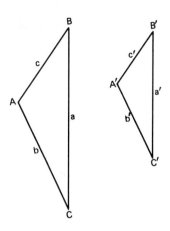

FIG. 16-13

$$\angle A = \angle A', \qquad \angle B = \angle B', \qquad \text{and} \qquad \angle C = \angle C'$$

If the two triangles ABC and $A'B'C'$ are similar, then, according to Rule 2

$$\frac{a}{a'} = \frac{b}{b'} = \frac{c}{c'}$$

The two triangles of Fig. 16-13 are similar, according to rule 3, if

$$\angle A = \angle A' \qquad \text{and} \qquad \angle B = \angle B'$$

or if

$$\angle B = \angle B' \qquad \text{and} \qquad \angle C = \angle C'$$

or if

$$\angle C = \angle C' \qquad \text{and} \qquad \angle A = \angle A'$$

It should not be difficult to see in the two right triangles ABC and $A'B'C'$ of Fig. 16-14 that since $\angle C = \angle C'$, the two right triangles will be similar, according to rule 3, if

$$\angle A = \angle A' \qquad \text{or if} \qquad \angle B = \angle B'$$

From the above, rule 4 follows:

4. *Two right triangles are similar if an acute angle of one right triangle is equal to an acute angle of the other right triangle.*

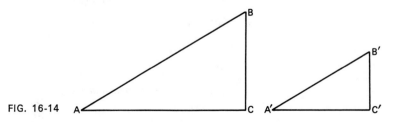

FIG. 16-14

Similar Right Triangles

EXAMPLE 16-6 Straight lines are intersecting as shown in Fig. 16-15. Find (a) side *a* and (b) ∠ *n'*.

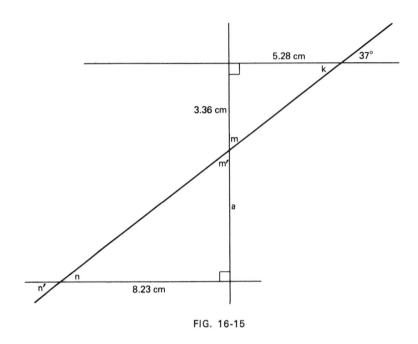

FIG. 16-15

SOLUTION

(a) The two right triangles of Fig. 16-15 are similar by rule 4, since their acute angles *m* and *m'* are equal as vertical angles. Since the two right triangles are similar, by rule 2,

$$\frac{5.28}{8.23} = \frac{3.36}{a}$$

Cross-multiplying and solving for *a*,

$$a = \frac{(8.23)(3.36)}{(5.28)}$$

$$= 5.24 \text{ cm}$$

(b) \qquad\qquad $\angle k = 37°$ \qquad\qquad *(by vertical angles)*

$\angle m = 90° - k$

$= 90° - 37°$

$= 53°$

$\angle m = \angle m'$ \qquad\qquad *(by vertical angles)*

Common Plane Figures and Common Solids

Thus $\angle m' = 53°$.

$$\angle n = 90° - \angle m'$$
$$= 90° - 53°$$
$$= 37°$$
$$\angle n = \angle n' \qquad \text{(by vertical angles)}$$

Thus $\angle n' = 37°$. Therefore, side $a = 5.24$ cm and $\angle n' = 37°$.

EXAMPLE 16-7 If the altitude, base, and hypotenuse of right triangle ABC of Fig. 16-16 are 3.68 in., 6.32 in., and 7.31 in., respectively, find the length of the hypotenuse and the base of the right triangle $A'B'C'$ if its altitude is 2.82 in.

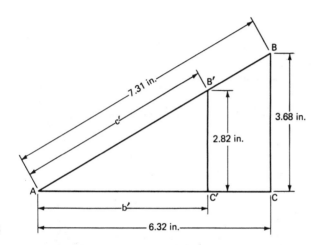

FIG. 16-16

SOLUTION Since $\angle A$ is a common angle for both right triangles ABC and $A'B'C'$, the two right triangles are similar by rule 4. Since triangles ABC and $A'B'C'$ are similar, their corresponding sides, by rule 2, are proportional. Thus,

$$\frac{3.68}{2.82} = \frac{6.32}{b'} \qquad (a)$$

and

$$\frac{3.68}{2.82} = \frac{7.31}{c'} \qquad (b)$$

From (a),

$$b' = \frac{(2.82)(6.32)}{(3.68)} = 4.84$$

and from (b),

$$c' = \frac{(2.82)(7.31)}{(3.68)} = 5.60$$

Therefore, $b' = 4.84$ in. and $c' = 5.60$ in.

Similar Right Triangles

EXERCISE 16-2

In each of exercises 1–3, two right triangles are given. State first whether the triangles in each case are similar, and if they are, find the missing side.

1.

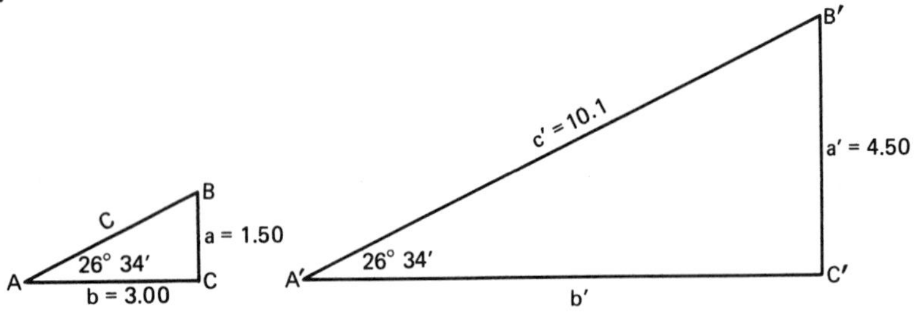

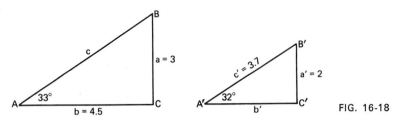

FIG. 16-17

2.

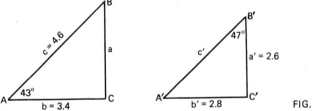

FIG. 16-18

3.

FIG. 16-19

4. Find distance BE and distance DE in Fig. 16-20.

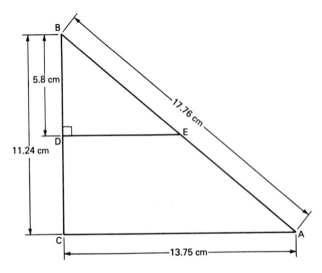

FIG. 16-20

5. Find the length of side *DE* of Fig. 16-21.

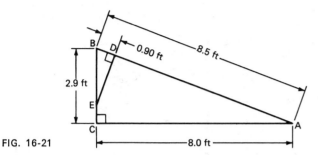

FIG. 16-21

6. Find the length of all the missing sides in the two triangles of Fig. 16-22.

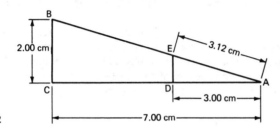

FIG. 16-22

7. Find ∠ *b*, ∠ *c*, ∠ *e*, and the length of side *f* of Fig. 16-23.

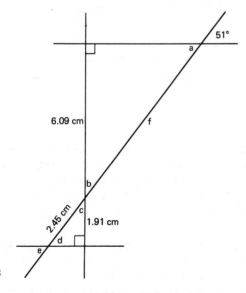

FIG. 16-23

8. At a certain moment of the day, the sun's rays have such an inclination that a telephone pole casts a shadow of 9.24 m. A nearby stick that is 1.35 m high and on the same level ground with it casts a shadow of 1.05 m (see Fig. 16.24). Find the height of the telephone pole. (Hint: Both the pole and the stick are perpendicular to the ground. Also, the rays

Similar Right Triangles

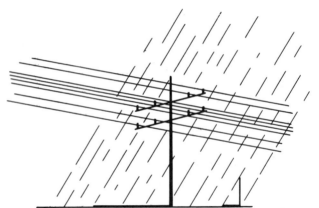

FIG. 16-24

of the sun are, for all practical purposes, assumed to be parallel, and form equal angles with the level ground.)

9. When the shadow that a tree casts is 28.6 ft, a 4.0-ft stick, near the tree and on the same level ground with it, casts a shadow of 6.6 ft. What is the height of the tree?

10. In an eye clinic, an examination booth is to be established in a 12-ft room (see Fig. 16-25). If the eye chart is 24-in. high, find the size of the smallest mirror which will show the entire chart when the patient is seated 8 ft from the mirror.

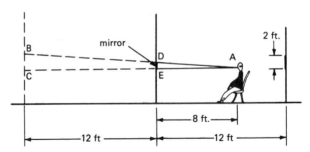

FIG. 16-25

16-4 Quadrilaterals

A **quadrilateral** is a polygon with four sides. There are several kinds of quadrilaterals. When there are no parallel lines in a quadrilateral, it is called a **trapezium** [Fig. 16-26(a)]. When there is one pair of parallel sides in a quadrilateral, it is called a **trapezoid** [Fig. 16-26(b)]. When both pairs of sides are parallel, the quadrilateral is called a **parallelogram** [Fig. 16-26(c)]. Opposite sides and opposite angles of a parallelogram are equal.

The straight line segment joining opposite angles is called a **diagonal**. The two parallel sides of a trapezoid are called its **bases**, and the perpendicular distance between them is called the **altitude** of the trapezoid [Fig. 16-26(b)].

When the two nonparallel sides of a trapezoid are equal in length, the trapezoid is called an **isosceles trapezoid** [Fig. 16-26(d)].

When the angles of a parallelogram are right angles, the parallelogram is called a **rectangle** [Fig. 16-26(e)].

Common Plane Figures and Common Solids

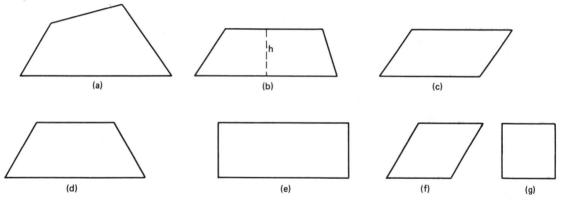

FIG. 16-26

The parallelogram whose sides are equal, but whose angles are not right angles is called a **rhombus** [Fig. 16-26(f)].

The rectangle whose sides are equal is called a **square** [Fig. 16-26(g)].

EXAMPLE 16-8 Find the altitude of the isosceles trapezoid of Fig. 16-27.

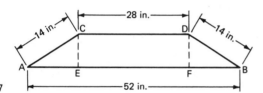

FIG. 16-27

SOLUTION The altitude of the trapezoid is $\overline{CE}$ or $\overline{DF}$. Since both $\overline{CE}$ and $\overline{DF}$ are perpendicular to $\overline{AB}$, and $\overline{CD}$ is parallel to $\overline{EF}$, the parallelogram $ECDF$ is a rectangle, and side $\overline{EF}$ is equal to 28 in. as it is opposite to side $\overline{CD}$. Then,

$$\overline{AE} = \overline{FB} = \frac{52 - 28}{2} = 12 \text{ in.}$$

Using the Pythagorean theorem, side $\overline{CE}$ of the right triangle ACE is

$$\overline{CE} = \sqrt{(14)^2 - (12)^2} = 7.2 \text{ in.}$$

Therefore, the altitude of the isosceles trapezoid is 7.2 in.

16-5 Perimeter and Area of Triangles and Quadrilaterals

The **perimeter** of a polygon is the sum of the lengths of its sides. Thus, the perimeter of a triangle is the sum of the lengths of its three sides, and the perimeter of a quadrilateral is the sum of the lengths of its four sides.

Since opposite sides in a rectangle are equal, the formula for the perimeter of a rectangle is

$$P = 2(L + W)$$

where L is the length of the rectangle and W is its width.

Perimeter and Area of Triangles and Quadrilaterals

The formula for the area of any kind of triangle is

$$A = \frac{bh}{2}$$

where h is the height or altitude of the triangle, and b is its base. The altitude is the perpendicular distance from a vertex of the triangle to the opposite side or an extension of that side [see Fig. 16-28(a) and (b)].

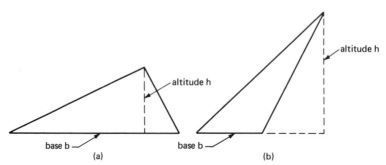

FIG. 16-28

EXAMPLE 16-9 Find the area of a triangle having a base of 14 in. and an altitude of 9 in.

SOLUTION Substituting in the formula $A = (bh)/2$, we have

$$A = \frac{(14)(9)}{2} = 63 \text{ in.}^2$$

Another formula for the area of a triangle whose three sides a, b, and c are known, but whose altitude is not given, or cannot be calculated, is

$$A = \sqrt{s(s - a)(s - b)(s - c)}$$

where $s = (a + b + c)/2$.

EXAMPLE 16-10 Find the area of a triangular lot, the sides of which are 112 m, 128 m, and 142 m, respectively.

SOLUTION

$$A = \sqrt{s(s - a)(s - b)(s - c)}$$

First we find s:

$$s = \frac{112 + 128 + 142}{2} = 191$$

and then we solve for A:

$$A = \sqrt{191(191 - 112)(191 - 128)(191 - 142)}$$
$$= 6830 \text{ m}^2$$

The formula for the area of a trapezoid is

$$A = \frac{(b_1 + b_2)h}{2}$$

Common Plane Figures and Common Solids

where b_1 and b_2 are the lengths of the two bases of the trapezoid, and h is the perpendicular distance between the two bases.

EXAMPLE 16-11 Find the area of the trapezoid having a lower base of 16.4 in., an upper base of 12.8 in. and an altitude of 5.2 in.

SOLUTION Substituting in the formula

$$A = \frac{(b_1 + b_2)h}{2}$$

we have

$$A = \frac{(16.4) + 12.8)(5.2)}{2}$$

$$= 76. \text{ in.}^2$$

The formula for the area of a parallelogram is

$$A = bh$$

where b is one of the two parallel sides and h is the perpendicular distance between them.

The formula for the area of a rectangle is

$$A = LW$$

where L is the length of the rectangle and W is the width.

The formula for the area of a square is

$$A = s^2$$

where s is the length of the side of the square.

EXAMPLE 16-12 Find the area of the parallelogram of Fig. 16-29.

SOLUTION Substituting in the formula $A = bh$, we have

$$A = (12)(7.0)$$

$$= 84 \text{ cm}^2$$

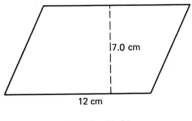

7.0 cm

12 cm

FIG. 16-29

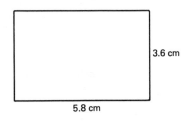

3.6 cm

5.8 cm

FIG. 16-30

EXAMPLE 16-13 Find the area of the rectangle of Fig. 16-30.

SOLUTION Substituting in the formula $A = LW$, we have

$$A = (5.8)(3.6)$$

$$= 21 \text{ cm}^2$$

Perimeter and Area of Triangles and Quadrilaterals

EXAMPLE 16-14   Find the area of a square whose side is 87.00 mm.

SOLUTION   Substituting in the formula $A = s^2$, we have

$$A = (87.00)^2$$
$$= 7569 \text{ mm}^2$$

EXAMPLE 16-15   Find the area of the composite figure of Fig. 16-31.

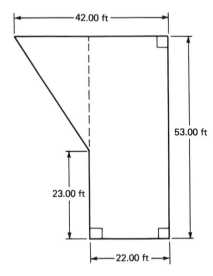

FIG. 16-31

SOLUTION   The dotted line divides the figure into a rectangle, the dimensions of which are 53.00 ft and 22.00 ft, and a right triangle, the base of which is 20.00 ft (42.00 − 22.00), and the altitude of which is 30.00 ft (53.00 − 23.00). The area of the rectangle is

$$A = (53.00)(22.00)$$
$$= 1166 \text{ ft}^2$$

The area of the triangle is

$$A = \frac{(20.00)(30.00)}{2}$$
$$= 300.0 \text{ ft}^2$$

The area of the composite figure is $1166 + 300.0 = 1466 \text{ ft}^2$.

EXERCISE 16-3

1. Find the altitude and the area of the isosceles trapezoid of Fig. 16-32.
2. A triangular piece of land was fenced, and then two strands of barbed wire were used above the fencing for extra protection. If the sides of the triangular lot are 218 ft, 123 ft, and 278 ft, respectively, how much barbed wire was used?
3. Find the perimeter and the area of a room 18.5-ft long, and 12.0-ft wide.
4. Find the area of a concrete driveway 64.0-ft long and 12-ft 9-in. wide.

Common Plane Figures and Common Solids

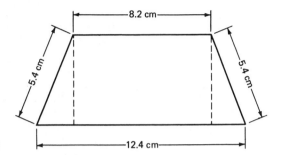

FIG. 16-32

5. Find the perimeter and the area of a square whose side is 17 mm.
6. Find the area of the L-shaped living room shown in Fig. 16-33.

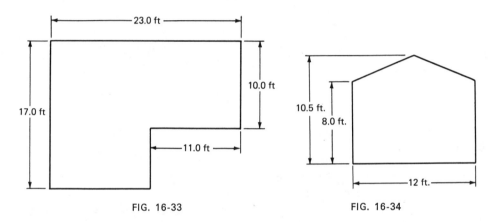

FIG. 16-33   FIG. 16-34

7. Find the area of the figure shown in Fig. 16-34.
8. Find the area of a triangle whose sides are 8.0 cm, 9.0 cm, and 11.0 cm.
9. Find the area of the figure of Fig. 16-35.

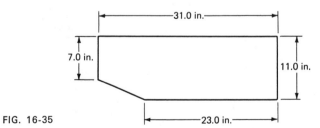

FIG. 16-35

10. A yacht has a triangular window whose sides are 14 in., 18 in., and 20 in., respectively.
   Find the area of the window.
11. Find the area of the two shaded triangles of Fig. 16-36. Can you do it in two different
   ways?

Perimeter and Area of Triangles and Quadrilaterals

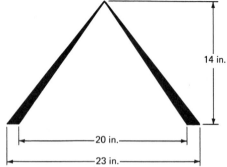

14 in.

20 in.

23 in.

FIG. 16-36

12. Find the area of the trapezoid of Fig. 16-37. Can you do it in two different ways?

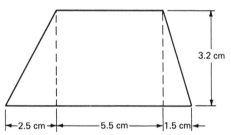

3.2 cm

2.5 cm   5.5 cm   1.5 cm

FIG. 16-37

16-6 Circles

The circle is a curved line, all points of which are in a plane and are a given distance from a point c, called the center (see Fig. 16-38).

A straight line segment from the center of a circle to any point on the circle, like $\overline{cR}$ of Fig. 16-38, is called a **radius** of the circle. The radius is represented by r.

A straight line segment which starts at any point on the circle, passes through the center, and is terminated by the circle, like $\overline{ST}$ of Fig. 16-38, is called a **diameter**. The diameter is represented by D. It must be evident that

$$D = 2r \quad \text{or} \quad r = \frac{D}{2}$$

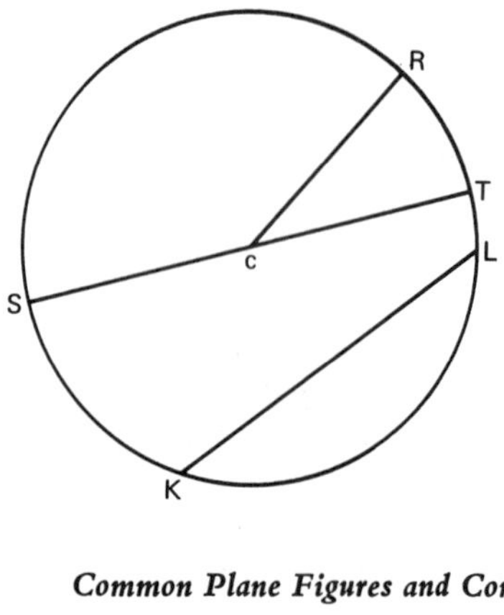

R

T

L

c

S

K

FIG. 16-38

Common Plane Figures and Common Solids

The length of the curved line is called the **circumference** of the circle. The circumference is represented by C.

The student should notice that the circumference is the length of the curved line, and not the curved line itself, which is the circle. The two things are often confused. We should remember that the circle is a shape, as a triangle is a shape; that the circumference is a length, as the perimeter is a length.

A part of a circle is called an **arc**. Thus, the part of the circle of Fig. 16-38 between the points K and L is an arc. In fact, there are two arcs between the points K and L, a small arc and a large arc.

A straight line segment, both ends of which are on the circle, like $\overline{KL}$ of Fig. 16-38 is called a **chord**. The diameter is a special chord which divides the circle into two equal arcs called **semicircles**. Any other chord divides the circle into a small and a large arc.

Two or more circles with the same center, but with different radii are called **concentric circles** (see Fig. 16-39).

An angle formed by two radii intercepting an arc, and having its vertex at the center of the circle, like $\angle$ AOB of Fig. 16-40 is called a **central angle**.

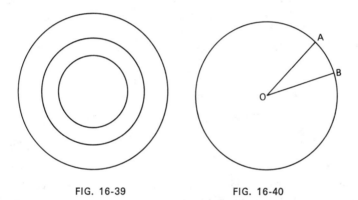

FIG. 16-39   FIG. 16-40

To measure a central angle, we use the intercepted arc. We have seen in the previous chapter that the central angle which intercepts an arc that is 1/360 of a circle is an angle of 1°. This is so because the circle is taken to be an arc of 360°. We can say now that *a given arc and the central angle, which intercepts the arc, have the same number of degrees.*

We have already seen that the ratio of the circumference of any circle to its diameter, that is the ratio C/D, is the constant π.

To the equation $C/D = \pi$, we can give three other forms which are useful for finding the circumference and the diameter of a circle. Thus,

$$C = \pi D$$

$$C = 2\pi r$$

$$D = \frac{C}{\pi}$$

Circles

EXAMPLE 16-16 Find the circumference of a circle whose diameter is $3\frac{7}{8}$ in.

SOLUTION Here, the diameter is $3\frac{7}{8}$, or 3.875 in. Substituting in the formula $C = \pi D$, we have

$$C = (3.875)(3.1416) = 12.17 \text{ in.}$$

EXAMPLE 16-17 The radius of a flywheel is 8.50 in. Find the diameter and the circumference of the wheel.

SOLUTION Substituting in the formula $D = 2r$, we have

$$D = 2(8.50)$$
$$= 17.0 \text{ in.}$$

Substituting in the formula $C = 2\pi r$, we have

$$C = 2(3.14)(8.50)$$
$$= 53.4 \text{ in.}$$

EXAMPLE 16-18 Find the diameter of a circle whose circumference is 23.1 cm.

SOLUTION Substituting in the formula $D = C/\pi$, we have

$$D = \frac{23.1}{3.142} = 7.35 \text{ cm}$$

The formula for the area of a circle is

$$A = \pi r^2$$

EXAMPLE 16-19 Find the area of a circle whose radius is 5.625 in.

SOLUTION Substituting in the formula $A = \pi r^2$, we have

$$A = (3.1416)(5.625)^2$$
$$= 99.40 \text{ in.}^2$$

The area of a ring (see Fig. 16-41) will be the difference between the areas of the two circles. Thus, if the radius of the large circle is R, and the radius of the small circle is r, then,

$$A = \pi R^2 - \pi r^2$$

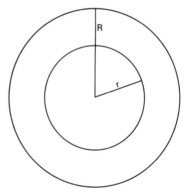

FIG. 16-41

Common Plane Figures and Common Solids

or

$$A = \pi(R^2 - r^2)$$

EXAMPLE 16-20 Find the area of a washer whose large diameter is 32 mm and small diameter is 12 mm.

SOLUTION

$$A = 3.14[(16)^2 - (6)^2]$$
$$= 3.14[(256 - 36)]$$
$$= 3.14(220)$$
$$= 690 \text{ mm}^2$$

EXERCISE 16-4

1. Find the radius of a circle whose circumference is 82.9 mm.
2. Find the radius of a circle whose area is 30.6 in.²
3. Find the circumference of a circle whose radius is 15.24 cm.
4. Find the circumference of a circle whose diameter is 32.7 m.
5. Find the diameter of a pulley whose circumference is 55.4 in.
6. Find the diameter of a circular mirror whose area is 468 cm².
7. The diameter of an automobile tire is 25.5 in. What is the circumference of that tire?
8. Two pulleys with 68.0-cm diameters are 2.35 m apart from center to center. Find the length of the belt connecting them.
9. The wheel of a certain car is 61.0 cm in diameter. Assuming that there was no slipping, how many turns did the wheel make in covering a distance of 2.3 km?
10. A circular swimming pool is 32 ft in diameter and has a 5-ft wide sidewalk around it. Find the area of the sidewalk.
11. The circle of Fig. 16-42 has a diameter of 38.0 cm. A chord of 32.0 cm intercepts the circle. Find the perpendicular distance d, from the center of the circle to the chord.
12. Find the area of the shaded region of the rectangle in Fig. 16-43.

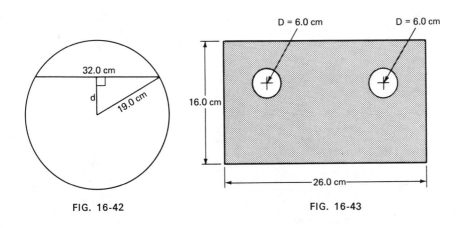

FIG. 16-42

FIG. 16-43

16-7 Solids, Volumes, and Surface Areas

A **rectangular solid** (see Fig. 16-44) has six plane sides called **faces** which form right angles with each other.

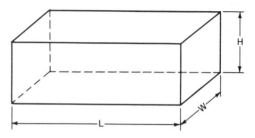

FIG. 16-44

The formula for finding the volume of a rectangular solid is

$$V = LWH$$

where L is the length of the solid, W is its width, and H is its height.

EXAMPLE 16-21   Find the volume in cubic feet of a rectangular box 44-in. long, 24-in. wide, and 18-in. high.

SOLUTION   Substituting in the formula $V = LWH$, we have

$$V = (44)(24)(18)$$
$$= 19{,}008 \text{ in.}^3$$
$$19{,}008 \text{ in.}^3 = 19{,}008 \text{ in.}^3 \left(\frac{1 \text{ ft}^3}{1728 \text{ in.}^3} \right)$$
$$= 11 \text{ ft}^3$$

The volume is 11 ft³.

Since opposite faces of a rectangular solid are equal in area, to find the total surface area, we can use the formula

$$\text{total area} = 2(\text{area of long face}) + 2(\text{area of end face})$$
$$+ 2(\text{area of bottom face})$$

EXAMPLE 16-22   Find the total surface area in square feet of the rectangular solid of Example 16-21.

SOLUTION

$$\text{Total area} = 2(44)(18) + 2(24)(18) + 2(44)(24)$$
$$= 4560 \text{ in.}^2$$
$$4560 \text{ in.}^2 = 4560 \text{ in.}^2 \left(\frac{1 \text{ ft}^2}{144 \text{ in.}^2} \right)$$
$$= 31.7 \text{ ft}^2$$

Thus, the total surface area is 31.7 ft².

Common Plane Figures and Common Solids

The cube (see Fig. 16-45) is a special case of the rectangular solid, because all three dimensions, L, W, and H, are equal. Therefore, the formula for the volume of a cube is

$$V = e^3$$

where e is the length of an edge of the cube.

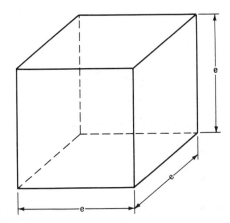

FIG. 16-45

The formula for the total surface area of a cube is

$$\text{total area} = 6e^2$$

EXAMPLE 16-23   Find the volume and the total surface area of the cube whose edge is 8.0 cm.

SOLUTION   Substituting in the formula $V = e^3$, we have

$$V = (8.0)^3$$
$$= 510 \text{ cm}^3$$

Substituting in the formula, total area $= 6e^2$, we have

$$\text{total area} = 6(8.0)^2 = 380 \text{ cm}^2$$

Therefore the volume is 510 cm³, and the total surface area is 380 cm².

The right circular cylinder (see Fig. 16-46) is a very common solid which is extensively used both in everyday life and in technology. This kind of cylinder has two circular bases (top and bottom) and a curved surface. It is called a **right circular cylinder** because the straight line segment $\overline{CD}$, connecting the centers of the two circular bases, is perpendicular to each base. The line $\overline{CD}$ is also the **altitude** (h) of the cylinder.
The formula for the volume of the cylinder is

$$V = \pi r^2 h$$

where r is the radius of the circular base and h is the height of the cylinder.

Solids, Volumes, and Surface Areas

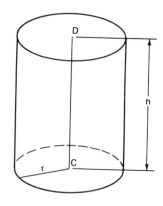

FIG. 16-46

The formula for the surface area, called **lateral area** (LA), of the cylinder is

$$LA = 2\pi rh$$

where r is the radius of the circular base and h is the altitude of the cylinder.

EXAMPLE 16-24 Find the volume in cubic feet and the lateral area in square feet of a right circular cylinder whose base has a radius of 7.50 in., and whose altitude is 20.0 in.

SOLUTION

$$V = (3.142)(7.50)^2(20.0)$$
$$= 3534.75 \text{ in.}^3$$

$$3534.75 \text{ in.}^3 = 3534.75 \text{ in.}^3 \left(\frac{1 \text{ ft}^3}{1728 \text{ in.}^3} \right)$$
$$= 2.05 \text{ ft}^3$$

$$LA = 2(3.142)(7.50)(20.0)$$
$$= 942.6 \text{ in.}^2$$

$$942.6 \text{ in.}^2 = 942.6 \text{ in.}^2 \left(\frac{1 \text{ ft}^2}{144 \text{ in.}^2} \right)$$
$$= 6.55 \text{ ft}^2$$

Thus, the volume of the right cylinder is 2.05 ft³ and the lateral area is 6.55 ft².

The formula for the volume of the **hollow cylinder** (see Fig. 16-47) is

$$V = V_L - V_S$$

where V_L is the volume of the larger cylinder and V_S is the volume of the smaller cylinder. If we represent the radius of the base of the large cylinder by R, and the radius of the base of the small cylinder by r, then the formula for the volume of the hollow cylinder becomes

$$V = \pi R^2 h - \pi r^2 h, \quad \text{or} \quad \pi h(R^2 - r^2)$$

Common Plane Figures and Common Solids

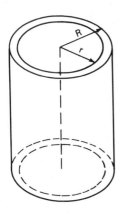

FIG. 16-47

EXAMPLE 16-25 Find the volume of a hollow cylinder in which $R = 12.0$ cm, $r = 7.50$ cm, and $h = 24.0$ cm.

SOLUTION

$$V = 3.142(24.0)[(12.0)^2 - (7.50)^2]$$
$$= 6620 \text{ cm}^3$$

A **right circular cone** (see Fig. 16-48) is a fairly common solid used both in everyday life and in technology. It is called a right circular cone because its base is a circle and because the straight line segment VC, from the point V of the cone, called the **vertex**, to the center of its circular base is perpendicular. The length of $\overline{VC}$ is the altitude (h) of the cone. A right circular cone is generated by rotating a right triangle about one of its perpendicular sides. The length of the hypotenuse of that right triangle is called the **slant height**. Thus, the line $\overline{VW}$ of Fig. 16-48 is the slant height of that cone.

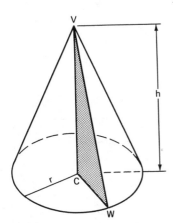

FIG. 16-48

The volume of a cone is one third that of a cylinder. Thus, the formula for the volume of the cone is

$$V = \frac{\pi r^2 h}{3}$$

where r is the radius of the circular base and h is the height.

Solids, Volumes, and Surface Areas

The formula for the lateral area of the cone is

$$LA = \frac{Cs}{2}$$

where C is the circumference of the circular base and s is the slant height.

EXAMPLE 16-26   A right circular cone has a radius of 5.0 in. and a height of 12.0 in. Find its volume and its lateral area.

SOLUTION

$$V = \frac{(3.14)(5.0)^2(12.0)}{3}$$

$$= 314 \text{ in.}^3$$

We use the Pythagorean theorem to find the length of the slant height

$$s = \sqrt{(5.0)^2 + (12.0)^2}$$

$$= \sqrt{169}$$

$$= 13$$

Therefore,

$$LA = \frac{2(3.14)(5.0)(13)}{2}$$

$$= 204 \text{ in.}^2$$

A **sphere** (see Fig. 16-49) is a closed curved surface, all points of which are the same distance from a given point called the center. A sphere will be generated if a circle is rotated about one of its diameters.

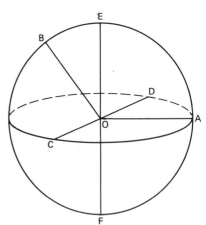

FIG. 16-49

The **radius** of a sphere is a straight line segment connecting any point of the sphere with its center. Thus, $\overline{OA}$ is a radius, and so is $\overline{OB}$.

A **diameter** of a sphere is a straight line segment which starts from a point of the sphere, passes through the center, and ends at a point on the sphere. Thus, $\overline{CD}$ is a diameter, and so is $\overline{EF}$.

Common Plane Figures and Common Solids

When a plane cuts through a sphere (see Fig. 16-50), the cut is a circular surface. If the cutting, or intersecting plane contains a diameter of the sphere, the circle formed is called a **great circle**. In all other cases, the circle formed is a **small circle**. The great circle cuts the sphere into two halves called **hemispheres**.

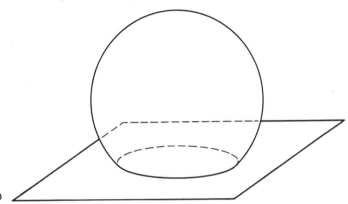

FIG. 16-50

The formula for the volume of a sphere is

$$V = \frac{4\pi r^3}{3}$$

where r is the radius of the sphere.

The formula for the surface area of a sphere is

$$A = 4\pi r^2$$

where r is the radius of the sphere.

EXAMPLE 16-27 How many gallons of water are there is a spherical container whose diameter is 44.0 in. ?

SOLUTION

$$V = \frac{4(3.142)(22.0)^3}{3}$$

$$= 44{,}608.021 \text{ in.}^3$$

$$44{,}608.021 \text{ in.}^3 = 44{,}608.021 \text{ in.}^3 \left(\frac{1 \text{ gal}}{231 \text{ in.}^3} \right)$$

$$= 193 \text{ gal}$$

EXAMPLE 16-28 Find the area of the surface of the sphere of Example 16-27.

SOLUTION

$$A = 4(3.142)(22.0)^2$$

$$= 6080 \text{ in.}^2$$

EXAMPLE 16-29 A sphere whose radius is 8.13 cm is ground and polished so that a flat circular area is formed. The radius of the small circle formed is 3.15 cm. Find the distance d from the center of the sphere to the center of the small circle (see Fig. 16-51).

Solids, Volumes, and Surface Areas

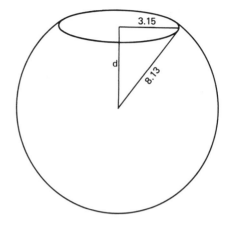

FIG. 16-51

SOLUTION

$$d = \sqrt{(8.13)^2 - (3.15)^2}$$
$$= \sqrt{56.2}$$
$$= 7.50 \text{ cm}$$

EXERCISE 16-5

1. How many cubic feet of concrete are there in a driveway 58-ft long, 16-ft wide, and 0.50- ft thick?
2. The cross section of a 5.00-m long brass bar is a square whose side is 3.00 cm. What is the weight of the bar if the density of brass is 8500 kg/m³?
3. An aquarium is 28.0-in. long, 16.0-in. wide, and 12.0-in. high. (a) How many gallons of water are there in the aquarium when it is filled to a level of 10.0 in.? (b) How many square inches of glass were used in making the aquarium? (Disregard the thickness of the glass.)
4. Find the weight of the compound rectangular solid of Fig. 16-52, if its dimensions

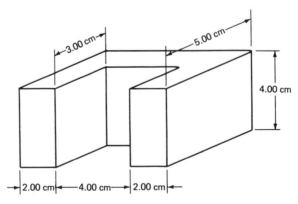

FIG. 16-52

Common Plane Figures and Common Solids

are as shown. The solid is made of aluminum, and the density of aluminum is 2.70 g/cm³.

5. The container of Fig. 16-53 is used to pour coolant into a lathe. Find the capacity of the container in cubic feet, if the radius of the cylinder and the cone is 7.0 in., the altitude of the cylinder is 16.0 in., and the altitude of the inverted cone is 6.0 in.

FIG. 16-53

6. Sand was dumped on the ground forming a conical pile having a radius of 24.0 ft, and an altitude of 12.0 ft. How many cubic yards of sand are there in the pile?

7. Find the capacity, in gallons, of a cylindrical tank having a diameter of 8.00 ft 8.00 in. and an altitude of 12.00 ft.

8. Find the number of cubic feet of steel in a solid steel cylinder that is 14 in. in diameter, and 32 ft 3 in. long.

9. How many cubic centimeters of steel were used in making a 4.25-m pipe whose outside diameter is 6.12 cm and whose inside diameter is 5.32 cm?

10. Find the lateral area of a right circular cylinder whose radius is 8 in., and whose altitude is 10 in.

11. Find the total area (lateral area and the area of the two bases) of a right circular cylinder with a radius of 8.2 ft, and an altitude of 10.4 ft.

12. Find the lateral area of a cone whose base is 8.0 in. in diameter, and whose altitude is 16 in.

13. A hollow spherical ball has an outside diameter of 15 in., and an inside diameter of 13 in. What is the volume of the metal used to construct the ball?

14. The hemispherical dome of a building has an outside diameter of 106.4 ft. Find the surface area of the dome.

15. To restore a medieval castle in Europe, the conical roof of one of its round towers had to be painted with a special protective silver paint. If the diameter of the conical roof base is 5.84 m, and its altitude is 9.76 m, what is the cost of gilding the roof if the rate of painting the roof is equivalent to $27.50/m²?

16. A steel sphere was ground and polished so that a flat circular surface was formed. If the radius r of the small circle is 3.8 cm, and the distance d from the center of the sphere to the center of the circle is 4.2 cm, what is the radius of the sphere? (See Fig. 16-54).

Solids, Volumes, and Surface Areas

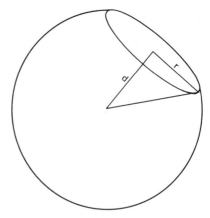

FIG. 16-54

REVIEW QUESTIONS

1. What is a polygon?
2. Are triangles polygons?
3. When are two triangles called congruent?
4. What is a perpendicular bisector, and what does it do to an isosceles triangle?
5. Which side of a right triangle is called the hypotenuse?
6. What is the common way of labeling a right triangle?
7. State the Pythagorean theorem, and write down the basic equation for it.
8. Write down the three forms of the Pythagorean theorem in which the basic equation is solved for a, b, and c, respectively.
9. When are two or more geometric figures called similar?
10. State the three rules about the similarity of two triangles.
11. What is the only requirement for two right triangles to be similar?
12. What is a quadrilateral?
13. When is a quadrilateral called a trapezoid?
14. When is a quadrilateral called a parallelogram?
15. When is a parallelogram called a rectangle?
16. When is a rectangle called a square?
17. What is the perimeter of a polygon?
18. Write down the formula for the area of a triangle whose base and altitude are known.
19. Write down the formula for the area of a triangle whose three sides are known.
20. Write down the formula for the area of a parallelogram.
21. Write down the formula for the area of a rectangle.
22. Write down the formula for the area of a square.
23. What is the radius of a circle?
24. What is the diameter of a circle?
25. What is the circumference of a circle?
26. What is an arc of a circle?
27. What is a central angle?

28. What is the ratio C/D equal to?
29. Give the formula for the volume of a rectangular solid.
30. Give the formula for the total surface area of a rectangular solid.
31. Give the formula for the volume of a right circular cylinder.
32. Give the formula for the lateral area of a right circular cylinder.
33. Give the formula for the volume of a hollow cylinder.
34. Give the formula for the volume of a right circular cone.
35. What is the slant height, and how can it be found?
36. Give the formula for the lateral area of a cone.
37. What is the radius of a sphere?
38. What is the diameter of a sphere?
39. What geometric figure is formed at the intersection of a sphere and a plane?
40. Give the formula for the volume of a sphere.
41. Give the formula for the surface area of a sphere.

REVIEW EXERCISES

1. Find the distance, d, across the two points A and B of the lake of Fig. 16-55.

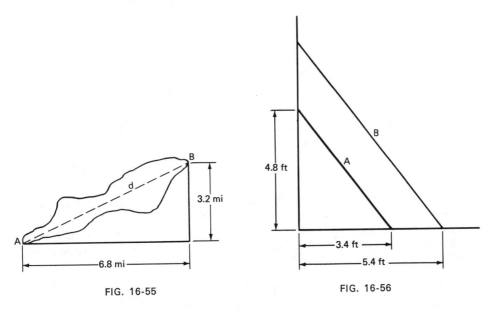

FIG. 16-55   FIG. 16-56

2. Two parallel bars were welded for reinforcement across the corner of a metal frame as shown in Fig. 16-56. Find the length of the longer bar, B.
3. Find the area of a triangular tract of land whose sides are 6.3 km, 4.8 km, and 5.1 km, respectively.
4. The radii of two concentric circles are $7\frac{1}{4}$ in. and $9\frac{3}{4}$ in., respectively. Find the area between the two concentric circles.
5. Find the volume of the compound figure of Fig. 16-57.

Review Exercises   291

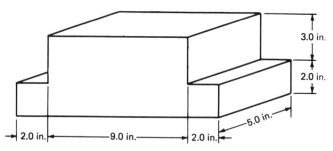

FIG. 16-57

6. Find the volume of the compound figure of Fig. 16-58.

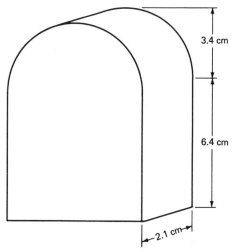

FIG. 16-58

7. An oil tank in the shape of an inverted cone has a base radius of 8.0 ft, and an altitude of 6.0 ft. If the tank is filled to the top, how many gallons will it hold?

8. A sphere with a 4.75-in. radius was machined so that a part of the sphere was removed and a flat circular surface was formed. The distance between the center of the circle formed and the center of the sphere is 3.50 in. Find the circumference of the circle formed.

BASIC TRIGONOMETRY

Trigonometry is the branch of mathematics that deals with finding the length of sides, and the size of angles of triangles. When we are finding the length of unknown sides, and the size of unknown angles of a triangle, we are said to be **solving** the triangle.

The basic trigonometry which we will study in the next three chapters developed from the need to determine, without actual measurement, the size of unknown parts of a triangle from the known parts of it. This kind of trigonometry, which is also known as numerical trigonometry, is used in the physical sciences, navigation, surveying, engineering, and other areas of technology. We will study first the trigonometry of right triangles, and then the trigonometry of oblique triangles. Of fundamental importance for the solution of both types of triangles are the **trigonometric ratios** and this is the topic we will start with.

Right Triangles

17-1 The Trigonometric Ratios

We have seen in Chapter 10 that a ratio is a comparison by division. Using this definition of a ratio, we say that the trigonometric ratios are comparisons of the lengths of the sides of a triangle. Since a triangle has three sides, and two sides can be taken at a time for a ratio, six trigonometric ratios are possible. Referring to the right triangle ABC of Fig. 17-1, it should be very easy for the student to see that any pair of sides of the triangle can form two different ratios. Thus,

—with the sides a and c, we can form the ratios a/c and c/a,
—with the sides b and c, we can form the ratios b/c and c/b, and
—with the sides a and b, we can form the ratios a/b and b/a.

NOTE: It is common practice in trigonometry, as in geometry, to label a right triangle with the method we used in the previous chapter, using the capital letters A and B for the acute angles and C for the right angle, and the lower case letters a opposite $\angle A$, b opposite $\angle B$, and c opposite the right angle C (see Fig. 17-1).

The trigonometric ratios always refer to one of the acute angles of the right triangle. We cannot talk about the trigonometric ratios of the triangle of Fig. 17-2, but only about the trigonometric ratios of one of the acute angles of the triangle, say angle A. If we refer to angle A, then side a is called **opposite** since it is opposite to angle A, and side b is called **adjacent** since it is "next to" angle A (see Fig. 17-2). Each

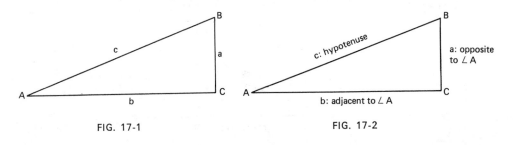

FIG. 17-1   FIG. 17-2

of the trigonometric ratios has a particular name. The definition, the name, and the abbreviation for each trigonometric ratio of angle A are given below:

—The ratio $\dfrac{\text{side opposite}}{\text{hypotenuse}}$, or $\dfrac{a}{c}$, is called **sine** of angle A, and is abbreviated sin A (read "sine A").

—The ratio $\dfrac{\text{side adjacent}}{\text{hypotenuse}}$, or $\dfrac{b}{c}$, is called **cosine** of angle A, and is abbreviated cos A (read "cosine A").

—The ratio $\dfrac{\text{side opposite}}{\text{side adjacent}}$, or $\dfrac{a}{b}$, is called **tangent** of angle A, and is abbreviated tan A (read "tangent A").

—The ratio $\dfrac{\text{side adjacent}}{\text{side opposite}}$, or $\dfrac{b}{a}$, is called **cotangent** of angle A, and is abbreviated cot A (read "cotangent A").

—The ratio $\dfrac{\text{hypotenuse}}{\text{side adjacent}}$, or $\dfrac{c}{b}$, is called **secant** of angle A, and is abbreviated sec A (read "secant A").

—The ratio $\dfrac{\text{hypotenuse}}{\text{side opposite}}$, $\dfrac{c}{a}$, is called **cosecant** of angle A, and is abbreviated csc A (read "cosecant A").

Using the above definitions, the trigonometric ratios for angle B, of the right triangle of Fig. 17-3 can be listed in the same way. The student should notice, however, that this time the "side opposite" is not side a as before, but side b, and the "side adjacent" is not side b, but side a (see Fig. 17-3). Thus, the trigonometric ratios of angle B of the right triangle of Fig. 17-3 are:

— $\dfrac{\text{side opposite}}{\text{hypotenuse}}$, or $\dfrac{b}{c}$, is sine of angle B; (sin B).

— $\dfrac{\text{side adjacent}}{\text{hypotenuse}}$, or $\dfrac{a}{c}$, is cosine of angle B; (cos B).

— $\dfrac{\text{side opposite}}{\text{side adjacent}}$, or $\dfrac{b}{a}$, is tangent of angle B; (tan B).

— $\dfrac{\text{side adjacent}}{\text{side opposite}}$, or $\dfrac{a}{b}$, is cotangent of angle B; (cot B).

— $\dfrac{\text{hypotenuse}}{\text{side adjacent}}$, or $\dfrac{c}{a}$, is secant of angle B; (sec B).

— $\dfrac{\text{hypotenuse}}{\text{side opposite}}$, or $\dfrac{c}{b}$, is cosecant of angle B; (csc B).

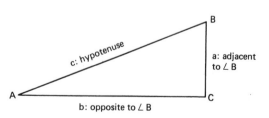

FIG. 17-3

Right Triangles

For the numerical trigonometry which we will study in this book, we do not need all six of the above trigonometric ratios; the first four ratios will be adequate. The student should memorize these four ratios and should be able to recognize them at once. To avoid confusion between the trigonometric ratios of angle A and angle B, the student should memorize them in a general form, that is in terms of side opposite, side adjacent, and hypotenuse, rather than in terms of side a, side b, and side c. Thus, for any acute angle R of a right triangle

$$\sin R = \frac{\text{side opposite}}{\text{hypotenuse}}$$

$$\cos R = \frac{\text{side adjacent}}{\text{hypotenuse}}$$

$$\tan R = \frac{\text{side opposite}}{\text{side adjacent}}$$

$$\cot R = \frac{\text{side adjacent}}{\text{side opposite}}$$

Let us now see these four ratios in specific numerical cases.

EXAMPLE 17-1 Find the four trigonometric ratios, sine, cosine, tangent, and cotangent, for angle A of the right triangle of Fig. 17-4.

SOLUTION

$$\sin A = \frac{\text{opposite}}{\text{hypotenuse}} = \frac{3 \text{ cm}}{5 \text{ cm}} = 0.6$$

$$\cos A = \frac{\text{adjacent}}{\text{hypotenuse}} = \frac{4 \text{ cm}}{5 \text{ cm}} = 0.8$$

$$\tan A = \frac{\text{opposite}}{\text{adjacent}} = \frac{3 \text{ cm}}{4 \text{ cm}} = 0.75$$

$$\cot A = \frac{\text{adjacent}}{\text{opposite}} = \frac{4 \text{ cm}}{3 \text{ cm}} = 1.3$$

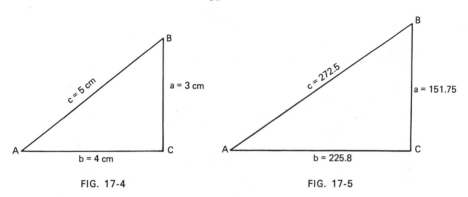

FIG. 17-4   FIG. 17-5

EXAMPLE 17-2 Given the right triangle of Fig. 17-5, find $\sin A$, $\cos A$, $\tan A$, and $\cot A$.

The Trigonometric Ratios

SOLUTION

$$\sin A = \frac{\text{opposite}}{\text{hypotenuse}} = \frac{151.75}{272.5} = 0.5569$$

$$\cos A = \frac{\text{adjacent}}{\text{hypotenuse}} = \frac{225.8}{272.5} = 0.8286$$

$$\tan A = \frac{\text{opposite}}{\text{adjacent}} = \frac{151.75}{225.8} = 0.6721$$

$$\cot A = \frac{\text{adjacent}}{\text{opposite}} = \frac{225.8}{151.75} = 1.4880$$

EXAMPLE 17-3 Given the right triangle of Fig. 17-6, find sin B, cos B, tan B, and cot B.

SOLUTION

$$\sin B = \frac{\text{opposite}}{\text{hypotenuse}} = \frac{6}{10} = 0.6$$

$$\cos B = \frac{\text{adjacent}}{\text{hypotenuse}} = \frac{8}{10} = 0.8$$

$$\tan B = \frac{\text{opposite}}{\text{adjacent}} = \frac{6}{8} = 0.75$$

$$\cot B = \frac{\text{adjacent}}{\text{opposite}} = \frac{8}{6} = 1.3$$

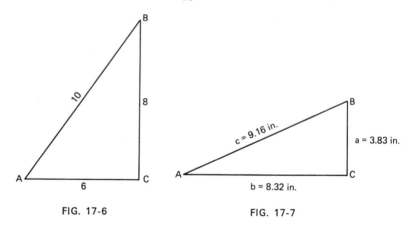

FIG. 17-6   FIG. 17-7

EXAMPLE 17-4 Given the right triangle of Fig. 17-7, find sin A, cos A, tan B, and cot B.

SOLUTION

$$\sin A = \frac{\text{opposite}}{\text{hypotenuse}} = \frac{3.83 \text{ in.}}{9.16 \text{ in.}} = 0.418$$

$$\cos A = \frac{\text{adjacent}}{\text{hypotenuse}} = \frac{8.32 \text{ in.}}{9.16 \text{ in.}} = 0.908$$

$$\tan B = \frac{\text{opposite}}{\text{adjacent}} = \frac{8.32 \text{ in.}}{3.83 \text{ in.}} = 2.172$$

$$\cot B = \frac{\text{adjacent}}{\text{opposite}} = \frac{3.83 \text{ in.}}{8.32 \text{ in.}} = 0.460$$

In each of the following exercises, the sides a, b, and c, of a right triangle ABC are given. Find in each case the trigonometric ratios as indicated. Make a sketch of the right triangle in each case.

1. $a = 16$, $b = 12$, $c = 20$. Find sin A, cos A, tan A, and cot A.
2. $a = 7.0$, $b = 24$, $c = 25$. Find sin B, cos B, tan B, and cot B.
3. $a = 7.02$, $b = 6.37$, $c = 9.48$. Find sin A, cos A, tan A, and cot A.
4. $a = 2.30$, $b = 7.70$, $c = 8.04$. Find sin A, cos A, tan A, and cot A.
5. $a = 2.12$, $b = 3.06$, $c = 3.72$. Find sin B, cos B, tan B, and cot B.
6. $a = 10.4$, $b = 5.57$, $c = 11.8$. Find sin A, cos A, tan A, and cot A.
7. $a = 3.17$, $b = 5.28$, $c = 6.17$. Find sin A, cos A, tan B, and cot B.
8. $a = 3.51$, $b = 6.31$, $c = 7.22$. Find sin B, cos B, tan A, and cot A.
9. $a = 6.21$, $b = 2.83$, $c = 6.82$. Find sin B, cos B, tan A, and cot A.
10. $a = 3.71$, $b = 9.44$, $c = 10.1$. Find sin A, cos A, tan B, and cot B.

17-2 The Trigonometric Ratios of an Angle. Trigonometric Tables

Consider the equilateral triangle whose sides are each 2.00 (see Fig. 17-8). We know from geometry that the bisector of one of the 60° angles bisects the opposite side also and forms two 30°–60° right triangles. Using the Pythagorean theorem, we find that the length of the bisector is $\sqrt{3}$, or 1.73 (approx.). Thus, the three sides of each of the two right triangles of Fig. 17-8 are 2.00, 1.00, and 1.73, respectively. We can now form the four trigonometric ratios of the specific 30° angle A of the right triangle ABC of Fig. 17-8. Thus,

$$\sin 30° = \frac{\text{opposite}}{\text{hypotenuse}} = \frac{1.00}{2.00} = 0.500$$

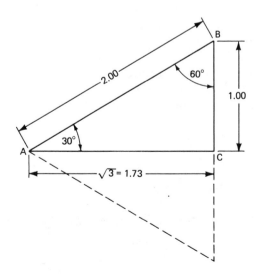

FIG. 17-8

The Trigonometric Ratios of an Angle. Trigonometric Tables

$$\cos 30° = \frac{\text{adjacent}}{\text{hypotenuse}} = \frac{1.73}{2.00} = 0.866$$

$$\tan 30° = \frac{\text{opposite}}{\text{adjacent}} = \frac{1.00}{1.73} = 0.577$$

$$\cot 30° = \frac{\text{adjacent}}{\text{opposite}} = \frac{1.73}{1.00} = 1.73$$

We can form the trigonometric ratios for the 60° angle of the above right triangle *ABC*. Thus,

$$\sin 60° = \frac{\text{opposite}}{\text{hypotenuse}} = \frac{1.73}{2.00} = 0.866$$

$$\cos 60° = \frac{\text{adjacent}}{\text{hypotenuse}} = \frac{1.00}{2.00} = 0.500$$

$$\tan 60° = \frac{\text{opposite}}{\text{adjacent}} = \frac{1.73}{1.00} = 1.73$$

$$\cot 60° = \frac{\text{adjacent}}{\text{opposite}} = \frac{1.00}{1.73} = 0.577$$

The student should notice in the above two cases of the ratios of the 30° and the 60° angle that the sine increased from 0.500 to 0.866 as the angle increased from 30° to 60°. The same happened to the tangent which increased from 0.577 to 1.73 as the angle increased from 30° to 60°. The cosine and the cotangent, however, decreased as the angle increased. Thus, the cosine decreased from 0.866 to 0.500 as the angle increased from 30° to 60°, and the cotangent decreased from 1.73 to 0.577 as the angle increased from 30° to 60°. We can say then that *the value of a trigonometric ratio depends on the size of the angle.* Because of this dependence of the variable ratio to the variable angle, we say that *the trigonometric ratio is a function of the angle.* In fact, the trigonometric ratios are more often called **trigonometric functions.**

Another important thing to notice in the trigonometric functions of the 30° angle and the 60° angle is that the value of sin 30°, which is 0.500, is the same as the value of cos 60°, which is also 0.500. Also, the value of tan 30°, which is 0.577, is the same as the value of cot 60°, which is also 0.577. Realizing that the 30° and the 60° angles are complementary angles, we can generalize the above fact by saying that *the function of any angle is equal to the cofunction of the complementary angle.* After the use of the trigonometric tables is introduced, the student will be able to see that

$$\sin 53° = \cos 37°$$

and

$$\tan 37° = \cot 53°$$

Let us now look at two more examples of the trigonometric functions of a 30° angle.

EXAMPLE 17-5 Find the four trigonometric functions of angle A in Fig. 17-9.

SOLUTION

$$\sin 30° = \frac{4.00}{8.00} = 0.500$$

$$\cos 30° = \frac{6.93}{8.00} = 0.866$$

$$\tan 30° = \frac{4.00}{6.93} = 0.577$$

$$\cot 30° = \frac{6.93}{4.00} = 1.73$$

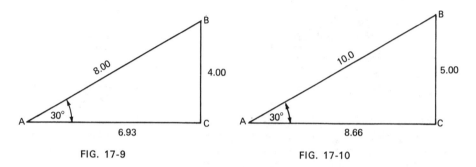

FIG. 17-9   FIG. 17-10

EXAMPLE 17-6 Find the four trigonometric functions of angle A in Fig. 17-10.

SOLUTION

$$\sin 30° = \frac{5.00}{10.00} = 0.500$$

$$\cos 30° = \frac{8.66}{10.00} = 0.866$$

$$\tan 30° = \frac{5.00}{8.66} = 0.577$$

$$\cot 30° = \frac{8.66}{5.00} = 1.73$$

The student must have noticed in the trigonometric functions of examples 17-5 and 17-6, that in both examples the sines have the same numerical value (0.500), and the same is true for the cosines (0.866), the tangents (0.577), and the cotangents (1.73) of both examples. In fact, this is the case for the trigonometric functions of the 30° angle of Fig. 17-8. This fact is of fundamental importance. Let us see first why this happens, and then why this fact is important.

The discussion will be simpler and more meaningful if we refer to the triple right triangle of Fig. 17-11. The three right triangles of Fig. 17-11 are the three right triangles of Figs. 17-8, 17-9, and 17-10, respectively.

The student should be able to recognize the three right triangles of Fig. 17-11 as similar right triangles, since $\angle$ A, the 30° angle is a common angle. The similarity

The Trigonometric Ratios of an Angle. Trigonometric Tables

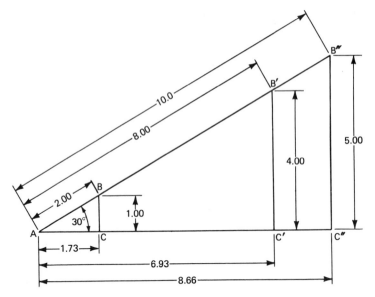

FIG. 17-11

of the three triangles explains why the sines of angle A of all three triangles have the same numerical value, and why the same is true for the values of the cosines, tangents, and cotangents. We know that because the right triangles ABC, $AB'C'$, and $AB''C''$ are similar triangles their sides are proportional. Therefore,

$$\sin 30° = \frac{\overline{BC}}{\overline{AB}} = \frac{\overline{B'C'}}{\overline{AB'}} = \frac{\overline{B''C''}}{\overline{AB''}} = \frac{1.00}{2.00} = \frac{4.00}{8.00} = \frac{5.00}{10.00} = 0.500$$

$$\cos 30° = \frac{\overline{AC}}{\overline{AB}} = \frac{\overline{AC'}}{\overline{AB'}} = \frac{\overline{AC''}}{\overline{AB''}} = \frac{1.73}{2.00} = \frac{6.93}{8.00} = \frac{8.66}{10.00} = 0.866$$

$$\tan 30° = \frac{\overline{BC}}{\overline{AC}} = \frac{\overline{B'C'}}{\overline{AC'}} = \frac{\overline{B''C''}}{\overline{AC''}} = \frac{1.00}{1.73} = \frac{4.00}{6.93} = \frac{5.00}{8.66} = 0.577$$

$$\cot 30° = \frac{\overline{AC}}{\overline{BC}} = \frac{\overline{AC'}}{\overline{B'C'}} = \frac{\overline{AC''}}{\overline{B''C''}} = \frac{1.73}{1.00} = \frac{6.93}{4.00} = \frac{8.66}{5.00} = 1.73$$

We are ready now to make a general and very important statement: *The lengths of the sides of a triangle do not affect the values of the trigonometric functions, and therefore for a given angle, the trigonometric functions of that angle are the same.*

Because the value of a trigonometric function for a given angle, sin 38°, for example, will be the same no matter what the lengths of the sides of the right triangles to which the given angle belongs, it becomes possible to construct trigonometric tables of angles. Such tables can be found in the appendices of this book. The values of the trigonometric functions in these tables have been computed for angles from 0° to 90°, to four decimal places.

To find the value of cos 34° 10′ in the Tables of Trigonometric Functions, follow down to 34° 10′ in the column to the extreme left with the heading "Degrees." Then follow across at this angle, and in the column under the heading "Cos," read 0.8274.

Right Triangles

The trigonometric functions in these tables are arranged so that the angles from 0° to 45° may be read downward in the extreme left column, and the angles from 45° to 90° may be read upward in the extreme right column. To locate the value of a given function, for angles from 0° to 45°, one should locate the angle and move across to the right using the name of the function which appears as heading. To locate the value of a given function for angles from 45° to 90°, one should locate the angle and move across to the left and use the name of the function which appears as the footing.

EXAMPLE 17-7 Find the value of sin 28° 40′.

SOLUTION Referring to the tables, locate the angle 28° 40′ in the left column. Then move to the right, and in the column under the heading "Sin," read 0.4797.

EXAMPLE 17-8 Find the value of tan 72° 50′.

SOLUTION Referring to the tables, locate the angle 72° 50′ in the right column. Then, move to the left, and in the column with the footing "Tan," read 3.2371.

The use of the tables is twofold:

1. Given an angle, we can find the value of any of its trigonometric functions.
2. Given the value of any trigonometric function of an angle, we can find the angle.

To find the angle when the value of a function is given, we simply reverse the procedure we have used.

EXAMPLE 17-9 Find A if sin A = 0.2840.

SOLUTION Referring to the tables, we first look for the value 0.2840 in the column having the heading "Sin." After we have located the value, we move to the left and read the angle 16° 30′. Thus, if sin A = 0.2840, A = 16° 30′.

There are cases (when the angle we are looking for is 45° or more), in which the correct name of the function will be the footing rather than the heading. In such cases, we should move across to the right rather than to the left to read the angle.

EXAMPLE 17-10 Find A if cot A = 0.4006.

SOLUTION Since the value 0.4006 is found in a column in which "Cot" is a footing, we move across to the right and read the angle 68° 10′. Thus, if cot A = 0.4006, then A = 68° 10′.

EXERCISE 17-2

Use the Tables of Trigonometric Functions to verify statements 1–12.

1. sin 31° 10′ = 0.5175
2. tan 43° 10′ = 0.9380
3. cos 20° 30′ = 0.9367
4. sin 68° 30′ = 0.9304
5. cot 77° 50′ = 0.2156
6. cos 21° 40′ = 0.9293
7. sin R = 0.3502, then R = 20° 30′
8. cot S = 0.9770, then S = 45° 40′
9. cos T = 0.6799, then T = 47° 10′
10. sin U = 0.9890, then U = 81° 30′
11. cos V = 0.9989, then V = 2° 40′
12. tan W = 2.0353, then W = 63° 50′

The Trigonometric Ratios of an Angle. Trigonometric Tables

Use the Tables of Trigonometric Functions to find the value of each of exercises 13–20.

13. cos 38° 40′   **14.** sin 53° 10′

15. tan 33° 50′   **16.** cos 69° 10′

17. cot 74° 10′   **18.** sin 72° 20′

19. tan 21° 30′   **20.** tan 44° 30′

Use the Tables of Trigonometric Functions to find angle R from the function of R given in each of exercises 21–26.

21. tan R = 0.4699   **22.** cos R = 0.3448

23. sin R = 0.6225   **24.** tan R = 0.9884

25. cot R = 1.0599   **26.** cos R = 0.7585

17-3 Interpolation

We must sometimes find the value to the nearest minute of a trigonometric function of an angle. We can use the same tables to do this. But, in such cases, the method of **interpolation** is used. Interpolation is a method of "reading between the lines." To interpolate, we use proportions.

EXAMPLE 17-11 Find the value of sin 37° 14′.

SOLUTION This angle is between 37° 10′ and 37° 20′. We can find from the tables that

$$\sin 37° 10′ = 0.6041$$

$$\sin 37° 20′ = 0.6065$$

Thus, the value for the sin 37° 14′ must be between 0.6041 and 0.6065. To show the differences we use the frame shown below.

$$10 \left[4 \left[\begin{array}{l} \sin 37° 10′ = 0.6041 \\ \sin 37° 14′ = \end{array} \right] x \\ \sin 37° 20′ = 0.6065 \end{array} \right] 0.0024$$

Now we set up the proportion

$$\frac{x}{0.0024} = \frac{4}{10}$$

and solving for x, we get

$$x = \frac{(4)(0.0024)}{10}$$

$$= 0.0010$$

Then

$$\sin 37° 14′ = 0.6041 + 0.0010$$

$$= 0.6051$$

EXAMPLE 17-12 Find the value of cos 73° 37′.

SOLUTION

$$10 \left[7 \left[\begin{array}{l} \cos 73° 30′ = 0.2840 \\ \cos 73° 37′ = \end{array} \right] x \\ \cos 73° 40′ = 0.2812 \end{array} \right] 0.0028$$

$$\frac{x}{0.0028} = \frac{7}{10}$$

$$x = \frac{(7)\,(0.0028)}{10}$$

$$= 0.0020$$

Then

$$\cos 73° 37' = 0.2840 - 0.0020$$

$$= 0.2820$$

The student must have noticed in the above two examples that in the case of the sine, the value of the function *increased* as the angle increased. This is why the value of x was added to the value of the function of the smaller angle. In the case of the cofunction cosine, however, the value of the function *decreased* as the angle increased. This is why the value of x was subtracted from the value of the function of the smaller angle.

EXAMPLE 17-13   Find the value of tan 66° 23′.

SOLUTION

$$10 \left[\begin{array}{l} 3 \left[\begin{array}{l} \text{tan } 66° 20' = 2.2817 \\ \text{tan } 66° 23' = \end{array} \right] x \\ \text{tan } 66° 30' = 2.2998 \end{array} \right] 0.0181$$

$$\frac{x}{0.0181} = \frac{3}{10}$$

$$x = 0.0054$$

Then

$$\text{tan } 66° 23' = 2.2817 + 0.0054$$

$$= 2.2871$$

EXAMPLE 17-14   Find the value of cot 27° 48′.

SOLUTION

$$10 \left[\begin{array}{l} 8 \left[\begin{array}{l} \text{cot } 27° 40' = 1.9074 \\ \text{cot } 27° 48' = \end{array} \right] x \\ \text{cot } 27° 50' = 1.8940 \end{array} \right] 0.0134$$

$$\frac{x}{0.0134} = \frac{8}{10}$$

$$x = 0.0107$$

Then

$$\text{cot } 27° 48' = 1.9074 - 0.0107$$

$$= 1.8967$$

What was said after example 17-12 about the sine and the cosine is also true about the tangent and the cotangent. Thus, in examples 17-13 and 17-14, the value of the function increased for the tangent, but decreased for the cotangent, as the angle increased. This is why the value of x was added to the value of the function of the smaller angle in the case of the tangent, but was subtracted from the value of the function of the smaller angle in the case of the cotangent.

EXAMPLE 17-15 Find A if $\sin A = 0.5341$.

SOLUTION

$$10 \left[x \left[\begin{matrix} \sin 32° \ 10' = 0.5324 \\ \sin A \qquad = 0.5341 \end{matrix} \right] 0.0017 \\ \quad\ \sin 32° \ 20' = 0.5348 \end{matrix} \right] 0.0024$$

$$\frac{x}{10} = \frac{0.0017}{0.0024}$$

$$x = 7$$

Then

$$A = 32° \ 10' + 7'$$

$$= 32° \ 17'$$

EXERCISE 17-3

Use interpolation to find the value of the function in exercises 1–8.

1. $\tan 28° \ 34'$ 2. $\sin 56° \ 27'$
3. $\cot 41° \ 48'$ 4. $\cos 49° \ 57'$
5. $\tan 39° \ 43'$ 6. $\sin 57° \ 38'$
7. $\cos 67° \ 14'$ 8. $\cot 17° \ 16'$

Use interpolation to find angle R in exercises 9–16.

9. $\cos R = 0.9940$ 10. $\tan R = 6.0080$
11. $\sin R = 0.2583$ 12. $\sin R = 0.8695$
13. $\cot R = 0.7893$ 14. $\cot R = 1.2153$
15. $\tan R = 0.9850$ 16. $\cos R = 0.6955$

17-4 Accuracy of Computed Results. Decimal Angles

Because the entries of the trigonometric tables, as well as the sides and angles of triangles in trigonometric problems, are approximate numbers, all computed results must be rounded according to Table 17-1.

TABLE 17-1   Corresponding accuracy of sides and angles

Accuracy of Side	Accuracy of Angle
1. Two significant digits	Nearest degree
2. Three significant digits	Nearest multiple of ten minutes
3. Four significant digits	Nearest minute

Trigonometric tables in some books, as well as calculators, express the angles in degrees and tenths of a degree instead of degrees and multiples of 10 minutes. For instance, 23.4°, 78.3°, 37.8°, and 81.6° are examples of this method of expressing the measure of an angle.

To change angles expressed in decimal degrees into degrees and minutes, multiply the decimal part of the degrees by 60. Thus, 5.3° = 5° 18′, since (0.3)(60) = 18. Also, 23.7° = 23° 42′, since (0.7)(60) = 42.

EXAMPLE 17-16 Change 58.9° to degrees and minutes.

SOLUTION Since (0.9)(60) = 54, 58.9° = 58° 54′.

To change angles expressed in degrees and minutes into decimal degrees, divide the minute part of the angle by 60. Thus, 23° 24′ = 23° + 24/60 = 23.4°, and 48° 38′ = 48° + 38/60 = 48.6333333 = 48.63°.

EXAMPLE 17-17 Change 31° 54′ to decimal degrees.

SOLUTION

$$31° \ 54′ = 31 + \frac{54}{60} = 31.9°$$

EXAMPLE 17-18 Change 83° 47′ to decimal degrees.

SOLUTION

$$83° \ 47′ = 83 + \frac{47}{60} = 83.78333333 = 83.78°$$

The value of a trigonometric function of an angle expressed in decimal degrees can be found by using either the trigonometric tables or by using a calculator.

To find the value of a trigonometric function of an angle expressed in decimal degrees using the tables, first change the decimal degrees into degrees and minutes, and then use the tables. Interpolate if necessary.

$$\sin 27.8° = \sin 27° \ 48′$$
$$= 0.4664$$

To find the value of a trigonometric function of an angle expressed in decimal degrees using a calculator, enter the numerical value of the angle, press the key of the function wanted, and read the answer. Thus, if you use a calculator,

$$\sin 27.8° = 0.46638664$$

This answer, of course, should be rounded to four significant digits, since the angle is accurate to the nearest minute. Then,

$$\sin 27.8° = 0.4664$$

EXAMPLE 17-19 Find cos 71.7° (a) using the Tables of Trigonometric Functions, and (b) using a calculator.

Accuracy of Computed Results. Decimal Angles

SOLUTION

(a) First we change the angle from decimal degrees to degrees and minutes. Thus, $71.7° = 71° \; 42'$. Then, using the tables and interpolation, $\cos 71° \; 42' = 0.3140$.

(b) First we enter 71.7 and press the "cos" key to get 0.31399246. Then rounding to four significant digits, we get $\cos 71.7° = 0.3140$.

EXERCISE 17-4

Change the angle measure in exercises 1–6 to degrees and minutes.

1. 17.3°	**2.** 67.1°	**3.** 84.2°
4. 33.6°	**5.** 2.8°	**6.** 11.5°

Use the Tables of Trigonometric Functions and interpolation or a calculator to find the value of the function in exercises 7–12. In rounding your results, refer to the Table 17-1.

7. tan 34.4°	**8.** sin 88.7°	**9.** sin 48.2°
10. cot 33.6°	**11.** cos 55.8°	**12.** tan 77.9°

Change the angle measure in exercises 13–20 to decimal degrees.

13. 71° 18′	**14.** 43° 42′	**15.** 73° 33′	**16.** 18° 21′
17. 66° 27′	**18.** 84° 38′	**19.** 31° 41′	**20.** 57° 17′

17-5 Solution of Right Triangles

A right triangle is said to be solved when all of its parts are determined. Using the trigonometric tables, *we can solve for the missing parts of a right triangle when either one of the acute angles and a side, or two sides are given.* Consider, for instance, the right triangle of Fig. 17-12. Here one angle and one side are given. Therefore, we can solve for the missing angle and the missing sides of this right triangle.

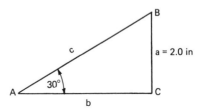

FIG. 17-12

We know from geometry that

$$B = 90° - A$$
$$= 90° - 30°$$
$$= 60°$$

To find c, we think as follows:
We know that

$$\sin A = \frac{a}{c} \qquad \textit{(definition of sine)}$$

$$\text{or} \quad \sin 30° = \frac{2.0}{c} \qquad \textit{(substituting 30° for A, and}$$
$$\textit{2.0 for a)}$$

$$0.500 = \frac{2.0}{c} \qquad \textit{(referring to the tables for}$$
$$\textit{the value of sin 30°)}$$

We recall at this point that the information of the tables, $\sin 30° = 0.500$, means that for any right triangle, the ratio of the side opposite a 30° angle to the side adjacent is 0.500, for any lengths of these two sides. Since this ratio and its value in this case form the equation

$$0.500 = \frac{2.0}{c}$$

side c has to be such that the equation is satisfied. It is very simple to solve the above equation, and $c = 4.0$ in.

To find b, we proceed as follows:

We know that

$$\cot A = \frac{b}{a}$$

$$\cot 30° = \frac{b}{2.0}$$

$$1.7321 = \frac{b}{2.0}$$

$$b = (2.0)(1.7321)$$

$$= 3.4642$$

$$= 3.5 \text{ in.}$$

Thus, $B = 60°$, $b = 3.5$ in., and $c = 4.0$ in.

The above discussion suggests the following steps when solving right triangle problems:

1. *Make a sketch, reasonably to scale, of the right triangle to be solved, and label it with the letters A, B, C, and a, b, c.*
2. *Show the known as well as the missing parts of the triangle.*
3. *Use the equation defining a trigonometric function which will include two known parts as well as the unknown part to be found.*
4. *Solve this simple equation for the unknown.*
5. *Find as many as possible of the missing parts of the triangle from the given parts rather than from the calculated parts.*

The student is reminded at this point that the table of corresponding accuracy of sides and angles (Table 17-1) should be kept in mind when solving triangles.

EXAMPLE 17-20 In a right triangle $A = 27°$, and $c = 7.0$ cm. Find a.

SOLUTION Figure 17-13 is a sketch of the right triangle of the problem. The triangle is labeled and the known parts as well as the side which is to be found are shown.

Solution of Right Triangles

The definition of the function involving the two known parts, angle A, and side c, as well as the unknown side a, is

$$\sin A = \frac{a}{c}$$

Substituting from the given data, and from the table, we can solve the equation for a. Thus,

$$0.4540 = \frac{a}{7.0}$$

$$a = (7.0)(0.4540)$$

$$= 3.2 \text{ cm}$$

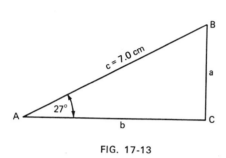

FIG. 17-13   FIG. 17-14

EXAMPLE 17-21   In a right triangle, $A = 38° 20'$, and $b = 8.00$ cm. Find c.

SOLUTION   Figure 17-14 is a sketch of the right triangle of the problem. The definition of the function involving the two known parts as well as the unknown part which is to be found is that of the cosine. Thus,

$$\cos A = \frac{b}{c}$$

$$\cos 38° 20' = \frac{8.00}{c}$$

$$c = \frac{8.00}{0.7844}$$

$$= 10.2 \text{ cm}$$

EXAMPLE 17-22   In a right triangle, $A = 77° 30'$, and $a = 19.4$ ft. Find b.

SOLUTION   It is evident, from Fig. 17-15, that we have two choices:

$$\tan 77° 30' = \frac{19.4}{b} \quad \text{or}$$

$$b = \frac{19.4}{\tan 77° 30'}$$

and

$$\cot 77° 30' = \frac{b}{19.4} \quad \text{or}$$

$$b = (19.4)(\cot 77° 30')$$

Right Triangles

Here, and in all other similar cases, we should prefer the choice involving multiplication which is simpler than division. Thus,

$$b = (19.4)(\cot 77° \ 30')$$
$$= (19.4)(0.2217)$$
$$= 4.30 \text{ ft}$$

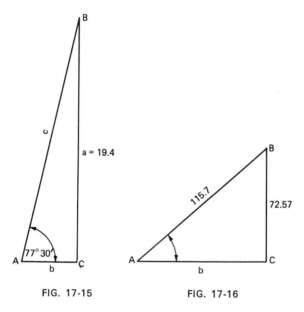

FIG. 17-15   FIG. 17-16

EXAMPLE 17-23   In a right triangle shown in Fig. 17-16, $a = 72.57$, and $c = 115.7$. Find A.

SOLUTION

$$\sin A = \frac{72.57}{115.7}$$
$$= 0.6271$$

Then we find from the tables the angle whose sine is 0.6271. We find it to be 38° 50′. Thus,

$$A = 38° \ 50'$$

EXAMPLE 17-24   Solve the right triangle in which $B = 61° \ 23'$ and $a = 31.17$ in.

SOLUTION   Figure 17-17 is a sketch of the right triangle of the problem. Since $B = 61° \ 23'$, $A = 90° - 61° \ 23'$, or $A = 28° \ 37'$. Solving for b, we have

$$\cot A = \frac{b}{a} \qquad \text{and}$$
$$b = (31.17)(\cot 28° \ 37')$$

By interpolation, $\cot 28° \ 37' = 1.8329$. Then,

$$b = (31.17)(1.8329)$$
$$= 57.13 \text{ in.}$$

Solution of Right Triangles

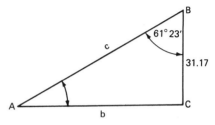

FIG. 17-17

Solving for c, we have

$$\sin A = \frac{a}{c} \quad \text{and}$$

$$c = \frac{31.17}{\sin 28° 37'}$$

By interpolation, $\sin 28° 37' = 0.4790$. Then,

$$c = \frac{31.17}{0.4790}$$

$$= 65.07 \text{ in.}$$

EXAMPLE 17-25 Solve the right triangle in which $a = 21.83$ and $c = 55.42$.

SOLUTION Figure 17-18 is a sketch of the right triangle of the problem. Solving for A, we have

$$\sin A = \frac{a}{c}$$

$$\sin A = \frac{21.83}{55.42}$$

$$= 0.3939$$

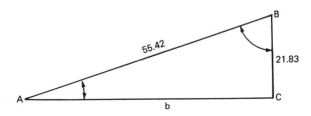

FIG. 17-18

By interpolation, $A = 23° 12'$. Since $A = 23° 12'$, $B = 90° - 23° 12'$ or $B = 66° 48'$. Solving for b, we have

$$\cot A = \frac{b}{a} \quad \text{and}$$

$$b = (21.83)(\cot 23° 12')$$

By interpolation, $\cot 23° 12' = 2.3332$. Then,

$$b = (21.83)(2.3332)$$

$$= 50.93.$$

EXAMPLE 17-26 Use the Tables of Trigonometric Functions or a calculator to solve the right triangle in which $A = 72.6°$ and $a = 11.21$.

SOLUTION   Figure 17-19 is a sketch of the right triangle of the problem.

$$B = 90° - 72.6°$$

$$= 17.4°$$

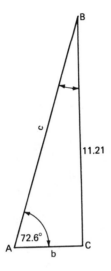

FIG. 17-19

Using tables:

$$\sin 72.6° = \sin 72° 36' = \frac{11.21}{c} \quad \text{and} \quad c = \frac{11.21}{\sin 72° 36'}$$

By interpolation, $\sin 72° 36' = 0.9542$. Then,

$$c = \frac{11.21}{0.9542}$$

$$= 11.75$$

$$\tan 72.6° = \tan 72° 36' = \frac{11.21}{b} \quad \text{and} \quad b = \frac{11.21}{\tan 72° 36'}$$

By interpolation, $\tan 72° 36' = 3.1910$. Then,

$$b = \frac{11.21}{3.1910}$$

$$= 3.513$$

Using a calculator:

$$\sin 72.6° = \frac{11.21}{c}$$

$$c = \frac{11.21}{\sin 72.6°}$$

$$= \frac{11.21}{0.9542}$$

$$= 11.75$$

Solution of Right Triangles

$$\tan 72.6° = \frac{11.21}{b}$$

$$b = \frac{11.21}{\tan 72.6°}$$

$$= \frac{11.21}{3.1910}$$

$$= 3.513$$

EXERCISE 17-5

Use the Tables of Trigonometric Functions to solve the right triangle in exercises 1–22. Interpolate wherever needed. Round answers to the accuracy justified by the data.

1. $A = 18°$, $b = 23$
2. $A = 64°$, $a = 8.2$
3. $B = 22°$, $c = 0.92$
4. $A = 38° \ 20'$, $a = 2.18$
5. $A = 43° \ 30'$, $b = 376$
6. $B = 24° \ 10'$, $c = 28.6$
7. $A = 47° \ 50'$, $a = 42.3$
8. $B = 57° \ 40'$, $b = 1.72$
9. $B = 51° \ 10'$, $c = 11.2$
10. $A = 78° \ 40'$, $b = 0.917$
11. $A = 28° \ 18'$, $a = 42.33$
12. $B = 13° \ 29'$, $c = 16.25$
13. $B = 47° \ 52'$, $a = 71.26$
14. $A = 43° \ 37'$, $b = 3142$
15. $B = 61° \ 43'$, $b = 6047$
16. $a = 41.38$, $c = 62.65$
17. $b = 4.12$, $c = 5.11$
18. $a = 0.723$, $b = 0.888$
19. $a = 1.293$, $b = 2.084$
20. $a = 0.5148$, $b = 0.4037$
21. $b = 13.17$, $c = 16.06$
22. $a = 7.018$, $b = 5.176$

Use either the Tables of Trigonometric Functions or a calculator to solve each of right triangles 23–32. Round answers to the accuracy justified by the data.

23. $A = 23.2°$, $c = 48.72$
24. $A = 67.7°$, $b = 3.142$
25. $B = 74.6°$, $c = 700.0$
26. $B = 18.1°$, $a = 1.214$
27. $A = 34.9°$, $a = 6127$
28. $B = 9.7°$, $c = 31.16$
29. $A = 26.3°$, $a = 16.81$
30. $B = 31.3°$, $c = 7234$
31. $A = 28.4°$, $b = 1.444$
32. $B = 57.8°$, $b = 0.6032$

17-6 Angles of Elevation and Depression. Applications

The straight line (or ray) from an observer O, to a point P, is called the **line of sight** (see Fig. 17-20). *The angle formed between the line of sight and the horizontal is called the* **angle of elevation**, *if the line of sight is above the horizontal, and the* **angle of depression**, *if the line of sight is below the horizontal.* Thus, $\angle e$ of Fig. 17-20(a) is an angle of elevation, and $\angle d$ of Fig. 17-20(b) is an angle of depression.

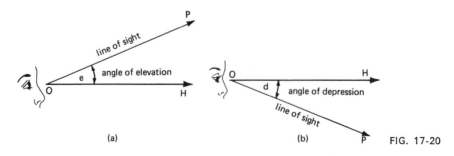

(a)   (b)   FIG. 17-20

Right Triangles

EXAMPLE 17-27 The angle of elevation of the balloon shown in Fig. 17-21 from a point O on the ground is $57°\ 20'$. How high is the balloon if the point directly under it is 143 m from point O?

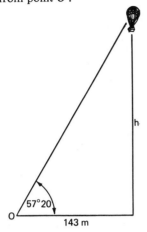

FIG. 17-21

SOLUTION

$$\tan 57°\ 20' = \frac{h}{143}$$

$$h = (143)(1.5597)$$

$$= 223\ \text{m}$$

EXAMPLE 17-28 A lighthouse keeper, 352 ft above sea level, is looking from the window of the lighthouse at a distant boat. The angle of depression d of the boat is $22°\ 10'$. The situation is shown in Fig. 17-22. How far is the boat from the lighthouse keeper?

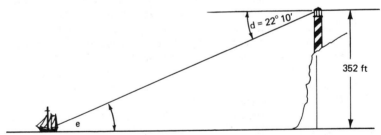

FIG. 17-22

SOLUTION Since $\angle\ d$ and $\angle\ e$ are alternate interior angles, $\angle\ e$ is also $22°\ 10'$. Then,

$$\sin 22°\ 10' = \frac{352}{c}$$

$$c = \frac{352}{0.3773}$$

$$= 933\ \text{ft}$$

EXAMPLE 17-29 To find the width $\overline{PQ}$ of a river, a civil engineer measured a distance $\overline{QR}$, perpendicular to $\overline{PQ}$, of 110.0 ft along the bank of the river. Then he found with a transit that $\angle\ R = 38°\ 52'$. What is the width of the river? (See Fig. 17-23.)

Angles of Elevation and Depression. Applications

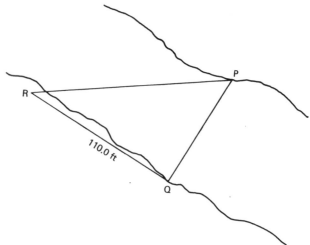

R

P

Q

110.0 ft

FIG. 17-23

SOLUTION

$$\tan R = \frac{\overline{PQ}}{110.0}$$

$$\tan 38° \ 52' = \frac{\overline{PQ}}{110.0}$$

By interpolation, tan 38° 52′ = 0.8059. Then

$$\overline{PQ} = (110.0)(0.8059)$$

$$= 88.65 \text{ ft}$$

EXERCISE 17-6

Use the Tables of Trigonometric Functions to solve each of the right triangle problems 1–13. Interpolate when needed. Round answers according to Table 17-1.

1. A boy flying a kite let out 85 m of string. Determine the height of the kite if the angle of elevation of the kite is 41° 10′, and the string is assumed to be a straight line.

2. From a point on the ground 130 ft from the foot of a flagpole, the angle of elevation of the top of the flagpole is 38° 20′. Find the height of the flagpole.

3. Each of the two equal sides of an isosceles triangle is 78.26 in. The base is 123.42 in. Determine the angles of the triangle.

4. The base of an isosceles triangle is 20.84 cm, and a base angle is 48° 17′. Find the length of each equal side, and the altitude of the triangle.

5. A ladder leans against the side of a building. The foot of the ladder is 9.3 ft away from the foot of the wall, and the ladder makes an angle of 68° with the ground. Assuming that the foot of the ladder and the foot of the wall are on the same level, find (a) how high from the ground is the top of the ladder, and (b) how long is the ladder?

6. A ladder is leaning against the side of a building. The ladder is 32.2-ft long, and reaches a height of 28.6 ft. What is the angle formed by the ladder and the wall?

7. A radio tower is supported at a height of 86 m by two cables of equal length which are anchored at the two points *C* and *D* (see Fig. 17-24). The distances $\overline{CB}$ and $\overline{BD}$ are equal, and points *C* and *D* are on the same level with the base of the tower. The angles the two cables form with the ground at *C* and *D* are 43° each. Find the lengths $\overline{AC}$ and $\overline{BC}$.

Right Triangles

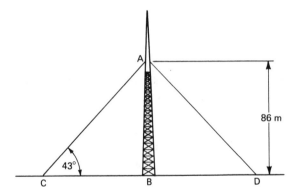

FIG. 17-24

8. The angle of depression to the airport, from an airplane flying at 14,800 ft is 19.6°. Find the distance D between the point below the airplane and the airport, assuming that the ground is level. The situation is shown in Fig. 17-25.

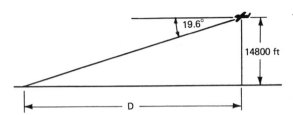

FIG. 17-25

9. From a balloon 1230 ft above the ground, the angles of depression to points on the near and the far banks of a river are 43° 40′ and 37° 20′, respectively. Determine the width of the river between these two points.

10. A radio tower is 623-ft high. At what distance from the center of the base of the tower will the angle of elevation to the top of the tower be 38° 40′?

11. A chord of a circle is 64.26 cm. If the central angle intercepting the chord is 63° 22′, find the radius of the circle, and the perpendicular distance from the center of the circle to the chord.

12. Find the height of a man who casts a shadow of 7.32 ft, when the angle of elevation of the sun is 39.2°.

13. The angle of elevation of the top of a building from a point 216.3 ft from the base of the building is 38.6°. There is a flagpole on the edge of the roof, and the angle of elevation of the top of the flagpole from the same point, 216.3 ft from the base of the building, is 44.5°. What is the height of the flagpole?

REVIEW QUESTIONS

1. If ratios are comparisons by division, what do the trigonometric ratios compare?

2. How many trigonometric ratios are possible?

3. Is it incorrect to talk about a certain trigonometric ratio of a right triangle, or about a certain trigonometric ratio of an acute angle of the right triangle?

4. For the right triangle ABC, labeled with the method adopted in this book, define sin A, cos A, tan A and cot A in terms of a, b, and c.

5. For any right triangle ABC, define sin A, cos A, tan A, and cot A in terms of opposite, adjacent, and hypotenuse.

6. For any right triangle ABC, define sin B, cos B, tan B, and cot B in terms of a, b, and c.

7. State the three entries of the table of corresponding accuracy of sides and angles.

8. State the five steps suggested for the solution of right triangles.

9. Define angles of elevation and depression.

REVIEW EXERCISES

Use the Tables of Trigonometric Functions to verify the statements 1–8.

1. cot 81° 50′ = 0.1435
2. sin 47° 20′ = 0.7353
3. cos 24° 40′ = 0.9088
4. tan 8° 10′ = 0.1435
5. tan M = 0.2401, then M = 13° 30′
6. sin N = 0.9315, then N = 68° 40′
7. cot P = 0.2278, then P = 77° 10′
8. cos Q = 0.5471, then Q = 56° 50′

Use the Tables of Trigonometric Functions to find the value for exercises 9–12.

9. cos 17° 20′
10. tan 64° 30′
11. cot 48° 10′
12. sin 81° 40′

Use the Tables of Trigonometric Functions to find angle K from the function of K given in exercises 13–16.

13. cos K = 0.8704
14. tan K = 1.6212
15. cot K = 0.8050
16. sin K = 0.5995

In exercises 17–20, use interpolation to find the function value of the angle.

17. sin 46° 34′
18. cot 18° 12′
19. tan 69° 14′
20. cos 76° 47′

In exercises 21–24, use interpolation to find angle K.

21. cot K = 1.7954
22. cos K = 0.9764
23. tan K = 3.5776
24. sin K = 0.7705

Change the angle measure in exercises 25–28 to degrees and minutes.

25. 21.4°
26. 32.6°
27. 77.8°
28. 44.1°

Use the Tables of Trigonometric Functions and interpolation to find the value of the function in exercises 29–32. In rounding results, use Table 17-1.

29. cos 33.7°
30. sin 63.3°
31. tan 41.8°
32. cot 18.4°

Change the angle measure in exercises 33–36 to decimal degrees.

33. 72° 17′
34. 21° 34′
35. 56° 23′
36. 41° 29′

Use the Tables of Trigonometric Functions to solve the right triangle in exercises 37–44. Interpolate wherever needed. Round answers to the accuracy justified by the data.

37. $A = 29°$, $a = 32$
38. $B = 68° 20′$, $a = 417$
39. $A = 34° 18′$, $b = 31.23$
40. $B = 27.2°$, $c = 17.16$

41. $B = 52.6°$, $b = 1.814$ **42.** $A = 47.7°$, $c = 132.2$

43. $a = 2.214$, $c = 3.018$ **44.** $b = 731$, $c = 818$

Use the Tables of Trigonometric Functions to solve each of the right triangle problems 45–48. Interpolate when needed. Round answers according to Table 17-1.

45. Figure 17-26 shows a radio tower. Two cables, AC and AD, starting from a point A on the tower 700 ft above the ground are firmly anchored at the two points C and D forming angles with the ground of $46°$ and $56°$, respectively. The tower and the two cables are in the same vertical plane, and the base of the tower is in the same level with the points C and D. Find the lengths of the cables, and the distance between the points C and D.

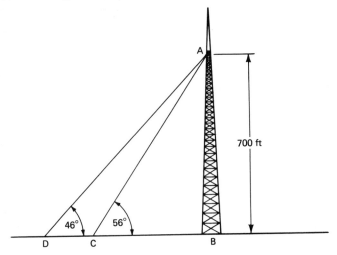

FIG. 17-26

46. The angles of elevation of the top of a radio tower for two observers C and D are $32° \ 20'$ and $38° \ 50'$, respectively (see Fig. 17-27). If the tower is 425-ft high, how far apart are the two observers? Assume that the two observers and the tower are in the same vertical plane, and on the same level ground.

47. Find the length l of the metal plate of Fig. 17-28.

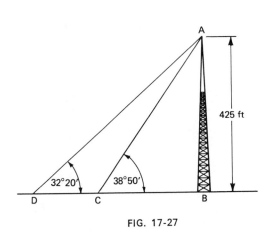

FIG. 17-27

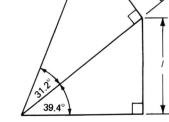

FIG. 17-28

Review Exercises

48. The angles of depression of two boats observed by a lighthouse keeper 184 ft above sea level are 23.8° and 27.7°, respectively. Assuming that the lighthouse and the two boats are in the same vertical plane, find the distance between the two boats (see Fig. 17-29).

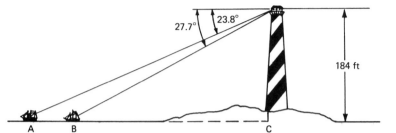

FIG. 17-29

18

The General
Trigonometric Angle.
Radian Measure

18-1 The Trigonometric Angle

We have defined the angle in geometry as the union of two rays with a common end point. *In trigonometry, the angle is defined as an amount of rotation of a ray.* Thus, if we rotate $\overrightarrow{AB}$ in Fig. 18-1 around point A, from its initial position along $\overrightarrow{AB}$ to its terminal position along $\overrightarrow{AC}$, the trigonometric angle t is generated. The side of a trigonometric angle from which the rotating ray starts is called the **initial side**, the side where the rotating ray stops is called the **terminal side**, and the fixed point around which the ray rotates is called the **vertex**. The curved arrow shows the direction in which the rotating ray moved.

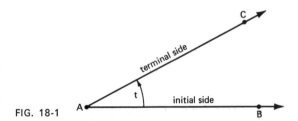

FIG. 18-1

When the rotating ray moves in a **counterclockwise** direction, the generated angle is a **positive** angle; when the rotating ray moves in a **clockwise** direction, the generated angle is a **negative** angle. Figure 18-2(a), (b), (c), and (d) show a positive angle of 120°, a negative angle of 160°, a positive angle of 530°, and a negative angle of 380°, respectively.

When the vertex of a trigonometric angle is at the origin of the rectangular coordinate system, and its initial side coincides with the positive x-axis, the angle is said to be in **standard position**. Thus, angles s and t of Fig. 18-3 are in standard position.

When the terminal sides of two trigonometric angles coincide, the angles are said to be **coterminal angles**. Thus, the 230° and the −130° angles of Fig. 18-4 are coterminal angles. Also, a 75° angle and a 435° angle are coterminal angles.

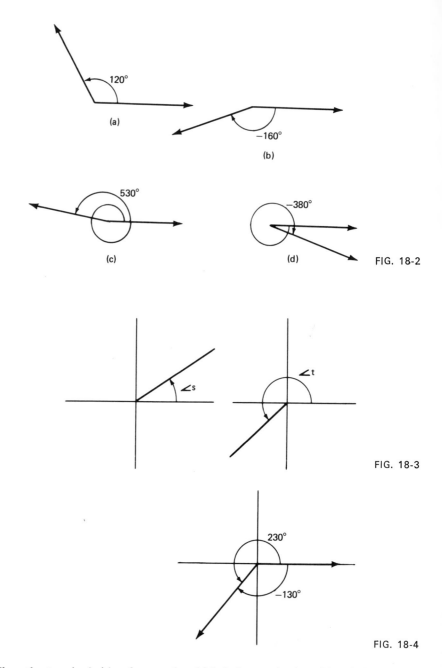

(a)

(b)

(c)

(d)

FIG. 18-2

FIG. 18-3

230°

-130°

FIG. 18-4

When the terminal side of an angle which is in standard position is in the first quadrant, the angle is said to be a **first quadrant angle**. When the terminal side of an angle in standard position is in the second quadrant, the angle is said to be a **second quadrant angle**, and so on for a **third** and a **fourth quadrant angle**. Figure 18-5(a), (b), (c), and (d), show first, second, third, and fourth quadrant angles, respectively.

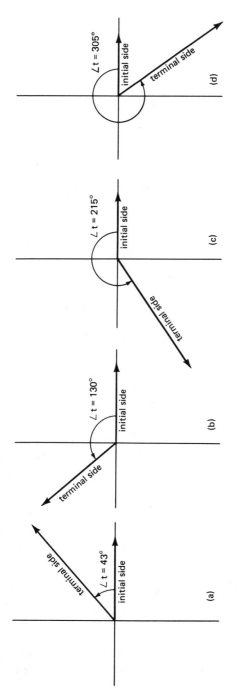

FIG. 18-5

When the terminal side of an angle in standard position coincides with one of the axes, the angle is called a **quadrantal** angle. A 180° angle, for instance, is a quadrantal angle.

EXERCISE 18-1

1. Which of the two angles of Fig. 18-6, $\angle a$ or $\angle b$, is a trigonometric angle?

FIG. 18-6

2. Which of the two angles of Fig. 18-7 is positive, and which is negative?

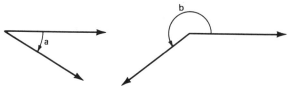

FIG. 18-7

3. Referring to Fig. 18-8, state which of the angles $\angle AOB$, $\angle BOC$, $\angle COD$, and $\angle EFG$ are in standard position?

4. Which quadrant angles are $\angle s$ and $\angle t$ of Fig. 18-9?

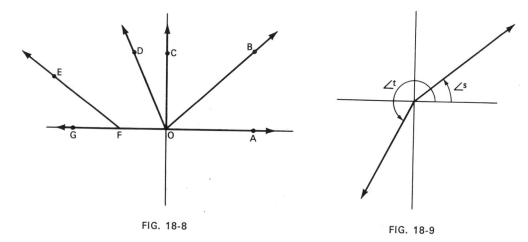

FIG. 18-8

FIG. 18-9

18-2 The Reference Angle

The Tables of Trigonometric Functions we used in solving right triangles give values for angles between 0°–90°. Soon, however, in solving oblique triangles and vector problems, we will have to deal with angles which are larger than 90°, and the question is how do we find the values of the functions of such angles. We will see in this section

The General Trigonometric Angle. Radian Measure

that any function of an angle from 0°–360° can be expressed as a function of an angle between 0°–90°, and that the value of any function of such an angle can be determined from the Tables of Trigonometric Functions.

Before we can use our tables to find the value of a given function of an angle larger than 90°, the concept of the **reference angle** must be introduced, and the functions of a trigonometric angle of any quadrant should be defined.

The reference angle a of a trigonometric angle t in standard position, is the positive acute angle between the terminal side of angle t and the x-axis.

According to this definition, the reference angle of the first quadrant angle t of Fig. 18-10(a) is the angle t itself. The reference angles for a second quadrant angle t, a third quadrant angle t, and a fourth quadrant angle t, are shown in Fig. 18-10(b), (c), and (d), respectively.

Referring to Fig. 18-10, it should be easy to see that to find the reference angle of an angle t,

1. *subtract t from 180° if angle t is in the second quadrant,*
2. *subtract 180° from t if angle t is in the third quadrant,*
3. *subtract t from 360° if angle t is in the fourth quadrant.*

EXAMPLE 18-1 Find the reference angle of each of the following angles:

(a) 260°   (b) 173°   (c) 327° 20′

SOLUTION

(a) A 260° angle is a third quadrant angle. The reference angle of a 260° angle is

$$260° - 180° = 80°$$

an 80° angle.

(b) A 173° angle is a second quadrant angle. The reference angle of a 173° angle is

$$180° - 173° = 7°$$

a 7° angle.

(c) A 327° 20′ angle is a fourth quadrant angle. The reference angle of a 327° 20′ angle is

$$360° - 327° 20′ = 32° 40′$$

a 32° 40′ angle.

In Fig. 18-11, four trigonometric angles t_1, t_2, t_3, and t_4, and their reference angles a_1, a_2, a_3, and a_4 are shown. In all four cases, a perpendicular line from a point P of the terminal side of the angle to the x-axis is dropped, forming a **reference triangle** for the reference angle in that triangle.

It should be evident from the reference triangles of Fig. 18-11, that for any of the four trigonometric angles t, the distance of point P from the origin is $r = \sqrt{x^2 + y^2}$. It must be evident also that r will always be positive while the signs of x and y depend on the quadrant in which P lies.

The functions of any trigonometric angle t, in terms of x, y, and r are defined as follows:

$$\sin t = \frac{y}{r} \qquad \tan t = \frac{y}{x}$$

$$\cos t = \frac{x}{r} \qquad \cot t = \frac{x}{y}$$

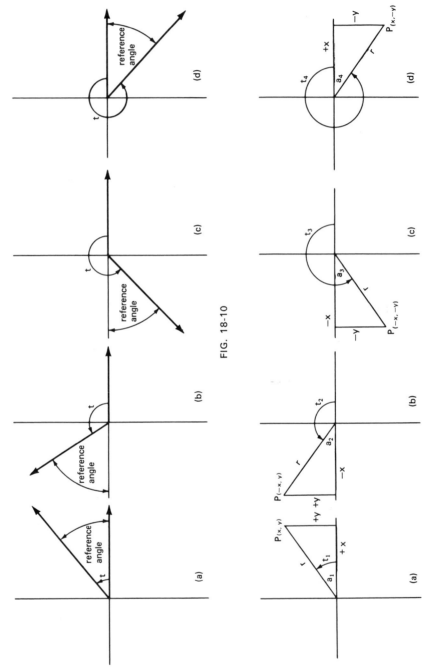

FIG. 18-10

FIG. 18-11

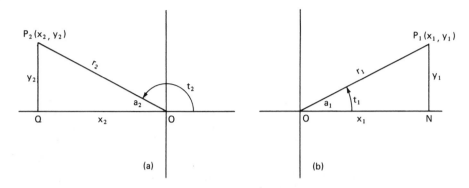

FIG. 18-12

Consider now in Fig. 18-12(a) the second quadrant angle t_2, its reference angle a_2, and its reference triangle OP_2Q. Then, consider the first quadrant angle t_1 of Fig. 18-12(b). Its reference angle a_1 has been constructed so that $\angle a_1 = \angle a_2$, and $r_1 = r_2$. Then the reference triangles OP_1N and OP_2Q are congruent and

$$x_1 = x_2, \qquad y_1 = y_2, \qquad \text{and} \qquad r_1 = r_2$$

Because, however, the value of x in the second quadrant is negative, x_2 of Fig. 18-12(a) is negative.

Expressing angle t_2 as $180° - a_2$, using the definitions of the functions for the trigonometric angle t, and remembering that $\angle a_1 = \angle a_2$, we see from Fig. 18-12 that

$$\sin t_2 = \sin (180° - a_2) = \frac{y_2}{r_2} = \frac{y}{r} = \sin a_1$$

$$\cos t_2 = \cos (180° - a_2) = \frac{x_2}{r_2} = \frac{-x}{r} = -\cos a_1$$

$$\tan t_2 = \tan (180° - a_2) = \frac{y_2}{x_2} = \frac{y}{-x} = -\tan a_1$$

$$\cot t_2 = \cot (180° - a_2) = \frac{x_2}{y_2} = \frac{-x}{y} = -\cot a_1$$

Because similar arguments can be made for a third quadrant angle t, and a fourth quadrant angle t, we can make the following general statement:

The value of any trigonometric function of an angle is numerically equal to the same function of its reference angle. The sign of the value of a function for a given angle is determined by the signs of x and y of the quadrant in which the terminal side of the given angle lies.

Since side r of a reference triangle is always positive, and side x and side y are positive, or negative according to the quadrant in which the reference triangle is (see

The Reference Angle

327

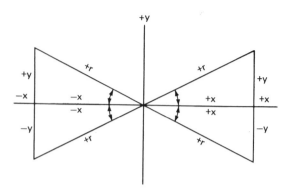

FIG. 18-13

TABLE 18-1 Algebraic signs of the functions of the general angle in the four quadrants

Function	First Quadrant	Second Quadrant	Third Quadrant	Fourth Quadrant
$\sin t = \dfrac{y}{r}$	$\dfrac{+y}{+r} = +$	$\dfrac{+y}{+r} = +$	$\dfrac{-y}{+r} = -$	$\dfrac{-y}{+r} = -$
$\cos t = \dfrac{x}{r}$	$\dfrac{+x}{+r} = +$	$\dfrac{-x}{+r} = -$	$\dfrac{-x}{+r} = -$	$\dfrac{+x}{+r} = +$
$\tan t = \dfrac{y}{x}$	$\dfrac{+y}{+x} = +$	$\dfrac{+y}{-x} = -$	$\dfrac{-y}{-x} = +$	$\dfrac{-y}{+x} = -$
$\cot t = \dfrac{x}{y}$	$\dfrac{+x}{+y} = +$	$\dfrac{-x}{+y} = -$	$\dfrac{-x}{-y} = +$	$\dfrac{+x}{-y} = -$

Fig. 18-13), we can construct a table (see Table 18-1) showing the algebraic signs of the functions of any angle in the four quadrants.

The following is a three-step procedure for finding the value of a function of any angle:

To find the value of the function of a given angle,

1. *Find the quadrant and the size of the reference angle of the given angle.*
2. *Find the numerical value of the function of the given angle by looking at the tables for the value of the same function of its reference angle.*
3. *Attach to the numerical value obtained from the tables, the proper algebraic sign by referring to Table 18-1, or by considering the signs of x and y in the quadrant of the angle.*

EXAMPLE 18-2 Find cos 125°.

SOLUTION

1. The 125° angle is a second quadrant angle, and its reference angle is 180° − 125° = 55° (see Fig. 18-14).
2. From the tables,

$$\cos 55° = 0.5736$$

328 *The General Trigonometric Angle. Radian Measure*

3. Since $\cos t = x/r$, by definition, and x in the second quadrant is negative, $-x/r$ will give a negative sign. Therefore,

$$\cos 125° = -0.5736$$

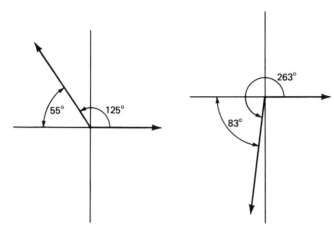

FIG. 18-14   FIG. 18-15

EXAMPLE 18-3 Find tan 263°.

SOLUTION

1. The 263° angle is a third quadrant angle, and its reference angle is $263° - 180° = 83°$ (see Fig. 18-15).
2. From the tables

$$\tan 83° = 8.1443$$

3. Since $\tan t = y/x$, by definition, and both x and y are negative in the third quadrant, $-y/-x$ will give a positive sign. Therefore,

$$\tan 263° = 8.1443$$

EXAMPLE 18-4 Find cot 314° 20′.

SOLUTION

1. The 314° 20′ angle is a fourth quadrant angle, and its reference angle is $360° - 314° 20′ = 45° 40′$ (see Fig. 18-16).

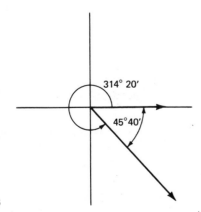

FIG. 18-16

2. From the tables

$$\cot 45° \, 40' = 0.9770$$

3. Since $\cot t = x/y$, by definition, and x is positive and y is negative in the fourth quadrant, $x/-y$ will give a negative sign. Therefore,

$$\cot 314° \, 20' = -0.9770$$

Now that reference angles are understood, it should be easy for the student to see that *the values of a given trigonometric function of two coterminal angles are the same*. The two angles have the same reference angle and they are angles of the same quadrant. Thus, 170° and 530° are coterminal angles, and

$$\sin 170° = \sin 530° = \sin 10° = 0.1736$$

We can find now the values of the functions of the quadrantal angles 0°, 90°, 180°, and 270°. (The values of the functions of the 0° and the 360° angle are the same.)

The 0° angle is shown in Fig. 18-17. Notice that when $\angle t = 0°$, the reference

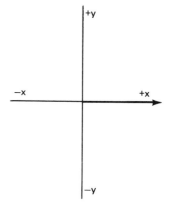

FIG. 18-17

triangle has no y side, because r and x coincide, and therefore, they have the same length. Thus, when $\angle t = 0°$, $x = r$, and $y = 0$. Then,

$$\sin 0° = \frac{y}{r} = \frac{0}{r} = 0$$

$$\cos 0° = \frac{x}{r} = \frac{r}{r} = 1$$

$$\tan 0° = \frac{y}{x} = \frac{0}{x} = 0$$

$$\cot 0° = \frac{x}{y} = \frac{x}{0} = \text{undefined}$$

Thinking in the same manner about the other three quadrantal angles, we can form Table 18-2.

The General Trigonometric Angle. Radian Measure

TABLE 18-2   Values of quadrantal angles

∠t	0°	90°	180°	270°
sin t	0	1	0	−1
cos t	1	0	−1	0
tan t	0	undefined	0	undefined
cot t	undefined	0	undefined	0

EXERCISE 18-2

For each of angles 1–14, find the reference angle.

1. 135°
2. 238°
3. 326°
4. 98°
5. 193° 20′
6. 217° 50′
7. 343° 40′
8. 111° 10′
9. 138° 17′
10. 253° 38′
11. 338° 53′
12. 172° 13′
13. 535°
14. 560°

Use the tables and find the values of the functions of the angles in exercises 15–26.

15. sin 160°
16. tan 210°
17. cos 295°
18. cot 263° 10′
19. cos 142° 40′
20. sin 184° 10′
21. tan 272° 30′
22. cos 164° 18′
23. cot 241° 36′
24. tan 323° 48′
25. sin 370°
26. tan 580°

18-3 Radian Measure

In this section, we will learn about another unit for angular measurement called the **radian**. The radian is useful for its properties, and more advanced mathematics as well as many technical problems are greatly simplified when the radian rather than the degree is used.

Consider the circle of Fig. 18-18 of radius r, and the central angle θ* which intercepts an arc AB. We say that the central angle θ is one radian if the length of arc AB is equal to the length r of the radius. Thus, *a radian is an angle which when made into a central angle by placing its vertex at the center of a circle will intercept an arc equal in length to the radius of the circle.*

It follows from the definition of the radian that the number of radians in a circle is equal to the quotient of the division of the circumference by the radius. Since $C = 2\pi r$,

$$\text{circumference} = \frac{2\pi r}{r} = 2\pi \text{ rad}$$

*The Greek letter θ (read "theta") is used in advanced mathematics as the symbol for angles.

Radian Measure

331

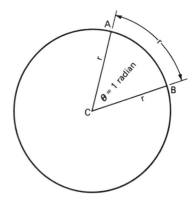

FIG. 18-18

But since circumference $= 360°$, we have the following relation of radians and degrees:

$$2\pi \text{ rad} = 360°$$

or

$$\pi \text{ rad} = 180° \qquad (a)$$

From relation (a), we obtain the following two forms:

$$1 \text{ rad} = \frac{180°}{\pi} = 57.3° \text{ (approx.)} \qquad (b)$$

$$1° = \frac{\pi}{180°} = 0.01745 \text{ rad (approx.)} \qquad (c)$$

EXAMPLE 18-5   Change 3.3 rad to degrees.

SOLUTION   Since by (b), 1 rad $= 57.3°$, 3.3 rad $= 3.3(57.3°) = 189.1°$.

EXAMPLE 18-6   Change 57.2° to radians.

SOLUTION   Since by (c), $1° = 0.01745$ rad, $57.2° = 57.2(0.01745) = 0.9981$ rad.

In converting the more common angles from degrees to radians, and from radians to degrees, the forms

$$1 \text{ rad} = \frac{180°}{\pi}$$

and

$$1° = \frac{\pi}{180°} \text{ rad}$$

of the equations (b) and (c) are used. Usually when converting degrees to radians, the result is left in terms of π. *When an angle like $\pi/3$ occurs without units, it is understood that the units are radians.*

Thus,

$$30° = 30°\left(\frac{\pi}{180°}\right)\text{rad}$$

$$= \frac{\pi}{6}\text{ rad or just }\frac{\pi}{6}$$

Also

$$\frac{\pi}{2} = \frac{\pi}{2}\left(\frac{180°}{\pi}\right)$$

$$= 90°$$

The above two examples suggest the following two guides in changing from degrees to radians, and from radians to degrees:

1. *To change degrees to radians in terms of π, multiply the number of degrees by π/180°.*
2. *To change radians in terms of π to degrees, multiply the number of radians by 180°/π.*

EXAMPLE 18-7   Change 150° to radians in terms of π.

SOLUTION

$$150° = 150°\left(\frac{\pi}{180°}\right) = \frac{5\pi}{6}\text{ rad or }\frac{5\pi}{6}$$

EXAMPLE 18-8   Change 3π/2 rad to degrees.

SOLUTION

$$\frac{3\pi}{2}\text{ rad} = \frac{3\pi}{2}\left(\frac{180°}{\pi}\right) = 270°$$

EXERCISE 18-3

In exercises 1–6, change the given angle from radians to degrees. Use 1 radian = 57.3°.
1. $\theta = 2$ **2.** $\theta = 4.3$ **3.** $\theta = 2.8$
4. $\theta = 5.7$ **5.** $\theta = 3.7$ **6.** $\theta = 6.8$
In exercises 7–12, change the given angle from degrees to radians. Use 1° = 0.01745 radians.
7. $\theta = 86.7°$ **8.** $\theta = 108.3°$ **9.** $\theta = 28.8°$
10. $\theta = 194°$ **11.** $\theta = 923.1°$ **12.** $\theta = 204.8°$
In exercises 13–20, change the given angle from radians to degrees.
13. $\theta = \frac{\pi}{4}$ **14.** $\theta = \frac{3\pi}{2}$ **15.** $\theta = \frac{2\pi}{3}$ **16.** $\theta = \frac{13\pi}{9}$

17. $\theta = 3\pi$ **18.** $\theta = \frac{5\pi}{6}$ **19.** $\theta = \frac{10\pi}{3}$ **20.** $\theta = \frac{5\pi}{2}$

In exercises 21–30, change the given angle to radian leaving your answers in terms of π.
21. $\theta = 60°$ **22.** $\theta = 45°$ **23.** $\theta = 270°$ **24.** $\theta = 150°$
25. $\theta = 120°$ **26.** $\theta = 170°$ **27.** $\theta = 85°$ **28.** $\theta = 36°$
29. $\theta = 75°$ **30.** $\theta = 12°$

18-4 Applications of Radian Measure

We have seen that a central angle of one radian intercepts an arc one radius long. Therefore, a central angle of two radians will intercept an arc two radii long, and so on. This fact makes the radian the quotient of two lengths measured in the same kind of unit, that is, a proper ratio. It also makes the length of an arc s of a circle (see Fig. 18-19) proportional to the central angle intercepting it, and we can write the equation

$$s = r\theta \qquad\qquad (d)$$

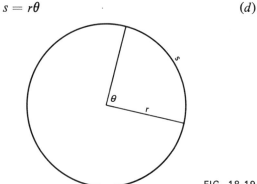

FIG. 18-19

Thus, the length s of an intercepted arc of a circle is equal to the product of the radius r of the circle and the number of radians in the intercepting angle θ.

EXAMPLE 18-9 If the radius of a circle is 25.0-ft long, find the length of the arc intercepted by a 54° central angle.

SOLUTION First we have to change the angle θ from degrees to radians. Thus,

$$\theta = 54° = 54°\left(\frac{\pi}{180°}\right) = \frac{3\pi}{10}$$

Then,

$$s = r\theta$$
$$= 25.0 \text{ ft}\left(\frac{3\pi}{10}\right)$$
$$= 7.5\pi \text{ ft}$$
$$= 23.6 \text{ ft} \qquad \textit{(substituting 3.14 for π and multiplying)}$$

The distance s which a point P on the rim of the rotating wheel of Fig. 18-20 covers is called **linear distance**.

The **linear speed**, v, of point P of the rotating wheel of Fig. 18-20 is the time rate of s; that is, s/t. The units of linear speed, v, are ft/sec, m/sec, mi/hr, etc. The formula for linear speed is

$$v = \frac{s}{t} \qquad\qquad (e)$$

The distance θ which the radius r of the rotating wheel of Fig. 18-20 covers is

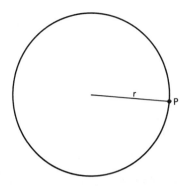

FIG. 18-20

called **angular distance** or **angular displacement**. Angular distance is measured in degrees, in revolutions, and in radians.

The **angular speed** ω* of the radius r of the rotating wheel of Fig. 18-20 is the time rate of θ, that is θ/t. The units of angular speed, ω, are radians/sec, revolutions/sec, and revolutions/min (RPM). The formula for angular speed is

$$\omega = \frac{\theta}{t} \qquad (f)$$

If we divide both sides of formula (d), $s = r\theta$, by t we get

$$\frac{s}{t} = \frac{r\theta}{t}$$

But $s/t = v$, by (e), and $\theta/t = \omega$, by (f).
Therefore, $s/t = r\theta/t$ becomes

$$v = r\omega \qquad (g)$$

Formula (g) says that *the linear speed v of a point on a revolving radius is equal to the product of the angular speed ω (in radians), and the distance r of the point from the center of rotation.*

In using formula (g), the linear units of v and r should be converted to the same kind of units if they are of different kinds.

EXAMPLE **18-10**   Convert the angular speed of 900 RPM into an angular speed of radians/min.

SOLUTION   Since there are 2π rad/rev,

$$\omega = \left(900 \ \frac{\text{rev}}{\text{min}}\right)\left(2\pi \ \frac{\text{rad}}{\text{rev}}\right)$$

$$= 1800 \ \pi \ \frac{\text{rad}}{\text{min}}$$

*The Greek letter ω (read "omega") is used in physics and in mathematics as the symbol for angular speed.

Applications of Radian Measure

335

EXAMPLE 18·11 A flywheel with a radius of 5.00 ft has an angular speed of 6 RPM. Find the linear speed of a point (a) on the rim of the flywheel; (b) 3.00 ft from the center of the wheel.

SOLUTION

(a)
$$\omega = \left(6\,\frac{\text{rev}}{\text{min}}\right)\left(2\pi\,\frac{\text{rad}}{\text{rev}}\right)$$

$$= 12\pi\,\frac{\text{rad}}{\text{min}}$$

Then,
$$v = r\omega$$

$$= 5\,\text{ft}(12\pi)$$

$$= 60\pi\,\frac{\text{ft}}{\text{min}}$$

$$= 188.5\,\frac{\text{ft}}{\text{min}}$$

(b)
$$\omega = 12\pi\,\frac{\text{rad}}{\text{min}}$$

Then,
$$v = r\omega$$

$$= 3\,\text{ft}(12\pi)$$

$$= 36\pi\,\frac{\text{ft}}{\text{min}}$$

$$= 113.1\,\frac{\text{ft}}{\text{min}}$$

EXAMPLE 18-12 A point on the rim of a wheel whose radius is 5 ft is moving with a linear speed of 25 ft/sec. Find the angular speed of the wheel.

SOLUTION From formula (g), $\omega = v/r$. Therefore,
$$\omega = \frac{25\,\text{ft/sec}}{5\,\text{ft}}$$

$$= 5\,\text{rad/sec}$$

The student should notice in the above example that length of arc in ft, over length of radius in ft, is radians, by definition. Because, however, radians are a proper ratio (feet over feet, in this case), they will not appear automatically as a unit next to the numerical result at the end of a problem like example 18-12, and they have to be put there. *Radians should be supplied as a unit at the end of a problem in which the formula* $\omega = v/r$ *and dimensional analysis are used. Also, radians should be removed from the units of the result of the problem in which either of the formulas* $s = r\theta$, *or* $v = \omega r$ *and dimensional analysis are used* (see examples 18-9 and 18-11).

EXAMPLE 18-13 The angular speed of the flywheel of a motor is 1500 RPM. Find its angular displacement in 3 sec.

SOLUTION First we change RPM to rad/sec.
$$\omega = 1500\,\frac{\text{rev}}{\text{min}}\left(\frac{2\pi\,\text{rad}}{\text{rev}}\right)\left(\frac{1\,\text{min}}{60\,\text{sec}}\right)$$

$$= 50\pi \frac{\text{rad}}{\text{sec}}$$

$$= 157 \frac{\text{rad}}{\text{sec}}$$

Then from formula (f), we get the formula for the angular displacement,

$$\theta = \omega t$$

Then,

$$\theta = \omega t$$

$$= 157 \frac{\text{rad}}{\text{sec}}(3 \text{ sec})$$

$$= 471 \text{ rad}$$

or

$$471 \text{ rad}\left(\frac{1 \text{ rev}}{2\pi \text{ rad}}\right) = 75 \text{ rev}$$

EXERCISE 18-4

In exercises 1–5, the radius r and the central angle θ of the circle are given. Find the length of the intercepted arc.
1. $r = 5.0$ in., $\theta = 240°$
2. $r = 3.0$ ft, $\theta = 108°$
3. $r = 2.0$ m, $\theta = 66°$
4. $r = 9$ in., $\theta = 75°$
5. $r = 15$ cm, $\theta = 36°$
6. Change 2400 RPM to rad/sec.
7. Change 8 radians to revolutions.
8. Change 420 revolutions to radians.
9. A grindstone makes 1200 RPM. The diameter of the grindstone is 8.0 in. Determine the linear speed of a point on the rim of the grindstone.
10. A point on the rim of a wheel is moving with a linear speed of 15 ft/sec. The diameter of the wheel is 4.0 ft. Find the angular speed of the wheel.
11. A point on the rim of a wheel moves with a linear speed of 18 m/sec. The diameter of the wheel is 4.0 m. Find the angular speed of the wheel in (a) rad/sec, (b) RPM.
12. A flywheel has an angular speed of 1200 RPM. Find its angular displacement in 5 sec.
13. The wheel of a bicycle is 28 in. in diameter. If the speed of the bicycle is 32 ft/sec, what is the angular speed of the wheel, (a) in rad/sec; (b) in RPM?
14. The spoke of a wheel of a bicycle makes 240 RPM. If the distance from the center of the wheel to the outer end of the spoke is 13.0 in., find the linear speed (a) of the outer end of the spoke, (b) of a point of the spoke 6.0 in. from the center of the wheel.

18-5 Variation of sin θ and cos θ

A circle with its center at the origin and with a radius OP equal to 1 is shown in Fig. 18-21(a). Such a circle is called a **unit circle**. For the triangle OPQ, $\sin \theta = y/1 = y$, and $\cos \theta = x/1 = x$. This means that $y = \sin \theta$, and $x = \cos \theta$.

When $\angle \theta = 0$ and the radius $\overline{OP}$ coincides with the x-axis, the length of $\overline{PQ} = 0$, $y = 0$, and $\sin \theta = 0$. As, however, $\angle \theta$ increases, the length of $\overline{PQ}$ increases, y

Variation of sin θ and cos θ

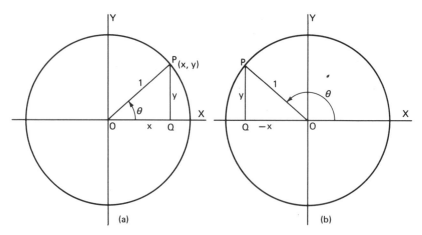

FIG. 18-21

increases, and sin θ increases. When $\angle\theta = 90°$, the radius $\overline{OP}$ coincides with the
y-axis, $\overline{PQ} = 1$, $y = 1$, and sin $\theta = 1$. We can say then that *the value of sin θ increases
as $\angle\theta$ increases from 0° to 90°, and that the maximum value sin θ can take is 1.* As the
radius $\overline{OP}$ continues to rotate from 90° to 180° [see Fig. 18-21(b)], $\angle\theta$ increases, $\overline{PQ}$
decreases, y decreases, and sin θ decreases from 1 to 0. In the same manner, we can
trace the variations of sin θ in the third and the fourth quadrant to find that sin θ
varies from 0 to −1 from 180° to 270°, and varies from −1 to 0 from 270° to 360°.

In a similar way we can trace the variations of cos θ by the variations of $\overline{OQ}$ as
radius $\overline{OP}$ rotates through the four quadrants and $\angle\theta$ increases from 0° to 360°. The
variations of sin θ and cos θ are shown in Table 18-3.

TABLE 18-3

Variation of θ in Degrees	Variation of θ in Radians	Variation of $y = \sin\theta$	Variation of $y = \cos\theta$
0° to 90°	0 to $\frac{\pi}{2}$	0 to 1	1 to 0
90° to 180°	$\frac{\pi}{2}$ to π	1 to 0	0 to −1
180° to 270°	π to $\frac{3\pi}{2}$	0 to −1	−1 to 0
270° to 360°	$\frac{3\pi}{2}$ to 2π	−1 to 0	0 to 1

Since, as we have seen, the values of the functions of coterminal angles are the
same, the value of sin θ will be equal to the value of sin $[\theta + n(360°)]$, where n is an
integer. Angle θ and angle $[\theta + n(360°)]$ are coterminal angles. The variations of
Table 18-3 then will be repeated as many times as the radius $\overline{OP}$ of Fig. 18-21 repeats
the 360° "trip." Since this repetition does not occur for an angle smaller than 360°,
we say that 360° or 2π is the **period** for the functions sine and cosine, and that sine and
cosine are **periodic functions**.

A "picture" of the variations of the sine can be obtained by drawing its curve. If we let x instead of θ stand for the angle in the relation $y = \sin x$, we will generate (x, y) ordered pairs of corresponding values. These pairs plotted in a rectangular coordinate system will produce the sine curve. First we make the table of corresponding values of x and y in intervals of 30°, or $\pi/6$ radians as shown in Table 18-4.

TABLE 18-4

x in degrees	0	30	60	90	120	150	180	210	240	270	300	330	360
x in radians	0	$\dfrac{\pi}{6}$	$\dfrac{\pi}{3}$	$\dfrac{\pi}{2}$	$\dfrac{2\pi}{3}$	$\dfrac{5\pi}{6}$	π	$\dfrac{7\pi}{6}$	$\dfrac{4\pi}{3}$	$\dfrac{3\pi}{2}$	$\dfrac{5\pi}{3}$	$\dfrac{11\pi}{6}$	2π
$y = \sin x$	0	0.50	0.87	1	0.87	0.5	0	−0.5	−0.87	−1	−0.87	−0.5	0

Notice that convenient values of x are shown in terms of both degrees and radians. The radian value of x is obtained by multiplying the corresponding value in degrees by $\pi/180$. Thus $60° = 60(\pi/180) = \pi/3$, and $270° = 270(\pi/180) = 3\pi/2$. The values of y are obtained from the Tables of Trigonometric Functions. Thus, $60° = \pi/3 = 1.0472$ rad, and $\sin 60°$, or the sine of 1.0472 rad, is 0.866 or 0.87. The ordered pairs of Table 18-4 are plotted to produce the curve shown in Fig. 18-22.

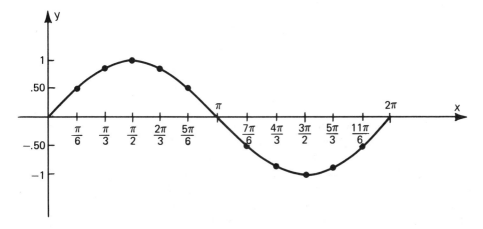

FIG. 18-22

As we expect, the period of the curve of $y = \sin x$ is 2π. This part of the curve is called the **cycle** of the curve. The maximum distance of the curve from the x-axis is called the **amplitude** of the curve, and this value for the curve $y = \sin x$ is 1.

Because the sine curve has a wave form, it is used extensively in electrical

The Curves of $y = \sin x$ *and* $y = \cos x$

engineering and in the study and description of wave motion of light and sound in physics.

In exactly the same manner, we can draw the curve of the cosine. The ordered pairs of corresponding values are shown in Table 18-5, and the curve in Fig. 18-23.

TABLE 18-5

x in degrees	0	30	60	90	120	150	180	210	240	270	300	330	360
x in radians	0	$\dfrac{\pi}{6}$	$\dfrac{\pi}{3}$	$\dfrac{\pi}{2}$	$\dfrac{2\pi}{3}$	$\dfrac{5\pi}{6}$	π	$\dfrac{7\pi}{6}$	$\dfrac{4\pi}{3}$	$\dfrac{3\pi}{2}$	$\dfrac{5\pi}{3}$	$\dfrac{11\pi}{6}$	2π
$y = \cos x$	1	0.87	0.50	0	−0.50	−0.87	−1	−0.87	−0.50	0	0.50	0.87	1

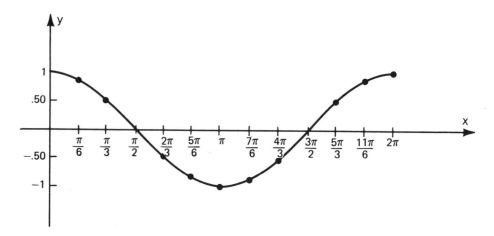

FIG. 18-23

The student should notice that the sine and cosine curves are of the same wave shape. They differ only by $\pi/2$ units, and if the sine curve is shifted $\pi/2$ units to the left, the two curves will coincide.

We will call the five points $(0, 0)$, $(\pi/2, 1)$, $(\pi, 0)$, $(3\pi/2, -1)$ and $(2\pi, 0)$ of the curve $y = \sin x$ *the five important points of the curve.* The five important points of the curve $y = \cos x$ are $(0, 1)$, $(\pi/2, 0)$, $(\pi, -1)$, $(3\pi/2, 0)$, and $(2\pi, 1)$. Notice first that the y-coordinates of the five important points for both curves are equal to either the amplitude ($+1$ or -1), or to zero. Notice also, that the x-coordinates of these five points for both curves are 0, $\pi/2$, π, $3\pi/2$, and 2π, and that they are located on the x-axis at intervals equal to $\frac{1}{4}$ of the period 2π. The expression "the five important points of the curve" will be used in the next two sections when we will be talking about the sketching of curves.

In the next two sections, we will be referring to the two equations $y = \sin x$, and $y = \cos x$ as the basic equations, to distinguish them from their variations.

The General Trigonometric Angle. Radian Measure

The curve of the equation $y = A \sin Bx$ is another wave shape curve different from that of the basic equation $y = \sin x$. Let us consider the three equations

$$y = 3 \sin x \qquad y = \sin 2x \qquad y = 3 \sin 2x$$

In $y = 3 \sin x$, $A = 3$ and $B = 1$. In $y = \sin 2x$, $A = 1$ and $B = 2$. In $y = 3 \sin 2x$, $A = 3$ and $B = 2$.

One cycle of the curves of each of the three equations $y = 3 \sin x$, $y = \sin 2x$, and $y = 3 \sin 2x$ is shown in Fig. 18-24, Fig. 18-25, and Fig. 18-26, respectively. In all the cases, the curve of the basic equation $y = \sin x$ is included for comparison.

Notice that the amplitude of the graph of $y = 3 \sin x$ (Fig. 18-24) is 3; that is

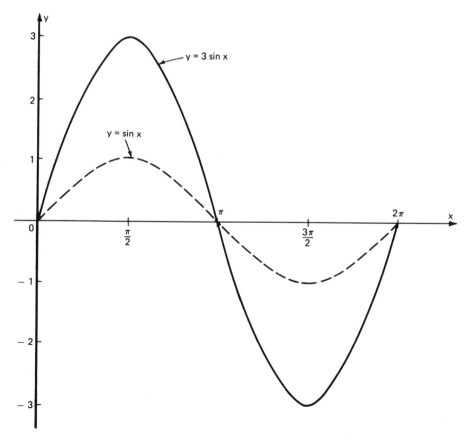

FIG. 18-24

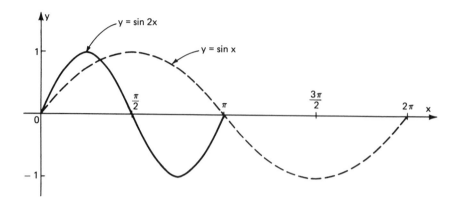

FIG. 18-25

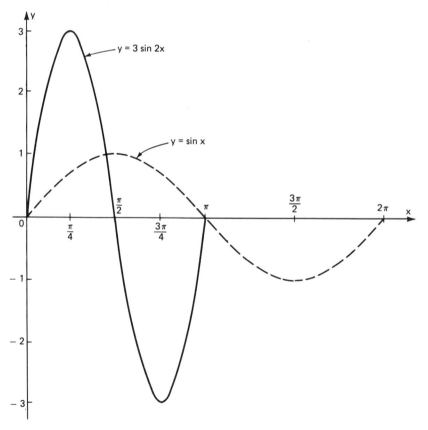

FIG. 18-26

three times that of the basic equation $y = \sin x$. The period for both $y = \sin x$ and $y = 3 \sin x$ is 2π.

Notice also that the period of the graph of $y = \sin 2x$ (Fig. 18-25) is π, that is half that of the curve of the basic equation $y = \sin x$ which is 2π.

In the curve of $y = 3 \sin 2x$ (Fig. 18-26), the amplitude is three times that of the basic equation $y = \sin x$, and the period is half that of the basic equation $y = \sin x$.

The above examples suggest that *the curve of an equation of the type* $y = A \sin Bx$ *has an amplitude equal to A and a period equal to* $2\pi/B$.

The above statement applies also for the curve of the equation of the type $y = A \cos Bx$, in which also the amplitude is A and the period is $2\pi/B$.

We can summarize now by giving the following *steps for sketching the curve for an equation of the type* $y = A \sin Bx$, *or* $y = A \cos Bx$.

1. *Complete the list:*
   a. *Amplitude:* $A =$
   b. *Period:* $2\pi/B =$
2. a. *Using the origin as the first point, mark four more points on the x-axis at intervals equal to $\frac{1}{4}$ of the period of step 1.b.*
   b. *Using information for the amplitude from step 1.a, plot the five important points of the curve corresponding to the five points of the x-axis.*
   c. *Connect these five points with a smooth curve to obtain the general wave shape of the curve of the basic sine or cosine equation.*

EXAMPLE **18-14**   Sketch one cycle of the curve $y = 3 \cos 2x$.

SOLUTION

1. a. Amplitude: $A = 3$
   b. Period: $2\pi/B = 2\pi/2 = \pi$
2. a. Starting with the point $(0, 0)$, the other four points on the x-axis, at intervals of $\pi/4$ are $\pi/4$, $\pi/2$, $3\pi/4$, π. They are shown in Fig. 18-27.
   b. The five important points of the curve $y = 3 \cos 2x$, corresponding to the five important points of the x-axis, are also shown in Fig. 18-27.
   c. The curve $y = 3 \cos 2x$ results when these five points are connected with a smooth curve (see Fig. 18-27).

EXAMPLE **18-15**   Sketch the curve $y = 2 \sin \frac{2}{3}x$.

SOLUTION

1. a. Amplitude: $A = 2$
   b. Period: $2\pi/B = 2\pi/\frac{2}{3} = 3\pi$.
2. a. The other four points on the x-axis, at intervals equal to $\frac{1}{4}$ of 3π (see step 1.b) are
   $$3\pi \cdot \frac{1}{4} = \frac{3\pi}{4}, \ 3\pi \cdot \frac{2}{4} = \frac{3\pi}{2}, \ 3\pi \cdot \frac{3}{4} = \frac{9\pi}{4}, \text{ and } 3\pi \cdot \frac{4}{4} = 3\pi.$$
   b. The five important points of the curve $y = 2 \sin \frac{2}{3}x$, corresponding to the five points of the x-axis, are shown in Fig. 18-28.
   c. The curve $y = 2 \sin \frac{2}{3}x$ results when these five points are connected with a smooth curve (see Fig. 18-28).

The Curve of **y = A sin Bx**

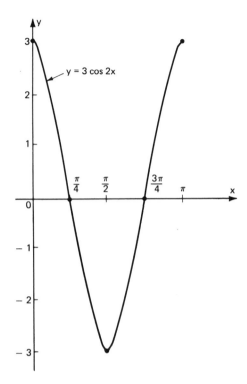

FIG. 18-27

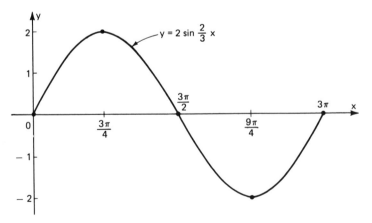

FIG. 18-28

To investigate the curve of an equation of the type $y = A \sin (Bx + C)$ we will consider first the equation $y = \sin (x + \pi/6)$, in which $A = 1$, $B = 1$ and $C = \pi/6$. We know that

when $x = 0$, $y = \sin (0 + \pi/6) = \sin \pi/6 = 0.50$
when $x = \pi/6$, $y = \sin (\pi/6 + \pi/6) = \sin \pi/3 = 0.87$
when $x = \pi/3$, $y = \sin (\pi/3 + \pi/6) = \sin \pi/2 = 1$

and so on. We can complete in this fashion the table of corresponding values, part of which is shown in Table 18-6.

TABLE 18-6

x	0	$\dfrac{\pi}{6}$	$\dfrac{\pi}{3}$	$\dfrac{\pi}{2}$	$\dfrac{2\pi}{3}$	$\dfrac{5\pi}{6}$
$x + \dfrac{\pi}{6}$		$\dfrac{\pi}{6}$	$\dfrac{\pi}{3}$	$\dfrac{\pi}{2}$	$\dfrac{2\pi}{3}$	$\dfrac{5\pi}{6}$
$y = \sin \left(x + \dfrac{\pi}{6} \right)$		0.50	0.87	1	0.87	

We notice very soon, however, that the values of y are in exactly the same order like those for the basic equation $y = \sin x$, shown in Table 18-4, but they have been *shifted* by an interval of $\pi/6$ to the left. This suggests that the curves of the basic equation $y = \sin x$, and the equation $y = \sin (x + \pi/6)$ are the same, with the only difference being that the curve of the $y = \sin (x + \pi/6)$ is shifted to the left by an amount equal to C; that is, $\pi/6$ (see Fig. 18-29). For this reason, *the amount by which the curve of the basic equation $y = \sin x$ is shifted is called the* **phase shift**.

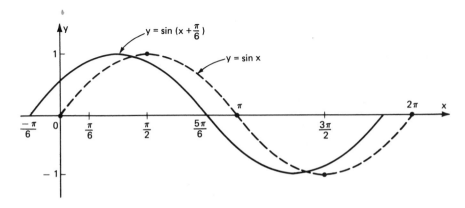

FIG. 18-29

The student should be careful at this point. The amount of the phase shift is not always equal to C. The numerical value of the phase shift is given by the equation

$$\text{Phase shift} = \frac{|C|}{B}$$

In the above example, the value of the phase shift is equal to C, that is $\pi/6$, because $B = 1$.

The above example suggests that *the curve of an equation of the type $y = A \sin(Bx + C)$ may be obtained by sketching first the curve of $y = A \sin Bx$, and then shifting it by an amount equal to $|C|/B$. The curve of $y = A \sin Bx$ should be shifted to the left if C is positive and to the right if C is negative.*

The above statement applies also for the curve of the equation of the type $y = A \cos(Bx + C)$.

We can summarize now by giving the following *steps for sketching the curve of an equation of the type $y = A \sin(Bx + C)$:*

1. *Complete the list:*
   a. *Amplitude: $A =$*
   b. *Period: $2\pi/B =$*
   c. *Phase Shift: $|C|/B =$*
   d. *Algebraic sign of C:*
2. *Sketch first the curve $y = A \sin Bx$.*
   a. *Using the origin as the first point, mark four more points on the x-axis at intervals equal to $\frac{1}{4}$ of the period of step 1.b.*
   b. *Corresponding to the five points of the x-axis, plot the five important points of the curve. Use the information of step 1.a.*
   c. *Connect these five points with a smooth curve.*
3. *Shift the curve $y = A \sin B$, of step 2, a number of units equal to information from step 1.c, to the left or to the right according to the information from step 1.d.*

EXAMPLE 18-16 Sketch one cycle of $y = 3 \sin(2x - \pi/3)$.

SOLUTION

1. a. Amplitude: $A = 3$
   b. Period: $2\pi/B = 2\pi/2 = \pi$
   c. Phase Shift: $|C|/B = (\pi/3)/2 = \pi/6$
   d. Algebraic sign of C: $(-)$
2. Sketch first $y = 3 \sin 2x$.
   a. Starting with point $(0, 0)$, the other four points on the x-axis, at intervals of $\pi/4$ are $\pi/4$, $\pi/2$, $3\pi/4$, and π. They are shown in Fig. 18-30.
   b. The five important points of the curve $y = 3 \sin 2x$ corresponding to the five points of the x-axis are shown in Fig. 18-30.
   c. The curve $y = 3 \sin 2x$ results when these five points are connected with a smooth curve (see the dotted-line curve of Fig. 18-30).
3. The curve $y = 3 \sin 2x$ is shifted $\pi/6$ units to the right (C is negative) to produce the curve $y = 3 \sin(2x - \pi/3)$ of Fig. 18-30.

The General Trigonometric Angle. Radian Measure

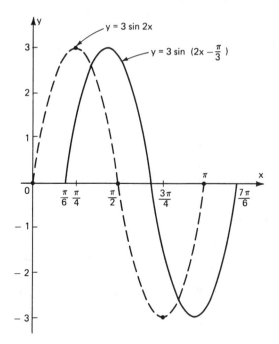

$y = 3 \sin 2x$

$y = 3 \sin \left(2x - \frac{\pi}{3}\right)$

FIG. 18-30

1. Sketch the graph of $y = \sin x$ and $y = \cos x$ of Fig. 18-26, and Fig. 18-28. Notice carefully how the values of these functions change between 0, $\pi/2$, π, $3\pi/2$, and 2π.

2. Sketch the sine graph for a complete cycle. Locate points every 15°, or $\pi/12$ radians.

3. Sketch the cosine graph for a complete cycle. Locate points every 15°, or $\pi/12$ radians.

In exercises 4–7 find the amplitude and the period of each equation.

4. $y = \sin 3x$

5. $y = 3 \cos \frac{2}{3}x$

6. $y = 8 \sin \frac{4}{5}x$

7. $y = \cos 2x$

In exercises 8–15 sketch one cycle of the curve of the equation.

8. $y = \sin 3x$

9. $y = \frac{1}{2} \sin x$

10. $y = \cos 2x$

11. $y = 2 \sin x$

12. $y = \cos \frac{1}{4}x$

13. $y = \frac{3}{4} \sin x$

14. $y = -2 \sin x$

15. $y = 3 \sin \frac{1}{2}x$

In exercises 16–22 sketch one cycle of the curve of the equation.

16. $y = \sin \left(x + \frac{\pi}{4}\right)$

17. $y = \sin \left(x - \frac{\pi}{2}\right)$

18. $y = \sin \left(2x - \frac{\pi}{2}\right)$

19. $y = 5 \cos \left(x - \frac{\pi}{4}\right)$

The Curve of $y = A \sin (Bx + C)$

347

20. $y = 3 \sin \left(\frac{1}{3}x + \pi \right)$ **21.** $y = 3 \sin \left(3x + \frac{\pi}{3} \right)$

22. $y = 2 \sin \left(2x - \frac{\pi}{3} \right)$

REVIEW QUESTIONS

1. State the definition of the angle in trigonometry.
2. Explain what is meant by the initial and the terminal side of an angle.
3. When is a trigonometric angle positive, when is it negative?
4. When is a trigonometric angle said to be in standard position?
5. When are two or more angles called coterminal?
6. When is an angle called a quadrantal angle?
7. What is the reference angle of a given angle?
8. State the way of determining the reference angle of a second, third, and fourth quadrant angle.
9. State in your own words the procedure for determining the value of a function of an angle which is more than 90°.
10. How does one decide on the algebraic sign of the value of a function of an angle which is more than 90°?
11. State the definition of a radian.
12. When an expression occurs in terms of π, and there are no units, what are the units understood to be?
13. How does one change degrees to radians in terms of π?
14. How does one change radians in terms of π to degrees?
15. State the formula for angular speed. What are the units for angular speed?
16. State the formula for changing angular speed to linear speed.
17. What happens to the value of the sine as the angle increases from 0° to 90°, from 90° to 180°?
18. What happens to the value of the cosine as the angle increases from 0° to 90°, from 90° to 180°?
19. What is the period of the sine, of the cosine curve?
20. What is the amplitude of the sine curve, the cosine curve?
21. Which part of the sine curve is called the cycle of the curve?
22. Give the five important points of the sine curve.
23. Give the five important points of the cosine curve.
24. What part of the period is the interval between the five points of the x-axis, corresponding to the five important points of the curves $y = \sin x$, and $y = \cos x$?
25. What is the amplitude of the curve of an equation of the form $y = A \sin Bx$?
26. Give the formula for the numerical value of the period of the curve of an equation of the form $y = A \sin Bx$.

The General Trigonometric Angle. Radian Measure

27. Give in your own words the steps for the sketching of the curve of an equation of the form $y = A \sin Bx$.

28. What is phase shift?

29. Give the formula for the numerical value of the phase shift of the curve of an equation of the type $y = A \sin (Bx + C)$.

30. What is the importance of the algebraic sign of C for the curve of an equation of the form $y = A \sin (Bx + C)$?

31. Give in your own words the steps for the sketching of the curve of an equation of the form $y = A \sin (Bx + C)$.

REVIEW EXERCISES

Find the reference angle for each of the angles in exercises 1–4.

1. $242°$ **2.** $175° \, 40'$

3. $343° \, 50'$ **4.** $570°$

Find the values of the functions of the angles in exercises 5–8.

5. $\sin 270°$ **6.** $\cot 168° \, 10'$

7. $\cos 180°$ **8.** $\tan 193° \, 18'$

In exercises 9–12, convert the given angle from radians to degrees. Use 1 radian $= 57.3°$.

9. $\theta = 5.1$ **10.** $\theta = 7.3$

11. $\theta = 3.7$ **12.** $\theta = 1.2$

In exercises 13–16, convert the given angle from degrees to radians. Use $1° = 0.01745$ rad.

13. $\theta = 14.7°$ **14.** $\theta = 98.3°$

15. $\theta = 117.8°$ **16.** $\theta = 213.9°$

In exercises 17–20, convert the given angle to degrees.

17. $\theta = \dfrac{13\pi}{6}$ **18.** $\theta = \dfrac{3\pi}{5}$

19. $\theta = \dfrac{7\pi}{9}$ **20.** $\theta = \dfrac{11\pi}{18}$

In exercises 21–24, convert the given angle to radians leaving your answer in terms of π.

21. $\theta = 18°$ **22.** $\theta = 260°$

23. $\theta = 78°$ **24.** $\theta = 285°$

25. Change 240 RPM to rad/sec.

26. A point on the rim of a flywheel moves with a linear speed of 20 m/sec. If the radius of the wheel is 2 m, what is the angular speed of the wheel (a) in rad/sec, (b) in RPM?

27. The diameter of the tire of a car is 28 in. The car is traveling at the speed of 55 mph. Find the angular speed of the wheels of the car (a) in rad/sec, (b) in RPM.

28. A grindstone has a diameter of 20 cm and makes 1500 RPM. Find the linear speed of a point on the rim of the grindstone.

In exercises 29 and 30, find the amplitude and the period of each equation.

29. $y = 4 \sin 4x$

30. $y = \dfrac{2}{3} \cos \dfrac{1}{4} x$

In exercises 31 and 32, sketch one cycle of the curve of the equation.

31. $y = 2 \cos \dfrac{1}{2} x$

32. $y = \dfrac{5}{4} \sin \dfrac{1}{4} x$

In exercises 33–35, sketch one cycle of the curve of the equation.

33. $y = \sin \left(2x - \dfrac{\pi}{6} \right)$

34. $y = 2 \sin \left(x - \dfrac{\pi}{4} \right)$

35. $y = \sin \left(3x + \dfrac{\pi}{3} \right)$

19

Solution of Oblique
Triangles.
Vectors

19-1 Basic Ideas About Oblique Triangles

Triangles having no right angle are called **oblique triangles**. An oblique triangle can be solved when three parts of the triangle, of which at least one is a side, are known. Depending on which three parts of the triangle are known, we can classify oblique triangle problems into four cases.

Case 1. One side and two angles are known.
Case 2. Two sides and the angle opposite one of them are known.
Case 3. Two sides and the included angle are known.
Case 4. Three sides are known.

In solving oblique triangles, we will label the angles of the triangles with A, B, and C, and the corresponding opposite sides with a, b, and c, not exactly as we did with right triangle problems, but as shown in Fig. 19-1.

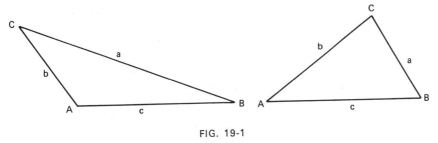

FIG. 19-1

19-2 The Law of Sines

It can be shown that *in any triangle, the sides are proportional to the sines of the opposite angles*. This statement is the **law of sines**, which when expressed mathematically will give

$$\frac{a}{\sin A} = \frac{b}{\sin B} = \frac{c}{\sin C}$$

In the above proportion, we have the three equations

$$\frac{a}{\sin A} = \frac{b}{\sin B}$$

$$\frac{a}{\sin A} = \frac{c}{\sin C}$$

$$\frac{b}{\sin B} = \frac{c}{\sin C}$$

In selecting the equation to use in a given oblique triangle problem, one should make sure that three of the four parts involved in the equation are known. Examining the above four cases of oblique triangle problems, we see that the law of sines can be used for cases 1 and 2.

EXAMPLE 19-1 Solve the oblique triangle in which $a = 23.1$, $A = 33°\ 10'$, and $B = 42°\ 40'$.

SOLUTION This is a case 1 problem. A sketch of the problem, reasonably to scale, is shown in Fig. 19-2. First

$$C = 180° - (A + B)$$

$$= 104°\ 10'$$

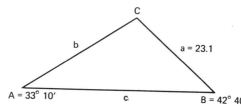

FIG. 19-2

Then, because of the given data, to solve for b we select the equation

$$\frac{a}{\sin A} = \frac{b}{\sin B}$$

Substituting, we have

$$\frac{23.1}{\sin 33°\ 10'} = \frac{b}{\sin 42°\ 40'}$$

and solving for b,

$$b = \frac{23.1(\sin 42°\ 40')}{\sin 33°\ 10'}$$

$$= 28.6$$

To solve for c, we substitute in the equation

$$\frac{a}{\sin A} = \frac{c}{\sin C}$$

Thus,

$$\frac{23.1}{\sin 33°\ 10'} = \frac{c}{\sin 104°\ 10'}$$

and

$$c = \frac{23.1(\sin 104° \, 10')}{\sin 33° \, 10'}$$

$$= 40.9$$

Solve each of the following triangles. Interpolate wherever needed. The method of labeling is according to the method adopted (see Fig. 19-1).

1. $A = 35° \, 20'$, $B = 71° \, 40'$, $a = 168$
2. $A = 58° \, 10'$, $B = 41° \, 40'$, $a = 367$
3. $B = 23° \, 50'$, $C = 58° \, 30'$, $b = 53.7$
4. $A = 44° \, 10'$, $C = 71° \, 20'$, $c = 3.41$
5. $A = 62° \, 20'$, $C = 74° \, 50'$, $a = 5.27$
6. $B = 37° \, 40'$, $C = 77° \, 30'$, $c = 0.606$
7. $A = 24° \, 40'$, $B = 75° \, 40'$, $a = 17.1$
8. $B = 67° \, 40'$, $C = 47° \, 10'$, $c = 1.89$
9. $A = 17° \, 50'$, $C = 107° \, 40'$, $c = 24.8$
10. $A = 101° \, 10'$, $B = 34° \, 40'$, $a = 31.6$
11. $A = 21° \, 17'$, $B = 63° \, 11'$, $a = 339$
12. $B = 34° \, 58'$, $C = 61° \, 33'$, $c = 1.23$
13. $A = 8° \, 48'$, $C = 127° \, 23'$, $a = 26.8$
14. $B = 51° \, 13'$, $C = 108° \, 14'$, $b = 12.8$
15. $A = 62° \, 42'$, $B = 53° \, 28'$, $b = 12.2$
16. Find length h in Fig. 19-3.

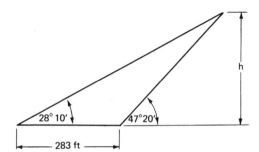

FIG. 19-3 |← — 283 ft — →|

28° 10'   47°20'

17. The angle of elevation of an airplane from a point A is $33° \, 20'$, and from a point B is $47° \, 30'$. The two points, A and B, are on level ground, and the distance between them is 2.82 km. What is the height of the airplane?

18. A and C are two points on the two banks of a river. From point A, $\overline{AB}$ is laid off equal to 258 ft (see Fig. 19-4). Angles A and B are measured and are found to be $123° \, 10'$ and $49° \, 40'$, respectively. Find distance AC.

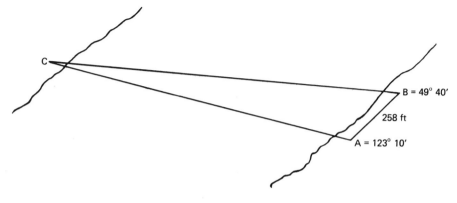

C
B = 49° 40'
258 ft
A = 123° 10'

FIG. 19-4

19-3 The Ambiguous Case

The solution of an oblique triangle is not as simple when two sides and an angle opposite one of them are given. This is case 2, and is known as the **ambiguous case** of the law of sines. The ambiguity of this case is in the fact that when the first of the two missing angles is found, it is possible that this angle is also the reference angle of another obtuse angle. It is possible, in other words, that the given data is common data for two different triangles. The procedure for determining how many solutions a given problem has—that is, whether the given data is for one or two triangles, or no triangle at all—is illustrated with the following three examples.

EXAMPLE 19-2 How many triangles are possible with $a = 168$, $b = 183$, and $A = 46° 10'$?

SOLUTION Using $\dfrac{a}{\sin A} = \dfrac{b}{\sin B}$, we have

$$\frac{168}{\sin 46° 10'} = \frac{183}{\sin B}$$

$$\sin B = \frac{183(0.7214)}{168}$$

$$= 0.7858$$

and

$$B = 51° 50'$$

Then, from the relation, $\angle A + \angle B + \angle C = 180°$, $\angle C = 82°$. Thus, $\angle A = 46° 10'$, $\angle B = 51° 50'$, $\angle C = 82°$.

To try to find out if a second triangle is involved, we assume that $\angle B = 51° 50'$ is the reference angle of another obtuse angle B'. Then,

$$\angle B' = 180° - \angle B = 180° - 51° 50' = 128° 10'$$

From the relation, $\angle A + \angle B' + \angle C' = 180°$, $\angle C' = 5° 40'$. Thus, the given data is common data for two different triangles. In the first triangle,

$$\angle A = 46° 10' \qquad \angle B = 51° 50' \qquad \angle C = 82°$$

and in the second triangle,

$$\angle A = 46° 10' \qquad \angle B' = 128° 10' \qquad \angle C' = 5° 40'$$

A sketch of the two triangles is shown in Fig. 19-5.

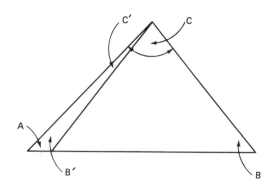

FIG. 19-5

Solution of Oblique Triangles. Vectors

EXAMPLE 19-3 How many triangles are possible with $a = 193$, $b = 143$, $A = 54° 50'$?

SOLUTION

$$\frac{a}{\sin A} = \frac{b}{\sin B}$$

$$\frac{193}{\sin 54° 50'} = \frac{143}{\sin B}$$

$$\sin B = 0.6057$$

$$B = 37° 20'$$

Then, $\angle A = 54° 50'$, $\angle B = 37° 20'$, and $\angle C = 87° 50'$. To test if a second triangle is involved, we assume that $\angle B$ is the reference angle of another obtuse angle B'. Then,

$$\angle B' = 180° - \angle B = 180° - 37° 20' = 142° 40'$$

But with $\angle A = 54° 50'$, and $\angle B' = 142° 40'$, no triangle is possible, since $\angle A + \angle B' = 197° 30'$. Therefore, the problem has only one solution.

EXAMPLE 19-4 How many triangles are possible with $a = 36.3$, $b = 40.6$, and $A = 108° 30'$?

SOLUTION

$$\frac{a}{\sin A} = \frac{b}{\sin B}$$

$$\sin B = \frac{b(\sin A)}{a}$$

$$= \frac{40.6(0.9483)}{36.3}$$

$$= 1.06$$

But no angle whose sine is greater than 1 exists. Therefore, there is no angle B, and there is no triangle.

The above three examples suggest that *to find the number of solutions which are possible when a, b, and A are given, find B and assume that B is the reference angle for an obtuse angle B'. If $A + B' < 180°$, a second triangle is possible. But if $A + B' > 180°$, the given data has only one solution. When $\sin B > 1$, no such angle exists, and there is no solution.*

EXERCISE 19-2

In exercises 1–10, solve all possible triangles.
  1. $a = 132$, $b = 153$, $A = 41° 30'$
  2. $a = 22.3$, $b = 44.1$, $A = 30° 20'$
  3. $a = 4.87$, $b = 2.31$, $A = 124° 50'$
  4. $b = 208$, $c = 267$, $B = 73° 10'$
  5. $a = 331$, $b = 237$, $A = 28° 20'$
  6. $a = 193$, $b = 158$, $A = 21° 10'$

The Ambiguous Case

7. $a = 33.3$, $c = 67.3$, $A = 38° 50'$
8. $a = 12.1$, $b = 14.9$, $A = 42° 50'$
9. $a = 4.39$, $b = 5.33$, $A = 51° 20'$
10. $b = 3.68$, $c = 2.34$, $B = 138° 10'$

19-4 The Law of Cosines

It can be shown that *in any triangle, the square of the length of any side is equal to the sum of the squares of the lengths of the other two sides minus twice the product of the lengths of those two sides and the cosine of the included angle.*

This statement is the **law of cosines**, which when expressed mathematically will give the following three equations:

$$a^2 = b^2 + c^2 - 2bc \cos A \qquad (a)$$

$$b^2 = a^2 + c^2 - 2ac \cos B \qquad (b)$$

$$c^2 = a^2 + b^2 - 2ab \cos C \qquad (c)$$

The law of cosines can be used for cases 3 and 4; that is, when two sides and the included angle or when three sides are known.

EXAMPLE 19-5   Find c if $a = 18.3$, $b = 23.4$, and $C = 46° 50'$.

SOLUTION   Using equation (c) of the law of cosines, and substituting from the given data, we have

$$c^2 = a^2 + b^2 - 2ab \cos C$$
$$= (18.3)^2 + (23.4)^2 - 2(18.3)(23.4)(\cos 46° 50')$$
$$= 334.9 + 547.6 - 856.4(0.6841)$$
$$= 296.6$$

and

$$c = \sqrt{296.6}$$
$$= 17.2$$

To find any other part of the triangle, or for a complete solution, the law of sines can now be used.

EXAMPLE 19-6   Find B if $a = 362$, $b = 468$, and $c = 398$.

SOLUTION   Using equation (b) of the law of cosines, and substituting from the given data, we have

$$b^2 = a^2 + c^2 - 2ac \cos B$$
$$(468)^2 = (362)^2 + (398)^2 - 2(362)(398)(\cos B)$$
$$2(362)(398)(\cos B) = (362)^2 + (398)^2 - (468)^2$$
$$\cos B = \frac{(362)^2 + (398)^2 - (468)^2}{2(362)(398)}$$
$$= 0.2444$$

Therefore, $B = 75° 50'$.

The student should remember that the cosine of an obtuse angle is negative, and that the proper algebraic sign should be attached any time an obtuse angle occurs.

EXAMPLE 19-7 Find c if $a = 2.76$, $b = 4.18$, and $C = 141°\ 10'$.

SOLUTION

$$c^2 = (2.76)^2 + (4.18)^2 - 2(2.76)(4.18)(-\cos 38°\ 50')$$
$$= 7.618 + 17.472 - 23.07(-0.7790)$$
$$= 25.09 + 17.97$$
$$= 43.06$$

and

$$c = \sqrt{43.06}$$
$$= 6.56$$

EXAMPLE 19-8 Find A if $a = 81.7$, $b = 49.8$, and $c = 40.3$.

SOLUTION Using equation (a), we have

$$(81.7)^2 = (49.8)^2 + (40.3)^2 - 2(49.8)(40.3)(\cos A)$$
$$\cos A = \frac{4104 - 6675}{4014}$$
$$= \frac{-2571}{4014}$$
$$= -0.6405$$

Because $\cos A$ is negative, A is a second quadrant angle. Its reference angle is $50°\ 10'$, and $A = 129°\ 50'$.

EXERCISE 19-3

In exercises 1–12, find the part of the triangle indicated using the law of cosines.
1. $a = 7.2$, $b = 5.1$, $c = 3.8$. Find A.
2. $a = 153$, $b = 172$, $c = 167$. Find B.
3. $b = 2.3$, $c = 1.6$, $A = 158°$. Find a.
4. $a = 372$, $c = 487$, $B = 141°\ 20'$. Find b.
5. $a = 3.32$, $b = 2.17$, $c = 2.87$. Find A.
6. $a = 272$, $b = 261$, $c = 418$. Find C.
7. $a = 53.7$, $c = 41.8$, $B = 158°\ 20'$. Find b.
8. $b = 8.34$, $c = 6.13$, $A = 138°\ 40'$. Find a.
9. $a = 14.2$, $b = 17.5$, $C = 28°\ 10'$. Find c.
10. $b = 29.3$, $c = 31.4$, $A = 53°\ 30'$. Find a.
11. $a = 154$, $b = 287$, $c = 248$. Find B.
12. $a = 47.3$, $b = 58.4$, $c = 36.3$. Find C.
13. The sides of a parallelogram are 43.6 cm and 54.3 cm. The angle between the two sides is $136°\ 20'$. Find the larger of the two diagonals.
14. A railroad track and a highway intersect and form an angle of $53°$. If the engine of a train is 85 ft from the intersection, and a car is 73 ft from the intersection, what is the distance between the car and the engine of the train?

The Law of Cosines

15. To find the length of a lake between points A and B, the distances AC and BC were measured and the size of angle C was determined with a transit. These measurements are shown in Fig. 19-6. What is the length of the lake?

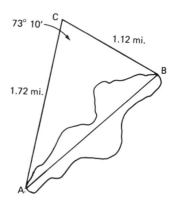

FIG. 19-6

19-5 Basic Ideas About Vectors

We have seen in chapter 9 that numbers representing a physical quantity are followed by the unit in which the quantity is measured, and that such numbers are called denominate numbers. Such quantities, as we have seen, are added by ordinary arithmetic. Thus,

$$4 \text{ km} + 3 \text{ km} = 7 \text{ km}$$

There are, however, certain other physical quantities for which not only the magnitude and the unit but the **direction** also should be stated for their complete specification. For instance, two physical quantities of this kind are 25 mph north and 2 kg downward. Quantities of this second kind are called **vector quantities**. Force, displacement, velocity, and acceleration are vector quantities.

A vector is represented graphically by a line segment, the length of which is proportional to the magnitude of the vector quantity. An arrowhead at the end of the line segment shows the direction, or **sense**, of the vector. The symbol for the vector represented by the line segment AB [see Fig. 19-7(a)] is $\overrightarrow{AB}$, and the symbol for vector A [see Fig. 19-7(b)] is $\vec{A}$.

FIG. 19-7

When the starting point of a vector is at the origin of the coordinate system, the vector is said to be in standard position. A vector in standard position is specified by its length and the counterclockwise angle θ [see Fig. 19-8(a)]. Thus, if vector A of Fig. 19-8(b) is 3 cm long we say $\vec{A} = 3$ cm at $135°$.

A vector is resolved when it is replaced by two or more vectors that will have the same net effect. These two or more vectors are called **components**. The sum of two or

Solution of Oblique Triangles. Vectors

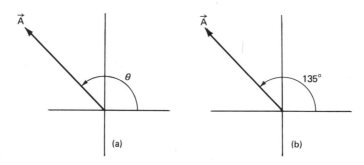

FIG. 19-8

(a) (b)

more vectors is called their **resultant**. Vector quantities do not add arithmetically, but geometrically. Geometric addition will be illustrated with a displacement example.

Suppose that a car, originally at point A, traveled 4 mi east to a point C, and then 3 mi north to a point B as shown in Fig. 19-9. The final displacement of the car from point A is the distance AB even though the car traveled first to C and then to B. The sum of the displacement $\vec{AC}$ and $\vec{CB}$ is $\vec{AB}$, and $\vec{AB}$ is 5 mi, not 4 mi $+$ 3 mi $=$ 7 mi. With our trigonometric background, we can find the direction of the displacement since it is simply the angle A.

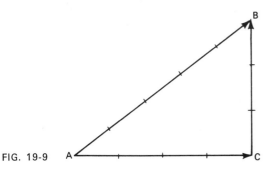

FIG. 19-9

There is a variety of methods of finding the resultant of two or more vectors. Here we will consider two methods: the **parallelogram method**, and the **method of components**.

19-6 The Parallelogram Method

This method can be used to resolve a vector into its two components, or to find the resultant of two vectors. *With the parallelogram method, the horizontal and the vertical components, V_h and V_v, of a given vector may be found by using the definitions of the sine and the cosine.* Thus,

$$V_h = V \cos \theta$$
$$V_v = V \sin \theta$$

EXAMPLE 19-9 A man is pulling a rope attached to a sled with a force of 35 lb. The rope makes an angle of 33° with the ground. Find the horizontal component of the pulling force tending to move the sled along the ground, and the vertical component of the pulling force tending to lift the sled upward.

SOLUTION The vector marked 35 (see Fig. 19-10) represents the pulling force. Two vectors $\vec{H}$ and $\vec{V}$ are drawn from the point of application of the pulling force, one parallel and the other perpendicular to the ground. Then the two lines parallel to the vectors $\vec{H}$ and $\vec{V}$ are drawn so that the pulling force is the diagonal of the parallelogram thus formed. It must be evident from the sketch of the parallelogram of Fig. 18-10 that

$$\vec{H} = 35 \cos 33° = 35(0.8387) = 29 \text{ lb}$$
$$\vec{V} = 35 \sin 33° = 35(0.5446) = 19 \text{ lb}$$

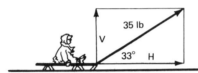

FIG. 19-10

The weight of an object placed on an inclined plane acts vertically downward and can be resolved into two components (see Fig. 19-11). One component, A, is parallel to the inclined plane, tending to roll the object down the plane, and the other component, B, is perpendicular to the plane pressing the object against the incline. Because vectors $\vec{B}$ and $\vec{W}$ of Fig. 19-11 are perpendicular to two sides of the inclined plane, $\angle K = \angle WOB$.

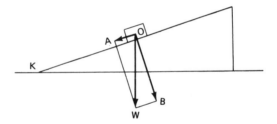

FIG. 19-11

EXAMPLE 19-10 A generator weighing 2880 lb (weight is force) is mounted on skids and is being pulled up an inclined plane making an angle of 18° 30′ with the horizontal. Ignoring friction, find the force tending to roll the generator down the incline, and the force with which the generator is pressed against the inclined plane. The situation is sketched in Fig. 19-12.

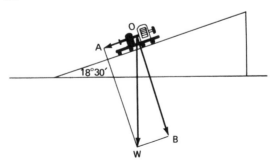

FIG. 19-12

Solution of Oblique Triangles. Vectors

SOLUTION   Since $\angle\, WOB = 18°\ 30'$, we have

$$\overrightarrow{OA} = \overrightarrow{BW} = 2880(\sin 18°\ 30')$$
$$= 2880(0.3173)$$
$$= 914\ \text{lb}$$
$$\overrightarrow{OB} = 2880(\cos 18°\ 30')$$
$$= 2880(0.9483)$$
$$= 2730\ \text{lb}$$

Therefore, the generator is tending to roll down the incline with a force of 914 lb, and is pressing against the inclined plane with a force of 2730 lb.

The resultant of two vectors forming any angle with each other may be found using the law of cosines and solving for the resultant represented by the diagonal of the parallelogram formed when two lines parallel to the given vectors are drawn.

Suppose, for instance, that two forces $\overrightarrow{A}$ and $\overrightarrow{B}$ are acting on the same object and that they form an angle k with each other (see Fig. 19-13). Their resultant $\overrightarrow{R}$ is the diagonal of the parallelogram which is formed when the dotted lines A' and B', parallel to the given vectors, are drawn.

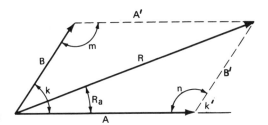

FIG. 19-13

In every parallelogram, opposite sides and opposite angles are equal, and the sum of the four angles is equal to 360°. The following relation of the angles of the parallelogram is evident:

$$\angle\, m = \angle\, n = \frac{360° - 2k}{2} = 180° - k$$

NOTE: We can arrive at the same value for $\angle\, n$ by using the fact that $\angle\, k$ and $\angle\, k'$ of Fig. 19-13 are equal as vertical angles to alternate interior angles, and the fact that $\angle\, n$ and $\angle\, k'$ are supplementary angles. Thus, $\angle\, n = 180° - \angle\, k' = 180° - \angle\, k$.

We can find the value of the resultant $\overrightarrow{R}$ of the two forces $\overrightarrow{A}$ and $\overrightarrow{B}$ using the law of cosines, since $\overrightarrow{R}$ is the side of an oblique triangle, two sides and the included angle of which are known. Then, using the law of sines, we can solve for $\angle\, R_a$, the angle the resultant $\overrightarrow{R}$ makes with the horizontal force, determining the direction of the resultant with respect to that force.

EXAMPLE 19-11   Two forces, $\overrightarrow{A}$ and $\overrightarrow{B}$, of 12 and 10 lb, respectively are acting on the same point and they form an angle of 30° with each other. Determine their vector sum, and $\angle\, R_a$, the direction of the resultant.

The Parallelogram Method

SOLUTION The two forces are represented by the two arrows $\vec{A}$ and $\vec{B}$ (see Fig. 19-14). Their resultant is represented by the diagonal $\vec{R}$ of the parallelogram which is formed when the dotted lines are drawn parallel to the given forces.

$$\angle\, n = 180° - 30° = 150°$$

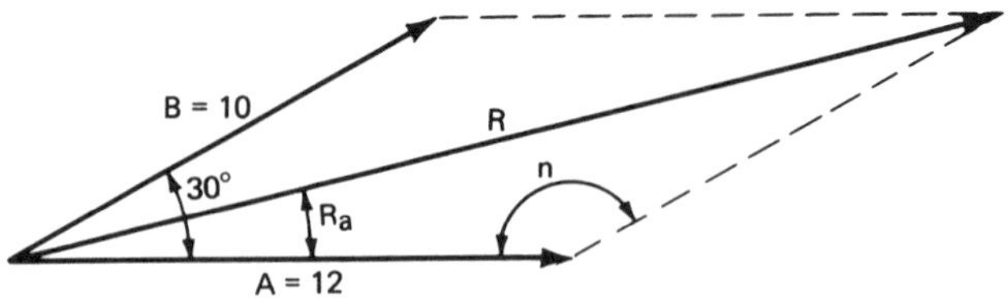

FIG. 19-14

Then,

$$\vec{R}^2 = \vec{A}^2 + \vec{B}^2 - 2AB\cos n$$

$$= (12)^2 + (10)^2 - 2(12)(10)\cos 150°$$

$$= 244 + 208$$

$$= 452$$

and

$$\vec{R} = 21 \text{ lb}$$

By the law of sines

$$\frac{B}{\sin R_a} = \frac{R}{\sin n}$$

$$\frac{10}{\sin R_a} = \frac{21}{\sin 150°}$$

$$\sin R_a = \frac{10(\sin 150°)}{21}$$

$$= 0.2381$$

and

$$\angle\, R_a = 14°$$

EXERCISE 19-4

1. A rope is attached to a box and a pulling force of 75 lb is applied to it. If the rope makes an angle of 28° with the ground, find the horizontal and the vertical components of the pulling force.
2. A force of 55 lb is applied on the handle of a lawn mower at an angle of 38° with the ground. Find the horizontal component of the applied force that moves the mower forward, and the vertical component of the applied force that pushes the mower against the ground.
3. The resultant of two forces, one vertical and the other horizontal, is 423 lb and makes an angle of 34° 20′ with the horizontal force. Find the two component forces.
4. The resultant of two vectors at right angles with each other is 524 lb. If the horizontal vector is 382 lb, how much is the other vector, and what is the angle that the horizontal vector makes with the resultant?
5. In loading a heavy safe onto a truck, an inclined plane is used making an angle of 34° 40′ with the ground. If the safe weighs 1250 lb, and friction is ignored, what is the force needed

Solution of Oblique Triangles. Vectors

to pull the safe up the incline? (Hint: The force needed to pull the safe up the incline is equal to the force tending to roll the safe down the incline.)

6. A force of 225 lb is needed to keep a loaded cart of 825 lb from rolling down a ramp. Find the angle that the ramp makes with the horizontal.
7. A force of 115 lb is required to prevent a cylindrical container from rolling down an inclined plane that makes an angle of 18° 20′ with the horizontal. What is the weight of the container?

Referring to Fig. 19-15, determine the resultant $\vec{R}$, of the two vectors $\overrightarrow{OA}$ and $\overrightarrow{OB}$, and $\angle R_a$ in exercises 8–10.

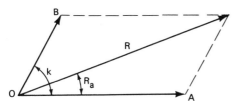

FIG. 19-15

8. $\overrightarrow{OA} = 21$, $\overrightarrow{OB} = 16$, $k = 60°$
9. $\overrightarrow{OA} = 218$, $\overrightarrow{OB} = 314$, $k = 23°\ 50'$
10. $\overrightarrow{OA} = 117$, $\overrightarrow{OB} = 187$, $k = 67°\ 40'$

19-7 The Method of Components

With this method, the size and the direction of the resultant of any number of vectors can be determined.

To find the resultant $\vec{R}$, and its direction $\angle R_a$ of, say, three forces $\vec{A}$, $\vec{B}$, and $\vec{C}$, acting on the same point [see Fig. 19-16(a)], first place the three vectors in standard

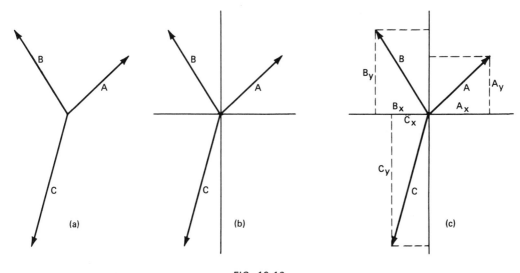

FIG. 19-16

The Method of Components

363

position [see Fig. 19-16(b)]. Resolve each vector into its x- and y-components by dropping a perpendicular from the arrow head to both the x- and y-axis [see Fig. 19-16(c)]. Notice the right triangles that are formed this way. Then add all the x-components, A_x, B_x, C_x, and call the sum ΣX*. Add in the same manner all the y-components, A_y, B_y, C_y, and call the sum ΣY. Using the Pythagorean theorem at this point, the value of $\vec{R}$ can be found (see Fig. 19-17) because

$$\vec{R}^2 = (\Sigma X)^2 + (\Sigma Y)^2$$
$$\vec{R} = \sqrt{(\Sigma X)^2 + (\Sigma Y)^2}$$

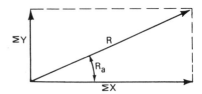

FIG. 19-17

Also, it must be evident from Fig. 19-17 that the direction of $\vec{R}$, the $\angle R_a$ the resultant makes with the x-axis, measured counterclockwise can be found by the relation

$$\tan R_a = \frac{\Sigma Y}{\Sigma X}$$

The method will be illustrated with the following example.

EXAMPLE 19-12 The following three vectors $\vec{A}, \vec{B}$, and $\vec{C}$ of Fig. 19-18(a), are in standard position.

$$\vec{A} = 9.0 \text{ units at } 40°$$
$$\vec{B} = 6.0 \text{ units at } 210°$$
$$\vec{C} = 8.0 \text{ units at } 270°$$

Find the resultant $\vec{R}$, and the direction of the resultant $\angle R_a$.

SOLUTION

$$A_x = 9.0(\cos 40°) = 6.9$$
$$B_x = 6.0(-\cos 30°) = -5.2$$
$$C_x = 8.0(\cos 90°) = 0.0$$
$$\overline{}$$
$$\Sigma X = 1.7$$
$$A_y = 9.0(\sin 40°) = 5.8$$
$$B_y = 6.0(-\sin 30°) = -3.0$$
$$C_y = 8.0(-\sin 90°) = -8.0$$
$$\overline{}$$
$$\Sigma Y = -5.2$$

*The Greek letter Σ (read "sigma") is used in advanced mathematics as a symbol for summation.

$$\vec{R} = \sqrt{(\Sigma X)^2 + (\Sigma Y)^2}$$
$$= \sqrt{(1.7)^2 + (-5.2)^2}$$
$$= 5.5$$

$$\tan R_a = \frac{\Sigma Y}{\Sigma X}$$
$$= \frac{-5.2}{1.7}$$
$$= -3.0588$$

The tangent is negative in the second and fourth quadrant, but since ΣX is positive, and ΣY is negative, $\angle\ R_a$ is a fourth quadrant angle, [see Fig. 19-18(b)]. Its reference angle is 72°, and $\angle\ R_a = 288°$.

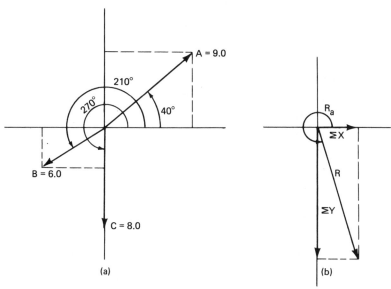

FIG. 19-18   (a)   (b)

EXERCISE 19-5

Use the method of components and find the resultant $\vec{R}$ and the direction of the resultant, $\angle\ R_a$, for exercises 1–5. All vectors are in standard position.

1. $\vec{A}$ = 3.0 units at 0°
   $\vec{B}$ = 12 units at 40°
   $\vec{C}$ = 7.0 units at 240°

2. $\vec{A}$ = 23 units at 0°
   $\vec{B}$ = 18 units at 104°
   $\vec{C}$ = 11 units at 223°

3. $\vec{A}$ = 227 units at 68° 20′
   $\vec{B}$ = 318 units at 114° 40′
   $\vec{C}$ = 206 units at 316° 50′

4. $\vec{A}$ = 8.2 units at 38°
   $\vec{B}$ = 5.4 units at 72°
   $\vec{C}$ = 4.6 units at 162°
   $\vec{D}$ = 7.1 units at 248°

5. $\vec{A}$ = 14.7 units at 23° 20′
   $\vec{B}$ = 24.6 units at 170° 40′
   $\vec{C}$ = 38.2 units at 222° 30′
   $\vec{D}$ = 16.3 units at 282° 40′

The Method of Components   365

REVIEW QUESTIONS

1. State the four cases of oblique triangle problems.

2. Give the mathematical form of the law of sines.

3. Which of the four cases of oblique triangle problems can be solved using the law of sines?

4. Which of the four cases of oblique triangle problems is the ambiguous case?

5. What is the ambiguity of the ambiguous case?

6. State the method of determining how many triangles are possible from given data.

7. Give all three equations of the mathematical form of the law of cosines.

8. Which cases of oblique triangle problems can be solved using the law of cosines?

9. Which quantities are vector quantities?

10. How is a vector represented graphically?

11. What is the symbol for a vector?

12. When is a vector in standard position?

13. How is a vector in standard position specified?

14. When is a vector said to be resolved?

15. What are the two vectors into which a given vector is resolved called?

16. What is the vector sum of two or more vectors called?

17. How are the horizontal and the vertical components of a given vector found with the parallelogram method?

18. State the procedure for finding the resultant of two vectors making any angle with each other, using the parallelogram method and the law of cosines.

19. Describe the procedure for finding the resultant of more than two vectors with the method of components. Give the formula for finding $\angle R_a$, the angle the resultant makes with the x-axis.

REVIEW EXERCISES

Solve each oblique triangle in exercises 1–4. Interpolate wherever needed.

1. $A = 25° 43'$, $B = 78° 52'$, $a = 18.60$

2. $b = 26$, $c = 18$, $A = 161°$

3. $a = 78$, $b = 53$, $c = 41$

4. $a = 146$, $b = 177$, $c = 153$

How many triangles are possible with the data given in exercises 5 and 6?

5. $b = 2.11$, $c = 2.73$, $B = 74° 40'$

6. $a = 24.3$, $b = 44.8$, $A = 28° 20'$

7. The chord of the circle shown in Fig. 19-19 is 8.23 m, and its central angle is 38° 20'. Find the radius r of the circle.

Solution of Oblique Triangles. Vectors

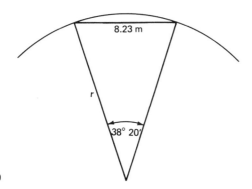

8.23 m

r

38° 20′

FIG. 19-19

8. From a certain point A, the angle of elevation of a windmill tower is 43°. From a point B, 23 ft closer to the tower, the angle of elevation is 53°. If points A and B and the tower are on the same level ground, find the height of the tower.

9. The distance between points A and B on the pond in Fig. 19-20 is needed. A distance AC equal to 638 ft was laid off. Then a transit was used and the angles A and C were found to be 112° 20′ and 28° 10′, respectively. Find the distance AB.

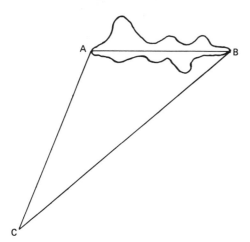

A   B

FIG. 19-20   C

10. The lower and the left sides of a parallelogram are 12.8 m and 10.3 m, and they make an angle of 53° 10′ with each other. Find the length of each diagonal.

11. A force of 115 lb makes an angle of 73° 20′ with another force of 83.6 lb. Find the magnitude and the direction of their resultant.

12. The following three forces act on the same point:

$$\vec{A} = 18.3 \text{ lb at } 60° \ 10′$$
$$\vec{B} = 20.8 \text{ lb at } 135° \ 40′$$
$$\vec{C} = 11.6 \text{ lb at } 210° \ 30′$$

Find the resultant $\vec{R}$, and the direction of the resultant $\angle R_a$.

20

Complex Numbers

20-1 Basic Ideas About Complex Numbers

The square root of a negative number is called an **imaginary number**. Thus, $\sqrt{-3}$, $\sqrt{-7}$, and $\sqrt{-16}$ are imaginary numbers.

We know from chapter 12 that $\sqrt{-3} = \sqrt{3} \cdot \sqrt{-1}$, $\sqrt{-7} = \sqrt{7} \cdot \sqrt{-1}$, and $\sqrt{-16} = \sqrt{16} \cdot \sqrt{-1} = 4\sqrt{-1}$. If we now agree to represent $\sqrt{-1}$ by the symbol i, we have

$$i = \sqrt{-1}$$

and $\sqrt{-3} = \sqrt{3} \cdot \sqrt{-1}$, $\sqrt{-7} = \sqrt{7} \cdot \sqrt{-1}$, and $\sqrt{-16} = \sqrt{16} \cdot \sqrt{-1} = 4\sqrt{-1}$ become $\sqrt{3}\,i$, $\sqrt{7}\,i$ and $4i$, respectively.

The symbol i, defined $i = \sqrt{-1}$, is used in mathematics books. In applied mathematics, however, and in technical problems, i is usually used to represent electric current. To avoid confusion it is customary to use the symbol j instead of the symbol i to represent $\sqrt{-1}$. Thus,

$$j = \sqrt{-1} \qquad j^2 = -1$$

The number j is usually called the **j-operator**.

EXAMPLE 20-1 Express each of the following imaginary numbers in terms of the j-operator:

$$\text{(a) } \sqrt{-9} \qquad \text{(b) } \sqrt{-\frac{1}{16}} \qquad \text{(c) } \sqrt{-48}$$

SOLUTION

(a) $\sqrt{-9} = \sqrt{9} \cdot \sqrt{-1} = 3\sqrt{-1} = 3j$

(b) $\sqrt{-\frac{1}{16}} = \sqrt{\frac{1}{16}} \cdot \sqrt{-1} = \frac{1}{4}\sqrt{-1} = \frac{1}{4}j$

(c) $\sqrt{-48} = \sqrt{(-3)(16)} = \sqrt{3} \cdot \sqrt{16} \cdot \sqrt{-1} = 4\sqrt{3}\,j$

The *j*-operator can be used along with the real numbers in the operations of addition, subtraction, multiplication, and division as if it were a real number. Its only difference from the real numbers is the consequence of its definition, $j^2 = -1$.

From the definition of *j*, the following powers of *j* easily follow:

$$j^1 = j$$
$$j^2 = -1$$
$$j^3 = j^2 j = (-1)j = -j$$
$$j^4 = j^3 j = (-j)j = -j^2 = +1$$
$$j^5 = j^4 j = (+1)j = j$$
$$j^6 = j^5 j = (j)j = j^2 = -1$$
$$j^7 = j^6 j = (-1)j = -j$$
$$j^8 = j^7 j = (-j)j = -j^2 = +1$$

Notice the cyclic property of the powers of *j*; after the fourth power the results are repeated. From the cyclic property of the powers of *j* it follows that to find the value of *j* raised to any given power *n*, where *n* is a positive integer, one should divide *n* by 4 and then raise *j* to the power of the remainder of that division. The remainder of the division of any number by 4 can only be 0, 1, 2, or 3, and therefore, the value of *j* for any exponent *n* will be either *j*, or −1, or −*j*, or 1.

EXAMPLE 20-2 Simplify:

(a) j^{27} (b) j^{52} (c) j^{106}

SOLUTION

(a) The remainder of $27 \div 4$ is 3, and $j^3 = -j$
(b) The remainder of $52 \div 4$ is 0, and $j^0 = +1$
(c) The remainder of $106 \div 4$ is 2, and $j^2 = -1$

A number of the form

$$a + bj$$

where *a* and *b* are real numbers, is called a **complex number**. Thus, $5 + 2j$ and $7 + 3j$ are complex numbers. A complex number of the form $a + bj$ is also called the **algebraic form** of a complex number.

Given the complex number $a + bj$,

1. If $b \neq 0$ the complex number $a + bj$ is called an **imaginary number**.
2. If $a = 0$ and $b \neq 0$, the complex number $a + bj$ is reduced to bj and is called a **pure imaginary number**.
3. If $b = 0$, the complex number $a + bj$ is reduced to *a*, which is a real number.

We see at this point that the complex numbers include all real numbers and all imaginary numbers.

Basic Ideas About Complex Numbers

The three kinds of complex numbers are illustrated below:

1. $2 + 3j$, $-7 + 4j$ and $3 - 2j$ are imaginary numbers.
2. $3j$, $-2j$ and $\sqrt{-5}$ are pure imaginary numbers.
3. 6, $-\dfrac{3}{7}$, and $-\sqrt{3}$ are real numbers.

The first term, a, of $a + bj$ is called the **real part** of the complex number, and the second term, bj, is called the **imaginary part**. Thus, in the complex number $2 + 4j$, 2 is the real part and $4j$ is the imaginary part.

> **EXAMPLE 20-3** Change each of the following numbers to the form $a + bj$:
> (a) $2 + \sqrt{-4}$ (b) $-5 - \sqrt{-16}$ (c) $2j^4 - j^2$

SOLUTION

(a) $2 + \sqrt{-4} = 2 + \sqrt{4} \cdot \sqrt{-1} = 2 + 2j$
(b) $-5 - \sqrt{-16} = -5 - \sqrt{16} \cdot \sqrt{-1} = -5 - 4j$
(c) $2j^4 - j^2 = 2(+1) - (-1) = 2 + 1 = 3$

Two complex numbers, $a + bj$ and $c + dj$, are said to be equal if and only if $a = c$ and $b = d$. Thus, $3 + 5j = 3 + 5j$, and $a + bj = 5 + 8j$ if $a = 5$ and $b = 8$.

> **EXAMPLE 20-4** What are the values of x and y which satisfy the equation $3 - 2j = -x + yj$?

SOLUTION By the definition of equality of complex numbers, $x = -3$ and $y = -2$.

The two complex numbers $a + bj$ and $a - bj$ are said to be the **conjugates** of each other. Also, the conjugate of $-3 - 5j$ is $-3 + 5j$, and the conjugate of $2j$ is $-2j$. Notice that *the conjugate of a complex number is the same complex number with the sign of its imaginary part changed.*

> **EXAMPLE 20-5** Write the conjugate for each of the following complex numbers:
> (a) $-4 + 2j$ (b) -5 (c) $-\sqrt{-9}$

SOLUTION

(a) Changing the sign of the imaginary part of $-4 + 2j$, its conjugate is $-4 - 2j$.
(b) Since the imaginary part of -5 is 0, its conjugate is -5.
(c) $-\sqrt{-9} = -3j$. Changing the sign of the imaginary part of $-3j$, its conjugate is $3j$.

EXERCISE 20-1

In exercises 1–9, express the imaginary number in terms of the j-operator.

1. $\sqrt{-25}$ 2. $\sqrt{-49}$ 3. $\sqrt{-169}$
4. $\sqrt{-\dfrac{1}{9}}$ 5. $\sqrt{-45}$ 6. $\sqrt{-24}$
7. $\sqrt{-\dfrac{4}{25}}$ 8. $\sqrt{-108}$ 9. $\sqrt{-80}$

In exercises 10–18, simplify the power of j.

10. j^9 **11.** j^{24} **12.** j^{67}

13. j^{18} **14.** j^{232} **15.** j^{149}

16. j^{44} **17.** j^{23} **18.** j^{266}

In exercises 19–27, change the number into the form $a + bj$.

19. $5 - \sqrt{-9}$ **20.** $-3 + \sqrt{-64}$ **21.** $-7 - \sqrt{-16}$

22. $2j^2 - 2j^3$ **23.** $j^3 - 6$ **24.** $-3j^5 + 2$

25. $3j - 5j^2$ **26.** $-2j^2 + j^6$ **27.** $5j^8 - 2j^2$

In exercises 28–31, state the values of x and y that will satisfy each equation.

28. $4 - 3j = x + yj$ **29.** $-2 - 6j = x - yj$

30. $-3 - 2j = -x + yj$ **31.** $7 - 3j = yj + x$

In exercises 32–40, name the conjugate of each complex number.

32. $-9 + 3j$ **33.** $-4 - 5j$ **34.** $2 + 5j$

35. -3 **36.** $-5j$ **37.** $-3j + 8$

38. $2j$ **39.** -11 **40.** $-\sqrt{-25}$

20-2 The Trigonometric Functions of 30°, 45°, and 60° in Terms of $\sqrt{2}$ and $\sqrt{3}$

It is simpler in many cases to express the functions of the 30°, 45°, and 60° angles using radicals. For this reason, and because these angles occur frequently, we will consider them in this section.

In Sect. 17-2 we have seen the 30°–60° triangle shown in Fig. 20-1. Notice that sides a and c are 1 and 2, respectively, and using the Pythagorean theorem, side b is found to be $\sqrt{3}$.

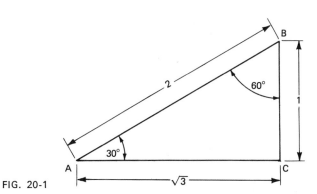

FIG. 20-1

Referring now to Fig. 20-1, and using the definitions of the trigonometric functions, the sine, cosine, and tangent of the 30° and 60° angles are as follows:

$$\sin 30° = \frac{1}{2} \qquad\qquad \sin 60° = \frac{\sqrt{3}}{2}$$

$$\cos 30° = \frac{\sqrt{3}}{2} \qquad\qquad \cos 60° = \frac{1}{2}$$

The Trigonometric Functions of 30°, 45° and 60° in Terms of $\sqrt{2}$ and $\sqrt{3}$

$$\tan 30° = \frac{1}{\sqrt{3}} = \frac{\sqrt{3}}{3} \qquad \tan 60° = \frac{\sqrt{3}}{1} = \sqrt{3}$$

Also, for the 45°–45° right triangle of Fig. 20-2, in which $a = 1$ and $b = 1$, using the Pythagorean theorem, $c = \sqrt{2}$.

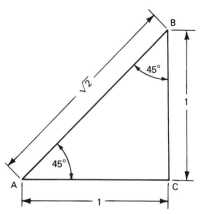

FIG. 20-2

Referring to Fig. 20-2, and using the definitions of the trigonometric functions, the sine, cosine, and tangent of the 45° angle are as follows:

$$\sin 45° = \frac{1}{\sqrt{2}} = \frac{\sqrt{2}}{2}$$

$$\cos 45° = \frac{1}{\sqrt{2}} = \frac{\sqrt{2}}{2}$$

$$\tan 45° = \frac{1}{1} = 1$$

For easy reference the trigonometric functions of the three angles considered above are given in Table 20-1.

TABLE 20-1

	30°	45°	60°
sin	$\frac{1}{2}$	$\frac{\sqrt{2}}{2}$	$\frac{\sqrt{3}}{2}$
cos	$\frac{\sqrt{3}}{2}$	$\frac{\sqrt{2}}{2}$	$\frac{1}{2}$
tan	$\frac{\sqrt{3}}{3}$	1	$\sqrt{3}$

The student should try to memorize the nine entries of this table. What is much easier and better, however, is for the student to form a mental picture of the 30°–60°, and 45°–45° right triangles of Figs. 20-1 and 20-2, and to use the definitions of the functions to produce mentally any of the entries of the table as the need arises.

Complex Numbers

Basic operations with complex numbers of the form $a + bj$ are performed in the same manner as they are with real numbers. *The square roots of negative numbers in such operations should be changed first to the bj form, and then the operations performed.* Below are the descriptions, definitions, and illustrations of these operations.

To add two complex numbers, add separately the real parts and the imaginary parts.

$$(a + bj) + (c + dj) = (a + c) + (b + d)j$$

Thus,

$$(3 + 5j) + (2 + 3j) = (3 + 2) + (5 + 3)j = 5 + 8j$$

and

$$(-4 + 2j) + (5 - 7j) = (-4 + 5) + (2 - 7)j = 1 - 5j$$

To subtract two complex numbers, subtract separately the real parts and the imaginary parts.

$$(a + bj) - (c + dj) = a + bj - c - dj = (a - c) + (b - d)j$$

Thus,

$$(3 + 4j) - (5 - 7j) = 3 + 4j - 5 + 7j = (3 - 5) + (4 + 7)j$$
$$= -2 + 11j$$

and

$$(-2 - 2j) - (-3 + 6j) = -2 - 2j + 3 - 6j$$
$$= (-2 + 3) + (-2 - 6)j$$
$$= 1 - 8j$$

To multiply two complex numbers, proceed as in the case of two ordinary binomials, and then replace j^2 by -1.

$$(a + bj)(c + dj) = ac + adj + bcj + bdj^2 = (ac - bd) + (ad + bc)j$$

Thus,

$$(3 + 2j)(1 - 4j) = 3 + (-12 + 2)j - 8j^2$$
$$= 3 - 10j - 8(-1)$$
$$= 11 - 10j$$

and

$$(-2 - 5j)(3 + 6j) = -6 + (-12 - 15)j - 30j^2$$
$$= -6 - 27j - 30(-1)$$
$$= 24 - 27j$$

To divide two complex numbers, multiply both numerator and denominator of the fraction denoting the division by the conjugate of the denominator, and then replace j^2 by -1.

Basic Operations with Complex Numbers   373

$$\frac{a+bj}{c+dj} = \frac{a+bj}{c+dj} \cdot \frac{c-dj}{c-dj} = \frac{ac-adj+bcj-bdj^2}{c^2-d^2j^2} = \frac{ac+bd+(bc-ad)j}{c^2+d^2}$$

Thus,

$$\frac{2+j}{3-2j} = \frac{2+j}{3-2j} \cdot \frac{3+2j}{3+2j} = \frac{6+7j-2}{9-4j^2} = \frac{4+7j}{13} = \frac{4}{13} + \frac{7}{13}j$$

and

$$\frac{4-2j}{3+5j} = \frac{4-2j}{3+5j} \cdot \frac{3-5j}{3-5j} = \frac{12-26j-10}{9-25j^2} = \frac{2-26j}{34} = \frac{1}{17} - \frac{13}{17}j$$

Notice that this procedure rationalizes the fractional expression, and changes the denominator into a real number.

EXAMPLE 20-6 Simplify $5 - 4j + 7j$.

SOLUTION

$$5 - 4j + 7j = 5 + 3j$$

EXAMPLE 20-7 Simplify $3j - (2j - 3) - 1$.

SOLUTION

$$3j - (2j - 3) - 1 = 3j - 2j + 3 - 1$$
$$= j + 2$$
$$= 2 + j$$

EXAMPLE 20-8 Multiply $3j(2 - 4j)$.

SOLUTION

$$3j(2 - 4j) = 6j - 12j^2$$
$$= 12 + 6j$$

EXAMPLE 20-9 Divide $\dfrac{3}{-5j}$.

SOLUTION

$$\frac{3}{-5j} = \frac{3}{-5j} \cdot \frac{5j}{5j} = \frac{15j}{-25j^2}$$
$$= \frac{15j}{25}$$
$$= \frac{3}{5}j$$

EXAMPLE 20-10 Multiply $(5 - \sqrt{-4})(6 + \sqrt{-9})$.

SOLUTION

$$(5 - \sqrt{-4})(6 + \sqrt{-9}) = (5 - 2j)(6 + 3j)$$
$$= 30 + (15 - 12)j - 6j^2$$
$$= 36 + 3j$$

EXAMPLE 20-11 Subtract $\sqrt{-8} - \sqrt{-2}$.

SOLUTION

$$\sqrt{-8} - \sqrt{-2} = \sqrt{8}\,j - \sqrt{2}\,j$$
$$= 2\sqrt{2}\,j - \sqrt{2}\,j$$
$$= \sqrt{2}\,j$$

Perform each of the following operations and express the result in the form $a + bj$.

1. $3 + 7j - 2j$

2. $(3 + 6j)(2 - 5j)$

3. $(4 + 3j)(4 - 3j)$

4. $\dfrac{7j}{4 - 5j}$

5. $3j - (8j + 3) - 5$

6. $(5 - 2j) - (3 + 4j)$

7. $(5 - j) + (2 - 9j)$

8. $\dfrac{2 - 3j}{3 + 6j}$

9. $(5 - 3j) + (7 + 2j)$

10. $(4 - 3j) - (3 + 5j)$

11. $7 - \sqrt{-72}$

12. $(\sqrt{-9} + 2)(\sqrt{-4})$

13. $(3 + \sqrt{-16}) - (\sqrt{-36} + 5)$

14. $\dfrac{1 - \sqrt{-4}}{3 + \sqrt{-9}}$

15. $(3 + 4j) + (3 - 4j)$

16. $(8 - 3j) - (8 + 3j)$

17. $(3 - \sqrt{-25}) - (9 - \sqrt{-4})$

18. $\dfrac{2 + j}{3 - 2j}$

19. $\dfrac{2}{j}$

20. $(7 - \sqrt{-25}) + (\sqrt{-49} - 2)$

21. $3\sqrt{-32} + 5\sqrt{-18}$

22. $(4 - 5\sqrt{-3})^2$

23. $\dfrac{2 - 5j}{4 + 3j}$

24. $\dfrac{3}{1 - \sqrt{-2}}$

25. $\dfrac{j}{j + 2}$

26. $(3 + 2j)^2$

27. $(4 - 3j)^2$

28. $\dfrac{6}{5j}$

29. $\dfrac{2 + 6j}{3 + j}$

30. $(2 + 7j) + (2 + j)$

31. $3(5 - 2j) + j(2 + j)$

32. $(3 + j) - (4 - 3j)$

33. $(1 - 2j) - (-5 + 3j)$

34. $\dfrac{1 + 3j}{2 + j}$

35. $(3 - 2j)(2 + j)$

36. $(-3 + 4j) + (5 + 6j)$

37. $(-1 - j) - (-3 + 2j)$

38. $(2 - 4j) + (-3 - 5j)$

39. $(1 + 3j) + (4 - 7j)$

40. $(-1 + 6j) - (3 - 5j)$

41. $\dfrac{2 + 2j}{3 - 3j}$

42. $(5 + j)^2$

43. $(3 + 2j)(1 - 4j)$

44. $(5 - 3j)(-2 + 4j)$

45. $\dfrac{3 - 2j}{5 + j}$

46. $\dfrac{3}{-8j}$

Basic Operations with Complex Numbers

20-4 Graphic Representation of Complex Numbers.
Rectangular Form

We can represent a complex number geometrically. The algebraic form $a + bj$ of a complex number is more often called the **rectangular form** of a complex number because it can be represented graphically in a rectangular coordinate system. In such a system we will call the horizontal axis RR' (see Fig. 20-3) the **real axis**, and the vertical axis II' the **imaginary axis**. Notice that in the rectangular coordinate system for complex numbers also, like in the coordinate system we have used before, there are positive and negative parts in the two axes. Thus, one unit of length in the direction OI will represent j, and one unit of length in the direction OI' will represent $-j$.

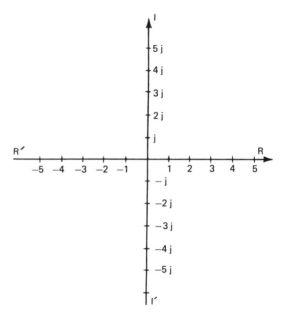

FIG. 20-3

The plane of the axes RR' and II' is called the **complex plane**.

Since the two axes RR' and II' are infinite in length, any complex number can be represented in this system. The two numbers $3 - 5j$ and $2 + 3j$ are shown in Fig. 20-4 as points A and B, respectively. The complex number $3 - 5j$ was plotted by locating the point which is 3 units to the right of the II' axis and 5 units below the RR' axis. Also, the complex number $2 + 3j$ was plotted by locating the point which is two units to the right of the II' axis, and 3 units above the RR' axis.

It must be evident now that to each point of the complex plane there corresponds one complex number, and conversely, to every complex number there corresponds one point of the complex plane.

The student should notice that since the complex number $5 + 0j$ is a real number it should be plotted as a point on the RR' axis. Also, since the complex

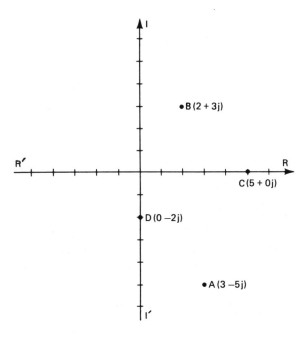

FIG. 20-4

number $0 - 2j$ is a pure imaginary number it should be plotted as a point on the II' axis (see points C and D of Fig. 20-4).

EXAMPLE 20-12 What complex numbers are represented by the points A, B, C, D, E, and F of Fig. 20-5?

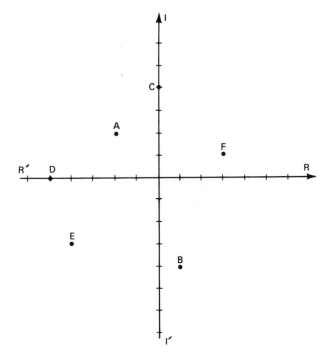

FIG. 20-5

SOLUTION

$$A = (-2 + 2j)$$
$$B = (1 - 4j)$$
$$C = (0 + 4j)$$
$$D = (-5 + 0j)$$
$$E = (-4 - 3j)$$
$$F = (3 + j)$$

A complex number is represented and is interpreted as a point in many applications. In other applications, electronic circuits for example, it is more convenient to interpret a complex number as a line segment which has both magnitude and direction, that is, as a **vector**.

The complex number $4 + 3j$ can be interpreted as a vector if a line segment is used to connect point $P(4 + 3j)$ with the origin, and an arrowhead is placed at $P(4 + 3j)$ (see Fig. 20-6). We say then that the complex number $4 + 3j$ is represented by $\overrightarrow{OP}$ of Fig. 20-6.

Similarly the complex numbers $0 + 3j$ and $5 + 0j$ are represented in Fig. 20-7 by $\overrightarrow{OP}$ and $\overrightarrow{OQ}$, respectively.

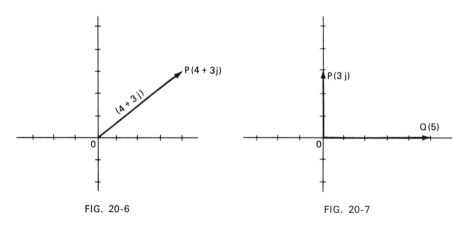

FIG. 20-6   FIG. 20-7

Like the vectors we have seen in the previous chapter, vectors representing two complex numbers can also be added. As we have already seen, the vector sum (resultant) of two given vectors is the diagonal of the parallelogram for which the two vectors (components) are adjacent sides. Thus, in Fig. 20-8 $\overrightarrow{OP} + \overrightarrow{OQ} = \overrightarrow{OR}$. Notice that graphically $\overrightarrow{OP} + \overrightarrow{OQ} = 5 + 4j$, and algebraically $(2 + 3j) + (3 + j) = 5 + 4j$.

EXAMPLE 20-13 Subtract $1 + 3j$ from $5 + 2j$ graphically and check the subtraction algebraically.

SOLUTION We know that

$$(5 + 2j) - (1 + 3j) = (5 + 2j) + (-1 - 3j)$$

378   *Complex Numbers*

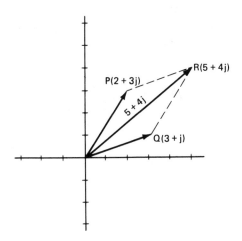

FIG. 20-8

Therefore, instead of subtracting $(1 + 3j)$ from $(5 + 2j)$ it will be the same to add $(-1 - 3j)$ to $(5 + 2j)$. To add $5 + 2j$ and $-1 - 3j$ graphically we proceed as before. The addition is shown in Fig. 20-9.

Graphically $\overrightarrow{OA} - \overrightarrow{OB} = 4 - j$, and algebraically $(5 + 2j) - (1 + 3j) = 4 - j$

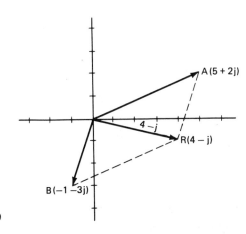

FIG. 20-9

In exercises 1–9, perform the operations graphically and check the result algebraically.

1. $(6 - 3j) + (2 + 4j)$
2. $(-2 + 3j) - (3 + j)$
3. $(2 - 3j) - (-5 + j)$
4. $(3 + 2j) + (5 - 3j)$
5. $(4 + 3j) - (-1 - 2j)$
6. $(-3 - 2j) + (-2 - 5j)$
7. $(3 - j) + 5j$
8. $(3 - 2j) - (2 + j)$
9. $(-2 + 2j) + (2 + 4j)$

Graphic Representation of Complex Numbers. Rectangular Form   379

20-5 Trigonometric Form of Complex Numbers

If we express the complex number $a + bj$ in terms of x and y we get $x + yj$. Representing $x + yj$ as a vector in the coordinate system of x- and y-axes we have the familiar case shown in Fig. 20-10, which we have already seen in Fig. 18-12(b) of Sect. 18-2. Notice that the line segment connecting the origin with point $P(x, y)$ is $\overrightarrow{OP}$ representing the complex number $x + yj$. Notice, also, that we let r stand for the length of $\overrightarrow{OP}$, and θ stand for the angle $\overrightarrow{OP}$ makes with the x-axis.

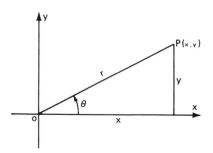

FIG. 20-10

From Fig. 20-10, the relationships (a), (b), (c), and (d) shown below easily follow when we use the Pythagorean theorem and the definitions of the trigonometric functions.

$$\overrightarrow{OP} = r = \sqrt{x^2 + y^2} \tag{a}$$

$$\sin \theta = \frac{y}{r} \quad \text{or} \quad y = r \sin \theta \tag{b}$$

$$\cos \theta = \frac{x}{r} \quad \text{or} \quad x = r \cos \theta \tag{c}$$

$$\tan \theta = \frac{y}{x} \tag{d}$$

The length of $\overrightarrow{OP}$, the number r, is called the **modulus**, or **absolute value** of the complex number $x + yj$, and it is always positive.

Angle θ, the angle $\overrightarrow{OP}$ makes with the positive x-axis of the real numbers, is called the **amplitude**, or **argument** of the complex number $x + yj$.

Substituting from the above relationships (a), (b), (c), and (d) in $x + yj$,

$$r \cos \theta \text{ for } x \text{ and } r \sin \theta \text{ for } y$$

we have

$$x + yj = r \cos \theta + j(r \sin \theta) \quad \text{or} \quad x + yj = r(\cos \theta + j \sin \theta) \tag{e}$$

Complex Numbers

The expression $r(\cos\theta + j\sin\theta)$ is called the **trigonometric form** or the **polar form** of a complex number. Thus, $5(\cos 25° + j\sin 25°)$, $2(\cos 225° + j\sin 225°)$, and $3(\cos 60° + j\sin 60°)$ are trigonometric forms of complex numbers.

EXAMPLE 20-14 Express $2 + 2j$ in trigonometric form.

SOLUTION In the rectangular form of the complex number $2 + 2j$,

$$x = 2, \qquad y = 2,$$

$$r = \sqrt{2^2 + 2^2} = \sqrt{8} = 2\sqrt{2}, \qquad \text{and} \qquad \tan\theta = \frac{y}{x} = \frac{2}{2} = 1$$

Since both x and y are positive, the vector representing $2 + 2j$ lies in the first quadrant, and the first quadrant angle whose tangent is 1 is the 45° angle.

Substituting in equation (e), 2 for x, 2 for y, $2\sqrt{2}$ for r, and 45° for θ, we have

$$2 + 2j = 2\sqrt{2}(\cos 45° + j\sin 45°)$$

The complex number $2 + 2j$ is shown graphically in Fig. 20-11.

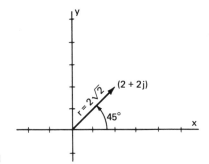

FIG. 20-11

EXAMPLE 20-15 Express $\sqrt{3} - j$ in trigonometric form.

SOLUTION

$$x = \sqrt{3}$$

$$y = -1$$

$$r = \sqrt{(\sqrt{3})^2 + (-1)^2} = \sqrt{4} = 2$$

$$\tan\theta = \frac{-1}{\sqrt{3}} = -\frac{\sqrt{3}}{3}$$

and θ is a 30° angle. Since x is positive and y is negative, the vector representing $\sqrt{3} - j$ lies in the fourth quadrant and $\theta = 360° - 30° = 330°$.

Substituting in equation (e) we have,

$$\sqrt{3} - j = 2(\cos 330° + j\sin 330°)$$

The complex number $\sqrt{3} - j$ is shown graphically in Fig. 20-12.

Trigonometric Form of Complex Numbers

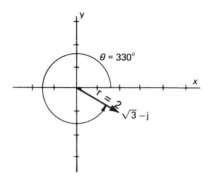

FIG. 20-12

EXAMPLE 20-16   Express $3 - 4j$ in trigonometric form.

SOLUTION

$$x = 3$$
$$y = -4$$
$$r = \sqrt{3^2 + (-4)^2} = 5$$
$$\tan \theta = \frac{-4}{3} = -1.33333$$

By interpolation, the angle whose tangent is 1.33333 is a 53° 8′ angle. Since x is positive and y is negative, θ is a fourth quadrant angle (see Fig. 20-13). Therefore $\theta = 360°-53° 8′ = 306° 52′$. Thus,

$$3 - 4j = 5(\cos 306° 52' + j \sin 306° 52')$$

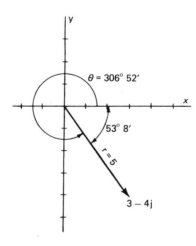

FIG. 20-13

EXAMPLE 20-17   Express $6(\cos 60° + j \sin 60°)$ in rectangular form.

SOLUTION   By equation (c) $x = r \cos \theta$, and by equation (b) $y = r \sin \theta$. Since $r = 6$ and $\theta = 60°$, we have

$$x = r \cos \theta$$
$$= 6 \cos 60°$$

Complex Numbers

$$= 6\left(\frac{1}{2}\right) \qquad (\textit{see Table 20-1})$$

$$= 3$$

$$y = r \sin \theta$$

$$= 6 \sin 60°$$

$$= 6\left(\frac{\sqrt{3}}{2}\right) \qquad (\textit{see Table 20-1})$$

$$= 3\sqrt{3}$$

Thus, $6(\cos 60° + j \sin 60°) = 3 + 3\sqrt{3}\,j$.

EXAMPLE 20-18 Express $2(\cos 315° + j \sin 315°)$ in rectangular form.

SOLUTION

$$r = 2$$

$$\theta = 315°$$

Since angle θ is a fourth quadrant angle, $\cos \theta$ is positive and $\sin \theta$ is negative.

$$x = 2[\cos(360° - 315°)]$$

$$= 2 \cos 45°$$

$$= 2\left(\frac{\sqrt{2}}{2}\right) \qquad (\textit{see Table 20-1})$$

$$= \sqrt{2}$$

$$y = 2[\sin(360° - 315°)]$$

$$= 2(-\sin 45°)$$

$$= 2\left(-\frac{\sqrt{2}}{2}\right) \qquad (\textit{see Table 20-1})$$

$$= -\sqrt{2}$$

Thus,

$$2(\cos 315° + j \sin 315°) = \sqrt{2} - \sqrt{2}\,j.$$

EXAMPLE 20-19 Express $3(\cos 305° + j \sin 305°)$ in rectangular form.

SOLUTION

$$r = 3$$

$$\theta = 305°$$

Since θ is a fourth quadrant angle, $\cos \theta$ is positive and $\sin \theta$ is negative.

$$x = 3[\cos(360° - 305°)]$$

$$= 3 \cos 55°$$

$$= 3(0.5736)$$

$$= 1.7208$$

$$y = 3[\sin(360° - 305°)]$$

$$= 3(-\sin 55°)$$

$$= 3(-0.8192)$$

$$= -2.4576.$$

Trigonometric Form of Complex Numbers

Thus,

$$3(\cos 305° + j \sin 305°) = 1.7208 - 2.4576j.$$

EXERCISE 20-4

Express each complex number in exercises 1–10 in trigonometric form.
1. $-5 + 5j$ 2. $6 - 6j$
3. $-1 + \sqrt{3}\,j$ 4. $-3 - 3j$
5. 9   6. $-2 - 2\sqrt{3}\,j$
7. $2\sqrt{3} - 2j$ 8. $-\sqrt{3} + j$
9. $3\sqrt{3} + 5j$ 10. $-2\sqrt{5} - 4j$

Express each complex number in exercises 11–20 in rectangular form.
11. $3(\cos 45° + j \sin 45°)$ 12. $6(\cos 240° + j \sin 240°)$
13. $4(\cos 315° + j \sin 315°)$ 14. $3(\cos 90° + j \sin 90°)$
15. $2(\cos 270° + j \sin 270°)$ 16. $6(\cos 180° + j \sin 180°)$
17. $4(\cos 150° + j \sin 150°)$ 18. $3(\cos 28° + j \sin 28°)$
19. $2(\cos 143° \, 10' + j \sin 143° \, 10')$ 20. $4(\cos 312° \, 17' + j \sin 312° \, 17')$

20-6 Multiplication and Division of Complex Numbers in Trigonometric Form

To multiply two complex numbers in trigonometric form, multiply their absolute values, and add their amplitudes.

$$[r_1(\cos \theta_1 + j \sin \theta_1)][r_2(\cos \theta_2 + j \sin \theta_2)] = r_1 r_2[\cos (\theta_1 + \theta_2) + j \sin (\theta_1 + \theta_2)]$$

Thus,

$$[3(\cos 230° + j \sin 230°)][5(\cos 50° + j \sin 50°)] = 15(\cos 280° + j \sin 280°)$$

To divide two complex numbers in trigonometric form, divide their absolute values, and subtract their amplitudes.

$$\frac{r_1(\cos \theta_1 + j \sin \theta_1)}{r_2(\cos \theta_2 + j \sin \theta_2)} = \frac{r_1}{r_2}[\cos (\theta_1 - \theta_2) + j \sin (\theta_1 - \theta_2)]$$

Thus,

$$6(\cos 240° + j \sin 240°) \div 2(\cos 50° + j \sin 50°) = 3(\cos 190° + j \sin 190°)$$

EXAMPLE 20-20   Multiply $5(\cos 50° + j \sin 50°)$ and $4(\cos 15° + j \sin 15°)$.

SOLUTION

$$[5(\cos 50° + j \sin 50°)][4(\cos 15° + j \sin 15°)]$$
$$= (5)(4)[\cos(50° + 15°) + j \sin(50° + 15°)]$$
$$= 20(\cos 65° + j \sin 65°)$$

EXAMPLE 20-21   Divide $12(\cos 315° + j \sin 315°)$ by $2(\cos 85° + j \sin 85°)$. Give your answer in rectangular form.

SOLUTION

$$\frac{12(\cos 315° + j \sin 315°)}{2(\cos 85° + j \sin 85°)} = \frac{12}{2}[\cos(315° - 85°) + j \sin(315° - 85°)]$$

$$= 6(\cos 230° + j \sin 230°)$$

$$r = 6$$

$$\theta = 230°$$

Since θ is a third quadrant angle, $\cos \theta$ is negative, and $\sin \theta$ is negative.

$$x = 6[\cos(230° - 180°)]$$

$$= 6(-\cos 50°)$$

$$= 6(-0.6428)$$

$$= -3.8568$$

and

$$y = 6[\sin(230° - 180°)]$$

$$= 6(-\sin 50°)$$

$$= 6(-0.7660)$$

$$= -4.5960$$

Thus,

$$6(\cos 230° + j \sin 230°) = -3.8568 - 4.5960j.$$

EXERCISE 20-5

In exercises 1–10 multiply or divide as indicated. Express answers in rectangular form.

1. $5(\cos 63° + j \sin 63°) \cdot 2(\cos 19° + j \sin 19°)$
2. $3(\cos 140° + j \sin 140°) \cdot 2(\cos 65° + j \sin 65°)$
3. $7(\cos 210° + j \sin 210°) \div (\cos 15° + j \sin 15°)$
4. $10(\cos 343° + j \sin 343°) \div 2(\cos 18° + j \sin 18°)$
5. $4(\cos 93° + j \sin 93°) \cdot 2(\cos 24° + j \sin 24°)$
6. $6(\cos 82° + j \sin 82°) \div 2(\cos 34° + j \sin 34°)$
7. $8(\cos 147° + j \sin 147°) \cdot \frac{1}{2}(\cos 88° + j \sin 88°)$
8. $16(\cos 122° + j \sin 122°) \div 8(\cos 43° + j \sin 43°)$
9. $9(\cos 347° + j \sin 347°) \div 3(\cos 219° + j \sin 219°)$
10. $3(\cos 357° + j \sin 357°) \div (\cos 110° + j \sin 110°)$

20-7 The *j*-Operator

In the first section of this chapter the symbol j was called the j-operator. We could not have given then any reason for this name; your background in complex numbers was not enough. Now that you know more about complex numbers, forms of, operations with, and graphic representation of complex numbers, the following explanation will be easy to understand. Here is why the symbol j is called an operator:

The effect of multipying a complex number, $a + bj$, by j is to rotate the vector representation of the number through an angle of 90° *in the positive, or counterclockwise direction, without any change of the modulus.* For instance, the vector of $5 + 2j$ is shown in Fig. 20-14. If we multiply $5 + 2j$ by j the product will be $5j + 2j^2$, or $-2 + 5j$. The

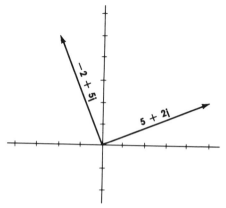

FIG. 20-14

graph of this product, that is, the vector $-2 + 5j$, is also shown in Fig. 20-14. The student should notice that the vector $5 + 2j$ was rotated by an angle of $90°$, and the length r of each of the vectors $5 + 2j$ and $-2 + 5j$ is the same.

EXAMPLE 20-22   Multiply the complex number $-7 + 3j$ by j, and graph the vectors of both the number and the product.

SOLUTION

$$j(-7 + 3j) = -7j + 3j^2 = -3 - 7j$$

The graphs of both $-7 + 3j$ and $-3 - 7j$ are shown in Fig. 20-15.

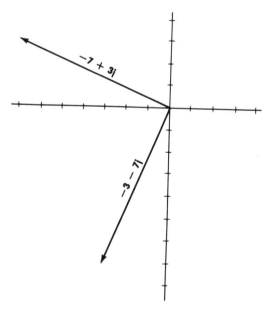

FIG. 20-15

Complex Numbers

In exercises 1–8 show graphically the effect of multiplying the given complex number by j.

1. $5 - 3j$
2. $-2 + 5j$
3. $-4 - 3j$
4. $3 + 5j$
5. $6 - 2j$
6. $5 - 5j$
7. $3 + 7j$
8. $-8 - 3j$

REVIEW QUESTIONS

1. What is the square root of a negative number called?
2. Which is the algebraic form of a complex number? What is another name for the algebraic form of a complex number?
3. What are the names of the two parts of the imaginary number $a + bj$?
4. What is the conjugate of a complex number?
5. How are two complex numbers of the form $a + bj$ added?
6. How are two complex numbers of the form $a + bj$ subtracted?
7. How are two complex numbers of the form $a + bj$ multiplied?
8. How are two complex numbers of the form $a + bj$ divided?
9. What are the axes, and the plane of the axes, called in which the complex number of the form $a + bj$ is represented graphically?
10. How is a complex number of the form $a + bj$ represented as a point?
11. How is a complex number of the form $a + bj$ represented as a vector?
12. How are two vectors representing two complex numbers of the form $a + bj$ added graphically?
13. How are two vectors representing two complex numbers of the form $a + bj$ subtracted graphically?
14. What is the modulus (or absolute value) and symbol of a complex number of the form $x + yj$?
15. What is the amplitude (or argument) and symbol of a complex number of the form $x + yj$?
16. What is a complex number of the form $r(\cos \theta + j \sin \theta)$ called?
17. What is the effect of multiplying a complex number by j?

REVIEW EXERCISES

In exercises 1–4, express the imaginary number in terms of the j-operator.

1. $\sqrt{-36}$
2. $\sqrt{-68}$
3. $\sqrt{-45}$
4. $\sqrt{-175}$

In exercises 5–8, simplify the power of j.

5. j^{13}
6. j^{71}
7. j^{208}
8. j^{42}

In exercises 9–12, change the number to the form $a + bj$.

9. $8 + \sqrt{-16}$
10. $-13 - \sqrt{-12}$

11. $-3j^3 + 5j^2$ **12.** $2j^6 - 3j^5$

In exercises 13–16, name the conjugate of each complex number.

13. $12 - 9j$ **14.** $7j - 11$

15. -19 **16.** $-\sqrt{-17}$

In exercises 17–34, perform the operation indicated. Express the result in the form $a + bj$.

17. $(3 - 2j)(3 + 2j)$ **18.** $(7 + 3j) + (-15 - 8j)$

19. $(5 + \sqrt{-9}) - (\sqrt{-81} - 2)$ **20.** $\dfrac{2 - \sqrt{-49}}{7 + \sqrt{-4}}$

21. $\dfrac{5 + 2j}{1 - 3j}$ **22.** $(8 - \sqrt{-4}) - (3j - 1)$

23. $(7j - 7) + (2 + 6j)$ **24.** $\dfrac{3j}{5 - 2j} \cdot \dfrac{6 + 2j}{5 + 2j}$

In exercises 25–27, perform the operation graphically, and check the result algebraically.

25. $(3 + 6j) - (-2 + 3j)$ **26.** $(2 + j) + (4 - 2j)$

27. $(4 + 3j) + (2 - j)$

Express each complex number in exercises 28–30 in trigonometric form.

28. $\sqrt{3} + j$ **29.** $-1 - 5j$

30. $-2 + 5j$

Express each complex number in exercises 31–33 in rectangular form.

31. $4(\cos 225° + j \sin 225°)$ **32.** $2(\cos 144° + j \sin 144°)$

33. $5(\cos 237° \, 20' + j \sin 237° \, 20')$

In exercise 34, perform the operation indicated. Express answer in rectangular form.

34. $6(\cos 253° + j \sin 253°) \cdot \frac{1}{2}(\cos 58° + j \sin 58°)$

In exercises 35 and 36, show graphically the effect of multiplying the complex number by j.

35. $2 + 3j$ **36.** $-3 - 3j$

APPENDICES

Tables of Trigonometric Functions

Degrees	Sin	Cos	Tan	Ctn	
0° 00'	0.0000	1.0000	0.0000		90° 00'
10	0.0029	1.0000	0.0029	343.77	50
20	0.0058	1.0000	0.0058	171.89	40
30	0.0087	1.0000	0.0087	114.59	30
40	0.0116	0.9999	0.0116	85.940	20
50	0.0145	0.9999	0.0145	68.750	10
1° 00'	0.0175	0.9998	0.0175	57.290	89° 00'
10	0.0204	0.9998	0.0204	49.104	50
20	0.0233	0.9997	0.0233	42.964	40
30	0.0262	0.9997	0.0262	38.188	30
40	0.0291	0.9996	0.0291	34.368	20
50	0.0320	0.9995	0.0320	31.242	10
2° 00'	0.0349	0.9994	0.0349	28.636	88° 00'
10	0.0378	0.9993	0.0378	26.432	50
20	0.0407	0.9992	0.0407	24.542	40
30	0.0436	0.9990	0.0437	22.904	30
40	0.0465	0.9989	0.0466	21.470	20
50	0.0494	0.9988	0.0495	20.206	10
3° 00'	0.0523	0.9986	0.0524	19.081	87° 00'
10	0.0552	0.9985	0.0553	18.075	50
20	0.0581	0.9983	0.0582	17.169	40
30	0.0610	0.9981	0.0612	16.350	30
40	0.0640	0.9980	0.0641	15.605	20
50	0.0669	0.9978	0.0670	14.924	10
4° 00'	0.0698	0.9976	0.0669	14.301	86° 00'
10	0.0727	0.9974	0.0729	13.727	50
20	0.0756	0.9971	0.0758	13.197	40
30	0.0785	0.9969	0.0787	12.706	30
40	0.0814	0.9967	0.0816	12.251	20
50	0.0843	0.9964	0.0846	11.826	10
5° 00'	0.0872	0.9962	0.0875	11.430	85° 00'
10	0.0901	0.9959	0.0904	11.059	50
20	0.0929	0.9957	0.0934	10.712	40
30	0.0958	0.9954	0.0963	10.385	30
40	0.0987	0.9951	0.0992	10.078	20
50	0.1016	0.9948	0.1022	9.7882	10
6° 00'	0.1045	0.9945	0.1051	9.5144	84° 00'
10	0.1074	0.9942	0.1080	9.2553	50
20	0.1103	0.9939	0.1110	9.0098	40
30	0.1132	0.9936	0.1139	8.7769	30
40	0.1161	0.9932	0.1169	8.5555	20
50	0.1190	0.9929	0.1198	8.3450	10
7° 00'	0.1219	0.9925	0.1228	8.1443	83° 00'
10	0.1248	0.9922	0.1257	7.9530	50
20	0.1276	0.9918	0.1287	7.7704	40
30	0.1305	0.9914	0.1317	7.5958	30
40	0.1334	0.9911	0.1436	7.4287	20
50	0.1363	0.9907	0.1376	7.2687	10
8° 00'	0.1392	0.9903	0.1405	7.1154	82° 00'
10	0.1421	0.9899	0.1435	6.9682	50
20	0.1449	0.9894	0.1465	6.8269	40
30	0.1478	0.9890	0.1495	6.6912	30
40	0.1507	0.9886	0.1524	6.6506	20
50	0.1536	0.9881	0.1554	6.4348	10
9° 00'	0.1564	0.9877	0.1584	6.3138	81° 00'
	Cos	Sin	Ctn	Tan	Degrees

Tables of Trigonometric Functions

Degrees	Sin	Cos	Tan	Ctn	
9° 00'	0.1564	0.9877	0.1584	6.3138	81° 00'
10	0.1593	0.9872	0.1614	6.1970	50
20	0.1622	0.9868	0.1644	6.0844	40
30	0.1650	0.9863	0.1673	5.9758	30
40	0.1679	0.9858	0.1703	5.8708	20
50	0.1708	0.9853	0.1733	5.7694	10
0°00'	0.1736	0.9848	0.1763	5.6713	80° 00'
10	0.1765	0.9843	0.1793	5.5764	50
20	0.1794	0.9838	0.1823	5.4845	40
30	0.1822	0.9833	0.1853	5.3955	30
40	0.1851	0.9827	0.1883	5.3093	20
50	0.1880	0.9822	0.1914	5.2257	10
11°00'	0.1908	0.9816	0.1944	5.1446	79° 00'
10	0.1937	0.9811	0.1974	5.0658	50
20	0.1965	0.9805	0.2004	4.9894	40
30	0.1994	0.9799	0.2035	4.9152	30
40	0.2022	0.9793	0.2065	4.8430	20
50	0.2051	0.9787	0.2095	4.7729	10
12°00'	0.2079	0.9781	0.2126	4.7046	78° 00'
10	0.2108	0.9775	0.2156	4.6382	50
20	0.2136	0.9769	0.2186	4.5736	40
30	0.2164	0.9763	0.2217	4.5107	30
40	0.2193	0.9757	0.2247	4.4494	20
50	0.2221	0.9750	0.2278	4.3897	10
13°00'	0.2250	0.9744	0.2309	4.3315	77° 00'
10	0.2278	0.9737	0.2339	4.2747	50
20	0.2306	0.9730	0.2370	4.2193	40
30	0.2334	0.9724	0.2401	4.1653	30
40	0.2363	0.9717	0.2432	4.1126	20
50	0.2391	0.9710	0.2462	4.0611	10
14°00'	0.2419	0.9703	0.2493	4.0108	76° 00'
10	0.2447	0.9696	0.2524	3.9617	50
20	0.2476	0.9689	0.2555	3.9136	40
30	0.2504	0.9681	0.2586	3.8667	30
40	0.2532	0.9674	0.2617	3.8208	20
50	0.2560	0.9667	0.2648	3.7760	10
15°00'	0.2588	0.9659	0.2679	3.7321	75° 00'
10	0.2616	0.9652	0.2711	3.6891	50
20	0.2644	0.9644	0.2742	3.6470	40
30	0.2672	0.9636	0.2778	3.6059	30
40	0.2700	0.9628	0.2805	3.5656	20
50	0.2728	0.9621	0.2836	3.5261	10
16°00'	0.2756	0.9613	0.2867	3.4874	74° 00'
10	0.2784	0.9605	0.2899	3.4495	50
20	0.2812	0.9596	0.2931	3.4124	40
30	0.2840	0.9588	0.2962	3.3759	30
40	0.2868	0.9580	0.2994	3.3402	20
50	0.2896	0.9572	0.3026	3.3052	10
17°00'	0.2924	0.9563	0.3057	3.2709	73° 00'
10	0.2952	0.9555	0.3089	3.2371	50
20	0.2979	0.9546	0.3121	3.2041	40
30	0.3007	0.9537	0.3153	3.1716	30
40	0.3035	0.9528	0.3185	3.1397	20
50	0.3062	0.9520	0.3217	3.1084	10
18°00'	0.3090	0.9511	0.3249	3.0777	72° 00'

	Cos	Sin	Ctn	Tan	Degrees

Tables of Trigonometric Functions

Degrees		Sin	Cos	Tan	Ctn		
18°	00'	0.3090	0.9511	0.3249	3.0777	72°	00'
	10	0.3118	0.9502	0.3281	3.0475		50
	20	0.3145	0.9492	0.3314	3.0178		40
	30	0.3173	0.9483	0.3346	2.9887		30
	40	0.3201	0.9474	0.3378	2.9600		20
	50	0.3228	0.9465	0.3411	2.9319		10
19°	00'	0.3256	0.9455	0.3443	2.9042	71°	00'
	10	0.3283	0.9446	0.3476	2.8770		50
	20	0.3311	0.9436	0.3508	2.8502		40
	30	0.3338	0.9426	0.3541	2.8239		30
	40	0.3365	0.9417	0.3574	2.7980		20
	50	0.3393	0.9407	0.3607	2.7725		10
20°	00'	0.3420	0.9379	0.3640	2.7475	70°	00'
	10	0.3448	0.9387	0.3673	2.7228		50
	20	0.3475	0.9377	0.3706	2.6985		40
	30	0.3502	0.9367	0.3739	2.6746		30
	40	0.3529	0.9356	0.3772	2.6511		20
	50	0.3557	0.9346	0.3805	2.6279		10
21°	00'	0.3584	0.9336	0.3839	2.6051	69°	00'
	10	0.3611	0.9325	0.3872	2.5826		50
	20	0.3638	0.9315	0.3906	2.5605		40
	30	0.3665	0.9304	0.3939	2.5386		30
	40	0.3692	0.9293	0.3973	2.5172		20
	50	0.3719	0.9283	0.4006	2.4960		10
22°	00'	0.3746	0.9272	0.4040	2.4751	68°	00'
	10	0.3773	0.9261	0.4074	2.4545		50
	20	0.3800	0.9250	0.4108	2.4342		40
	30	0.3827	0.9239	0.4142	2.4142		30
	40	0.3854	0.9228	0.4176	2.3945		20
	50	0.3881	0.9216	0.4210	2.3750		10
23°	00'	0.3907	0.9205	0.4245	2.3559	67°	00'
	10	0.3934	0.9194	0.4279	2.3369		50
	20	0.3961	0.9182	0.4314	2.3183		40
	30	0.3987	0.9171	0.4348	2.2998		30
	40	0.4014	0.9159	0.4383	2.2817		20
	50	0.4041	0.9147	0.4417	2.2637		10
24°	00'	0.4067	0.9135	0.4452	2.2460	66°	00'
	10	0.4094	0.9124	0.4487	2.2286		50
	20	0.4120	0.9112	0.4522	2.2113		40
	30	0.4147	0.9100	0.4557	2.1943		30
	40	0.4173	0.9088	0.4592	2.1775		20
	50	0.4200	0.9075	0.4628	2.1609		10
25°	00'	0.4226	0.9063	0.4663	2.1445	65°	00'
	10	0.4253	0.9051	0.4699	2.1283		50
	20	0.4279	0.9038	0.4734	2.1123		40
	30	0.4305	0.9026	0.4770	2.0965		30
	40	0.4331	0.9013	0.4806	2.0809		20
	50	0.4358	0.9001	0.4841	2.0665		10
26°	00'	0.4384	0.8988	0.4877	2.0503	64°	00'
	10	0.4410	0.8975	0.4913	2.0353		50
	20	0.4436	0.8962	0.4950	2.0204		40
	30	0.4462	0.8949	0.4986	2.0057		30
	40	0.4488	0.8936	0.5022	1.9912		20
	50	0.4514	0.8923	0.5059	1.9768		10
27°	00'	0.4540	0.8910	0.5095	1.9626	63°	00'
		Cos	Sin	Ctn	Tan		Degrees

Tables of Trigonometric Functions

Degrees		Sin	Cos	Tan	Ctn		
27°	00'	0.4540	0.8910	0.5095	1.9626	63°	00'
	10	0.4566	0.8897	0.5132	1.9486		50
	20	0.4592	0.8884	0.5169	1.9347		40
	30	0.4617	0.8870	0.5206	1.9210		30
	40	0.4643	0.8857	0.5243	1.9074		20
	50	0.4669	0.8843	0.5280	1.8940		10
28°	00'	0.4695	0.8829	0.5317	1.8807	62°	00'
	10	0.4720	0.8816	0.5354	1.8676		50
	20	0.4746	0.8802	0.5392	1.8546		40
	30	0.4772	0.8788	0.5430	1.8418		30
	40	0.4797	0.8774	0.5467	1.8291		20
	50	0.4823	0.8760	0.5505	1.8165		10
29°	00'	0.4848	0.8746	0.5543	1.8040	61°	00'
	10	0.4874	0.8732	0.5581	1.7917		50
	20	0.4899	0.8718	0.5619	1.7796		40
	30	0.4924	0.8704	0.5658	1.7675		30
	40	0.4950	0.8689	0.5696	1.7556		20
	50	0.4975	0.8675	0.5735	1.7437		10
30°	00'	0.5000	0.8660	0.5774	1.7321	60°	00'
	10	0.5025	0.8646	0.5812	1.7205		50
	20	0.5050	0.8631	0.5851	1.7090		40
	30	0.5075	0.8616	0.5890	1.6977		30
	40	0.5100	0.8601	0.5930	1.6864		20
	50	0.5125	0.8587	0.5969	1.6753		10
31°	00'	0.5150	0.8572	0.6009	1.6643	59°	00'
	10	0.5175	0.8557	0.6048	1.6534		50
	20	0.5200	0.8542	0.6088	1.6426		40
	30	0.5225	0.8526	0.6128	1.6319		30
	40	0.5250	0.8511	0.6168	1.6212		20
	50	0.5275	0.8496	0.6208	1.6107		10
32°	00'	0.5299	0.8480	0.6249	1.6003	58°	00'
	10	0.5324	0.8465	0.6289	1.5900		50
	20	0.5348	0.8450	0.6330	1.5798		40
	30	0.5373	0.8434	0.6371	1.5697		30
	40	0.5398	0.8418	0.6412	1.5597		20
	50	0.5422	0.8403	0.6453	1.5497		10
33°	00'	0.5446	0.8387	0.6494	1.5399	57°	00'
	10	0.5471	0.8371	0.6536	1.5301		50
	20	0.5495	0.8355	0.6577	1.5204		40
	30	0.5519	0.8339	0.6619	1.5108		30
	40	0.5544	0.8323	0.6661	1.5013		20
	50	0.5568	0.8307	0.6703	1.4919		10
34°	00'	0.5592	0.8290	0.6745	1.4826	56°	00'
	10	0.5616	0.8274	0.6787	1.4733		50
	20	0.5640	0.8258	0.6830	1.4641		40
	30	0.5664	0.8241	0.6873	1.4550		30
	40	0.5688	0.8225	0.6916	1.4460		20
	50	0.5712	0.8208	0.6959	1.4370		10
35°	00'	0.5736	0.8192	0.7002	1.4281	55°	00'
	10	0.5760	0.8175	0.7046	1.4193		50
	20	0.5783	0.8158	0.7089	1.4106		40
	30	0.5807	0.8141	0.7133	1.4019		30
	40	0.5831	0.8124	0.7177	1.3934		20
	50	0.5854	0.8107	0.7221	1.3848		10
36°	00'	0.5878	0.8090	0.7265	1.3764	54°	00'

		Cos	Sin	Ctn	Tan	Degrees	

Tables of Trigonometric Functions

Degrees		Sin	Cos	Tan	Ctn		
36°	00'	0.5878	0.8090	0.7265	1.3764	54°	00'
	10	0.5901	0.8073	0.7310	1.3680		50
	20	0.5925	0.8056	0.7355	1.3597		40
	30	0.5948	0.8039	0.7400	1.3514		30
	40	0.5972	0.8021	0.7445	1.3432		20
	50	0.5995	0.8004	0.7490	1.3351		10
37°	00'	0.6018	0.7986	0.7536	1.3270	53°	00'
	10	0.6041	0.7969	0.7581	1.3190		50
	20	0.6065	0.7951	0.7627	1.3111		40
	30	0.6088	0.7934	0.7673	1.3032		30
	40	0.6111	0.7916	0.7720	1.2954		20
	50	0.6134	0.7898	0.7766	1.2876		10
38°	00'	0.6157	0.7880	0.7813	1.2790	52°	00'
	10	0.6180	0.7862	0.7860	1.2723		50
	20	0.6202	0.7844	0.7907	1.2647		40
	30	0.6225	0.7826	0.7954	1.2572		30
	40	0.6248	0.7808	0.8002	1.2497		20
	50	0.6271	0.7790	0.8050	1.2423		10
39°	00'	0.6293	0.7771	0.8098	1.2349	51°	00'
	10	0.6316	0.7753	0.8146	1.2276		50
	20	0.6338	0.7735	0.8195	1.2203		40
	30	0.6361	0.7716	0.8243	1.2131		30
	40	0.6383	0.7698	0.8292	1.2059		20
	50	0.6406	0.7679	0.8342	1.1988		10
40°	00'	0.6428	0.7660	0.8391	1.1918	50°	00'
	10	0.6450	0.7642	0.8441	1.1847		50
	20	0.6472	0.7623	0.8491	1.1778		40
	30	0.6494	0.7604	0.8541	1.1708		30
	40	0.6517	0.7585	0.8591	1.1640		20
	50	0.6539	0.7566	0.8642	1.1571		10
41°	00'	0.6561	0.7547	0.8693	1.1504	49°	00'
	10	0.6583	0.7528	0.8744	1.1436		50
	20	0.6604	0.7509	0.8796	1.1369		40
	30	0.6626	0.7490	0.8847	1.1303		30
	40	0.6648	0.7470	0.8899	1.1237		20
	50	0.6670	0.7451	0.8952	1.1171		10
42°	00'	0.6691	0.7431	0.9004	1.1106	48°	00'
	10	0.6713	0.7412	0.9057	1.1041		50
	20	0.6734	0.7392	0.9110	1.0977		40
	30	0.6756	0.7373	0.9163	1.0913		30
	40	0.6777	0.7353	0.9217	1.0850		20
	50	0.6799	0.7333	0.9271	1.0786		10
43°	00'	0.6820	0.7314	0.9325	1.0724	47°	00'
	10	0.6841	0.7294	0.9380	1.0661		50
	20	0.6862	0.7274	0.9435	1.0599		40
	30	0.6884	0.7254	0.9490	1.0538		30
	40	0.6905	0.7234	0.9545	1.0477		20
	50	0.6926	0.7214	0.9601	1.0416		10
44°	00'	0.6947	0.7193	0.9657	1.0355	46°	00'
	10	0.6967	0.7173	0.9713	1.0295		50
	20	0.6988	0.7153	0.9770	1.0235		40
	30	0.7009	0.7133	0.9827	1.0176		30
	40	0.7030	0.7112	0.9884	1.0117		20
	50	0.7050	0.7092	0.9942	1.0058		10
45°	00'	0.7071	0.7071	1.000	1.000	45°	00'
		Cos	Sin	Ctn	Tan	Degrees	

Answers to Odd-Numbered Exercises

NOTE: Answers for simple cases in which the student can quickly check his results have been omitted.

EXERCISE 1-1

1.
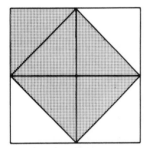

3.
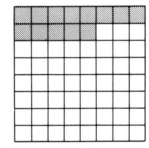

5.
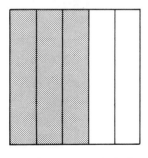

7.
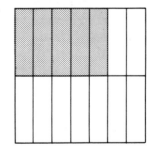

(Answers for exercises 1, 3, 5, and 7 may vary)

11. $\frac{5}{7}, \frac{28}{15}, \frac{8}{15}, \frac{1}{25}, \frac{14}{3}$ **13.** 1, 3, 23, 1, 213 **15.** $3\frac{3}{5}$ **17.** $1\frac{4}{7}$ **19.** $5\frac{1}{14}$ **21.** $6\frac{5}{13}$ **23.** $13\frac{2}{5}$ **25.** $9\frac{4}{9}$
27. $21\frac{14}{15}$ **29.** $29\frac{21}{29}$ **31.** $16\frac{15}{77}$ **33.** $47\frac{39}{87}$ **35.** $\frac{9}{3}, \frac{48}{16}$ **37.** $\frac{36}{3}, \frac{192}{16}$ **39.** $\frac{69}{3}, \frac{368}{16}$ **41.** $\frac{7}{2}$ **43.** $\frac{37}{8}$
45. $\frac{47}{7}$ **47.** $\frac{40}{3}$ **49.** $\frac{91}{4}$ **51.** $\frac{337}{24}$ **53.** $\frac{6137}{55}$ **55.** $\frac{1076}{27}$

EXERCISE 1-2

1.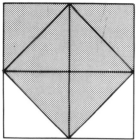

3.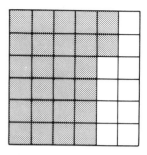

5.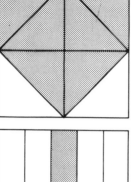

7.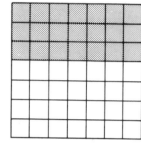

(Answers for exercises 1, 3, 5, and 7 may vary)

9. $\frac{15}{40}$ 11. $\frac{176}{77}$ 13. $\frac{34}{46}$ 15. $\frac{279}{63}$ 17. $\frac{56}{96}$ 19. $\frac{2}{5}$ 21. $\frac{14}{12}$ 23. $\frac{6}{8}$ 25. $\frac{2}{5}$ 27. 6, 30 29. 21, 42
31. 81, 4 33. not prime 35. prime 37. prime 39. $2 \times 2 \times 3$ 41. 2×13 43. $2 \times 3 \times 13$
45. $2 \times 5 \times 5$ 47. $2 \times 2 \times 2 \times 2 \times 2 \times 5$ 49. $3 \times 3 \times 17$ 51. $5 \times 5 \times 7$ 53. 11×11
55. $\frac{2}{3}$ 57. $\frac{1}{8}$ 59. $\frac{2}{3}$ 61. $\frac{5}{6}$ 63. $\frac{4}{5}$ 65. $\frac{1}{3}$ 67. $\frac{7}{8}$ 69. $\frac{1}{4}$ 71. $\frac{1}{3}$ 73. $\frac{3}{5}$ 75. $\frac{1}{6}$ 77. $\frac{4}{9}$ 79. yes 81. no

EXERCISE 1-3

1. no 3. no 5. 198 7. 396 9. 280 11. 3360 13. 1260 15. 2520 17. 520 19. $\frac{56}{126}, \frac{45}{126}, \frac{12}{126}$
21. $\frac{12}{72}, \frac{15}{72}$ 23. $\frac{81}{360}, \frac{56}{360}, \frac{320}{360}$ 25. $\frac{1089}{1584}, \frac{616}{1584}, \frac{216}{1584}$ 27. $\frac{216}{720}, \frac{135}{720}, \frac{80}{720}, \frac{288}{720}$ 29. $\frac{6}{7}$ 31. 2
33. $\frac{35}{44}$ 35. $\frac{707}{720}$ 37. $1\frac{257}{588}$ 39. $2\frac{61}{504}$ 41. $1\frac{409}{450}$ 43. $2\frac{1}{216}$ 45. $11\frac{43}{48}$ 47. $11\frac{71}{72}$ 49. $16\frac{103}{132}$
51. $8\frac{53}{315}$ 53. $15\frac{3}{8}$ hr 55. $155\frac{31}{36}$ lb

EXERCISE 1-4

1. $\frac{1}{3}$ 3. $\frac{1}{4}$ 5. $\frac{1}{12}$ 7. $\frac{11}{60}$ 9. $\frac{1}{112}$ 11. $8\frac{5}{6}$ 13. $2\frac{4}{5}$ 15. $8\frac{23}{24}$ 17. $5\frac{10}{17}$ 19. $\frac{16}{45}$ 21. $4\frac{17}{60}$ 23. $276\frac{47}{80}$
25. $221\frac{13}{36}$ lb

EXERCISE 1-5

1. $\frac{6}{35}$ 3. $\frac{5}{12}$ 5. $\frac{1}{8}$ 7. $\frac{3}{7}$ 9. $\frac{1}{49}$ 11. $\frac{10}{21}$ 13. 1 15. $2\frac{1}{2}$ 17. $\frac{1}{64}$ 19. $\frac{33}{98}$ 21. 4 23. $6\frac{1}{4}$ 25. $264\frac{3}{5}$
27. $1\frac{13}{42}$ 29. $7\frac{7}{9}$ 31. $5\frac{5}{8}$ 33. $18\frac{3}{8}$ 35. 12 37. $\frac{7}{20}$ 39. 184 mi 41. 33

EXERCISE 1-6

1. $1\frac{1}{15}$ 3. $\frac{7}{10}$ 5. $\frac{6}{7}$ 7. $\frac{1}{2}$ 9. $1\frac{7}{25}$ 11. $\frac{2}{5}$ 13. $14\frac{2}{5}$ 15. $6\frac{1}{2}$ 17. $\frac{3}{26}$ 19. $\frac{5}{12}$ 21. $1\frac{7}{11}$ 23. $1\frac{25}{31}$
25. $2\frac{1}{2}$ gal 27. $2\frac{3}{16}$ in. 29. $\frac{2}{11}$ in.

EXERCISE 1-7

3. $\frac{8}{75}$ 5. $\frac{21}{6}$ 7. $\frac{33}{8}$ 9. $\frac{119}{388}$ 11. $\frac{1}{24}$ 13. $\frac{1}{11}$ 15. $\frac{3}{2}$ 17. $\frac{3}{8}$

EXERCISE 1-8

1. 1027.979 3. 3763.023 5. 3019.9261 7. 0.073 9. 6.629 11. 199.13 13. 0.198 in.
15. 0.26 17. 0.4924

EXERCISE **1-9**

1. 0.06 **3.** 0.136 **5.** 0.00826 **7.** 12.282 **9.** 0.003483 **11.** 8.178 **13.** 0.000078 **15.** 3.65
17. 551.58 mi **19.** $23.22

EXERCISE **1-10**

1. 3.05 **3.** 0.478 **5.** 1.583 **7.** 0.333 **9.** 0.0625 **11.** 19.506 **13.** 1.471 **15.** 24.482
17. 3.4 cu ft **19.** $4.15

EXERCISE **1-11**

1. exact **3.** exact **5.** approximate **7.** approximate **9.** approximate
11. 4 s.d.; ten thousandth **13.** 1 s.d.; hundred **15.** 4 s.d.; thousandth
17. 1 s.d.; thousandth **19.** 2 s.d.; hundred **21.** 5 s.d.; ten **23.** 4 s.d.; tenth
25. 3 s.d.; hundredth

EXERCISE **1-12**

1. 162.1, 162, 160 **3.** 23.94, 23.9, 24 **5.** 21.13, 21.1, 21 **7.** 618.5, 619, 620
9. 9.324, 9.32, 9.3 **11.** 834.4, 834, 830

EXERCISE **1-13**

1. 40.95 **3.** 615.01 **5.** 287.12 **7.** 21.7 **9.** 218.39 **11.** 15.3 **13.** 16 **15.** 3.3 **17.** 23 **19.** 5

EXERCISE **1-14**

3. $\frac{17}{50}$ **5.** $\frac{67}{250}$ **7.** $\frac{27}{250}$ **9.** $\frac{5}{8}$ **11.** $3\frac{7}{50}$ **13.** $2\frac{1}{2}$ **15.** $81\frac{1}{4}$ **17.** $31\frac{19}{500}$ **19.** 0.25 **21.** 0.6 **23.** 0.05
25. 0.1875 **27.** 0.25 **29.** 0.3125 **31.** 0.56 **33.** 4.75 **35.** 8.43 **37.** 16.27

EXERCISE **1-16**

1. $6^2 = 6 \times 6 = 36$ **3.** $7^4 = 7 \times 7 \times 7 \times 7 = 2401$ **5.** $8^3 = 8 \times 8 \times 8 = 512$
7. 11 **9.** 14.731 **11.** 5.196 **13.** 1.808 **15.** 0.816 **17.** 1.5

REVIEW EXERCISES FOR CHAPTER 1

1. $3 \times 5 \times 7 \times 7$ **3.** $2 \times 2 \times 2 \times 3 \times 3 \times 5 \times 5$ **5.** $\frac{7}{15}$ **7.** $1\frac{107}{630}$ **9.** $5\frac{2683}{3969}$ **11.** $12\frac{37}{144}$
13. $9\frac{5}{7}$ **15.** $\frac{22}{25}$ **17.** $\frac{8}{7}$ **19.** 16 **21.** $\frac{1}{26}$ **23.** 1.9 **25.** 22.922 **27.** 0.0003 **29.** 2.8
31. approximate **33.** 1 s.d.; hundredth **35.** 3 s.d.; tenth **37.** 18 **39.** 6.5 **41.** 130 **43.** 6.2
45. 1.2 **47.** 2.125 **49.** 13 **51.** 0.41 **53.** 1.86 **55.** 7.75 min **57.** 87 hr **59.** 0.288 in.
61. 15.75 ft

EXERCISE **2-1**

3. to find the number of cm in 7 km easier. **5.** 0.38 m **7.** 1170 mm **9.** 47,100 cm **11.** 76 ft
13. 15.3 ft **15.** $132 **17.** 4.7 km **19.** 0.318 cm **21.** (a) 1.5 cm, (b) 15 mm
23. (a) 980 km per hr, (b) 980,000 m per hr, (c) 16,000 m per min **25.** 11,000

EXERCISE **2-2**

1. 470 ft² **3.** 6Ö0 cm² **5.** (a) 4.0 m³, (b) 140 ft³ **7.** 172 m² **9.** (a) 2.41 ft³, (b) 0.0681 m³
11. 1.3 m² **13.** (a) 328 ft, (b) 62Ö0 ft²

EXERCISE **2-3**

1. 7.3 lb **3.** 1.25 gal **5.** (a) 1 900 mℓ, (b) 1 900 cc **7.** 0.825 ℓ **9.** 1.83 m, 72.56 kg
11. 75.8 ℓ **13.** 3 300 t **15.** 1 300 kg

REVIEW EXERCISES FOR CHAPTER 2

1. 180 mm **3.** 2 318 m **5.** 0.398 km **7.** 2.4 m **9.** 957 km/hr **11.** 1.1 m² **13.** 210,000 m²
15. (a) 38.50 ft³, (b) 1.090 m³ **17.** 21 ℓ **19.** (a) 2 800 mℓ, (b) 2 800 cc

Answers to Odd–Numbered Exercises

EXERCISE 3-1
1. (a) Associative Law of Addition, (b) Commutative Law of Multiplication, (c) Distributive Law, (d) Commutative Law of Addition, (e) Associative Law of Multiplication 13. F
15. T 17. T 19. T 21. F 23. T 25. T 27. F 29. T

EXERCISE 3-2
1. $+3$ 3. -13 5. -28 7. -22 9. $+57$ 11. 10 13. -135 15. -7 17. -107 19. -182

EXERCISE 3-3
1. 15 3. -5 5. -22 7. -45 9. -1 11. 121 13. -673 15. -431 17. 202 19. 174

EXERCISE 3-4
1. -11 3. -30 5. 29 7. -112 9. 17 11. 127 13. -143 15. 273

EXERCISE 3-5
3. -12 5. -21 7. 16 9. -77 11. 19 13. -15 15. -9 17. 23 19. 12 21. -18 23. -12
25. -32 27. 6 29. -4 31. -3 33. 17 35. -11 37. 8 39. -1 41. 9 43. -3 45. 6
47. -11 49. -12

REVIEW EXERCISES FOR CHAPTER 3.
1. (a) Associative Law of Multiplication, (b) Associative Law of Addition, (c) Commutative Law of Addition, (d) Distributive Law, (e) Commutative Law of Multiplication 5. T 7. F
9. F 11. T 13. -1 15. 1 17. -15 19. -48 21. 24

EXERCISE 4-1
1. the division of the positive product $2 \cdot a$ by the positive product $3 \cdot m$
3. the positive product $5 \cdot x$ decreased by y 5. $3x + 2y$ 7. $\dfrac{x}{y} + 3xy$ 9. $\dfrac{k-3}{4}$ 17. two
19. two 21. $5 \cdot \dfrac{1}{a}$ 23. $15xy \cdot \dfrac{1}{7d}$ 25. (a) 3, (b) $+1$, (c) -1 27. F 29. T 31. T 33. c

EXERCISE 4-2
1. Associative Law of Addition 3. Distributive Law 5. Associative Law of Multiplication
7. $12 + 3v + 3x$ 9. $sr + st$ 11. 16 13. -64 15. -25 17. 1 19. 23 21. 30 23. 66
25. 105 27. 110

EXERCISE 4-3
1. $a \cdot a \cdot a$ 3. $y \cdot y$ 5. w^5 7. d^2k^3 9. 64 11. m^5h^8 13. 27 15. $\dfrac{1}{k}$ 17. $\dfrac{1}{V^5}$ 19. $\dfrac{k^2}{m^2}$ 21. $\dfrac{t^5w}{v}$
23. $\dfrac{wx}{u^3v^2}$ 25. 16 27. $m^{10}n^{15}k^5$ 29. 3 31. $\dfrac{81}{15,625}$ 33. $\dfrac{x^{20}}{z^{12}}$

EXERCISE 4-4
1. $5t$ 3. $3x$ 5. $7y$ 7. $9w$ 9. $2k^3$ 11. $6x^2y - xy^2$ 13. $d + e + 10f$ 15. $4vw^2z^3 + 7v^2wz^3$
17. $4gt^3 - 5k^2m - 1$ 19. $x^2 - 2xy - 5x - 2y^2 - 2$

EXERCISE 4-5
1. 3 3. 6 5. 65 7. $-5m$ 9. $-c^3 - 4c^2d + cd^2$ 11. $10m^2 - mt$ 13. -6

REVIEW EXERCISES FOR CHAPTER 4
1. $x + 5y$ 3. $x + 5(y - z)$ 5. Distributive Law 7. Commutative Law of Addition
9. Associative Law of Multiplication 11. -27 13. -1 15. 5 17. 128 19. a^7 21. 3^{10}
23. $\dfrac{s^2}{tv}$ 25. $\dfrac{g^6}{h^8}$ 27. $-x - y$ 29. $6k - 20$ 31. $-3d - 8$

EXERCISE 5-1

1. $8x - 5y$ **3.** $11t^2 - 10s$ **5.** $6g + 2h + 7k$ **7.** $5m + 2s - 4t$ **9.** $9a^2 + 12ab$
11. $6k^2 + 6m^2 - 4km + 4m^2$ **13.** $11a^2b - 2ab - ab^2$ **15.** $-ax^3 - 2by - 1$

EXERCISE 5-2

1. $d^2 + 5e^2$ **3.** $-5x^2 + 5y + 7$ **5.** $a - b + c$ **7.** $18h - 8k + 7$ **9.** $-x - 11y + 9$
11. $3a^2 - 4b - c$

EXERCISE 5-3

1. $6x^2y^2$ **3.** $3x^3y^3 + 3x^2y^2$ **5.** $21a^2b^2 - 14ab^3 - 21ab^2$ **7.** $m^4 + m^2t^2 + t^4$
9. $25x^2 - 10gx + g^2$ **11.** $6a^2 - 7ab + 2b^2$ **13.** $2a^3 - a^2 - 7a + 6$

EXERCISE 5-4

1. $m^2 - 4m + 4$ **3.** $c^2 + 6c + 9$ **5.** $t^2 - 121$ **7.** $3v + 15w - 3x + 6y$
9. $a^2 - 24a + 144$ **11.** $9 + 6b + b^2$ **13.** $x^2 + 10x + 21$ **15.** $36g^2 - h^2$
17. $4v^2 + 28v + 49$ **19.** $N^2 - 11N + 24$ **21.** $4d^2 - 32d + 64$ **23.** $g^2 + 2gR + R^2$
25. $h^2 - 6bh + 9b^2$ **27.** $16x^2 - 25y^2$ **29.** $49r^2 + 84rt + 36t^2$

EXERCISE 5-5

1. $18P^2 - 36P - 14$ **3.** $8z^2 - 10z + 3$ **5.** $2t^2 + 17t + 21$ **7.** $9y^2 + 35y - 50$
9. $6v^2 + 9v - 6$ **11.** $6y^2 - 15y - 9$ **13.** $42k^2 - 33k + 6$ **15.** $44t^2 - 100t + 24$
17. $378d^2 - 213d + 30$

EXERCISE 5-6

1. $3x^3$ **3.** $\dfrac{5k}{2m^2n}$ **5.** $\dfrac{x^4 + 2x^3}{3}$ **7.** $5a^3c + b^2c^2 + 2a$ **9.** $\dfrac{4x^2}{3m} - \dfrac{4m}{k} + \dfrac{3m^2}{k^2}$

EXERCISE 5-7

1. $3(x + 4)$ **3.** $2(R - 5)$ **5.** $2(2e - 1)$ **7.** $7(2cv^2 - dv - 3)$ **9.** $c(-4ab + 3a - 5b)$
11. $9gh(g^2h - 9h + 3g)$ **13.** $6a(ab + 4b + 11)$ **15.** $v(7uw - 3u + 5w)$
17. $3t^2s^2(s^2 + 5ts - 6t^2)$ **19.** $4(s + 4)(s + 3t)$ **21.** $(h - 3)(3 + 2g)$ **23.** $(x - 2)(5x + 3y)$
25. $(M + 3)(2M + 3d)$ **27.** $(t + 2)(t - 2)$ **29.** $16(2v + w)(2v - w)$ **31.** $(1 + D)(1 - D)$
33. $(4y + 5z)(4y - 5z)$ **35.** $(e + 1)(e - 1)$ **37.** $(F + 2)(F - 2)$ **39.** $(13g + 3h)(13g - 3h)$
41. $2(5 + A)(5 - A)$ **43.** $7(2R + 1)(2R - 1)$ **45.** $3(3x + 2y)(3x - 2y)$

EXERCISE 5-8

1. $(m + n)^2$ **3.** $(R - S)^2$ **5.** $(5L + 2M)^2$ **7.** $(3g - 1)^2$ **9.** $(2t + 5s)^2$ **11.** $(6g + 7h)^2$
13. $(k - 5m)^2$ **15.** $(3x - 2y)^2$ **17.** $(x + 6)(x - 4)$ **19.** $(x + 7)(x - 3)$ **21.** $(t + 7)(t + 1)$
23. $(7k - 3)(2k - 1)$ **25.** $3(3x - 7)(x + 1)$ **27.** $(3x - 2)(2x - 3)$ **29.** $(t - 5)(3t + 1)$
31. $2(d - 3)(3d + 5)$

EXERCISE 5-9

1. 29 **3.** $-\frac{4}{3}$ **5.** -4 **7.** 31 **9.** 31 **11.** 4 **13.** 21 **15.** 0 **17.** -72 **19.** -9

REVIEW EXERCISES FOR CHAPTER 5

1. $7d + 5e - 11g$ **3.** $-3h^2 - 2k - 15$ **5.** $-12x^3y^3$ **7.** $2t^4 - mt^3 + 3m^2t^2 + m^3t + 4m^4$
9. $r^2 - 2rq + q^2$ **11.** $M^2 - 15M + 56$ **13.** $6A^2 + 5AB - 25B^2$ **15.** $21h^2 + 32h - 5$
17. $\dfrac{4a^2}{bc}$ **19.** $(x - 3)(7x - 2t)$ **21.** $(8x - y)^2$ **23.** $(2x + 3)(x - 1)$ **25.** -20

EXERCISE 6-1

1. $\dfrac{x}{6y}$ **3.** $\dfrac{7}{8}$ **5.** $\dfrac{3}{5v}$ **7.** $\dfrac{9k^2}{4m}$ **9.** $\dfrac{3ACD}{4BE}$ **11.** $\dfrac{1}{3b + 1}$ **13.** $\dfrac{2x^2}{5}$ **15.** $\dfrac{1}{2x - y}$ **17.** $3t + 1$

19. $\dfrac{7k - 8}{k - 1}$ **21.** $\dfrac{3y - 2}{y + 3}$

EXERCISE 6-2

1. $12x^2y^2z^3$ **3.** $30u^3vw^3x$ **5.** $24x^5y^3$ **7.** $3(2x - y^3)$ **9.** $24(b + 2c)$
11. $(x + y)(x + y)(x - y)$ **13.** $k(2m + 5)(2m - 5)(2m - 5)$ **15.** $12(h - k)^2$

EXERCISE 6-3

1. $\dfrac{37v}{12}$ **3.** $\dfrac{7x + 12}{20x}$ **5.** $\dfrac{21 - 8c}{3c(c + 3)}$ **7.** $\dfrac{11m}{5t}$ **9.** $\dfrac{hw - x^2}{wx^2}$ **11.** $\dfrac{5g^2 + 5g + 6}{6g}$ **13.** $\dfrac{4 - k}{k - 1}$

15. $\dfrac{5x - 13}{(x - 3)^2}$ **17.** $\dfrac{k - mS}{mt}$ **19.** $\dfrac{x^2 - 3x + 5}{x - 3}$ **21.** $\dfrac{x^3 - 3x^2 - 3x}{(x + 1)(x - 1)}$ **23.** $\dfrac{19a + 8b}{21x}$

25. $\dfrac{x^3 - 4x^2 - 24x - 5}{(x + 3)(x + 3)(x - 1)}$ **27.** $\dfrac{6x^2 - 12x - 28}{(x + 2)(x + 2)(x - 2)}$ **29.** $\dfrac{-a^3 + 2a^2 + 4a}{(a + 3)(a - 3)}$

EXERCISE 6-4

1. $\dfrac{6}{c^2 + 2cd + d^2}$ **3.** 15 **5.** $\dfrac{x^2 - 1}{x^2 - 4}$ **7.** $\dfrac{5}{3s + 9}$ **9.** $\dfrac{3x + 8}{3 + x}$ **11.** $\dfrac{3}{a^2}$ **13.** $\dfrac{5dh}{9g^2}$ **15.** $\dfrac{9a}{40cd}$

17. $\dfrac{5x}{a - x}$ **19.** $\dfrac{2x + 10}{3x + 3}$

EXERCISE 6-5

1. $\frac{3}{2}$ **3.** $\dfrac{7t}{12t - 4}$ **5.** $\dfrac{g^2}{3h + 5g}$ **7.** $\dfrac{2x - 3}{5 - x}$ **9.** $\dfrac{x^2 + x^2y + xy^2 - y^2}{-x^2 + xy + y^2}$

REVIEW EXERCISES FOR CHAPTER 6

1. $\dfrac{d}{4h}$ **3.** $\dfrac{5kx}{4}$ **5.** $\frac{1}{7}$ **7.** $\dfrac{2m + 3}{1 - m}$ **9.** $7(3t - r^2)$ **11.** $3(x + 4)(x + 4)(x - 4)$ **13.** $\dfrac{53d}{24}$ **15.** $\dfrac{25k}{12b}$

17. $\dfrac{13 - t}{7 - t}$ **19.** $\dfrac{18 - 5x}{(3 - x)^2}$ **21.** $\dfrac{33mx - 23my}{12(x + y)(x - y)}$ **23.** $\dfrac{b - c}{1 - 6d}$ **25.** $\dfrac{5m}{42k}$ **27.** $\dfrac{7A}{6x - 5}$ **29.** $\frac{18}{5}$

31. $\dfrac{7c + cd}{c^2 - 2c - 15}$ **33.** $\dfrac{c^2d + 2}{1 + c^2}$

EXERCISE 7-1

1. 4 **3.** 5 **5.** 3 **7.** 4 **9.** 27 **11.** 63 **13.** 7 **15.** 14 **17.** 15 **19.** 22 **21.** 16 **23.** 9

EXERCISE 7-2

1. 3 **3.** 9 **5.** 2 **7.** 38 **9.** 7 **11.** 23 **13.** 33 **15.** 3 **17.** 4 **19.** 19

EXERCISE 7-3

1. 8 **3.** 17, 51 **5.** $W = 11$, $M = 17$ **7.** $W = 5$ m, $L = 14$ m
9. carpenter 42 hr, assistant 31 hr **11.** 45 tons Hematite, 75 tons Taconite
13. 75.0 kg of the first kind, 125.0 kg of the second kind
15. the 140-lb boy 3 ft, the 105-lb boy 4 ft **17.** 6 kg **19.** 60 mi **21.** 11:30 AM **23.** 3 hr
25. in $2\frac{1}{3}$ hr or 2 hr 20 min

REVIEW EXERCISES FOR CHAPTER 7

1. 9 **3.** 71 **5.** 14 **7.** 17 **9.** $21n$, $23d$, $7q$
11. central span is 115 m, extreme spans are 210 m each **13.** 4 ft

EXERCISE 8-1

1. 4 **3.** 5 **5.** $\frac{6}{7}$ **7.** 9 **9.** 5 **11.** 3 **13.** 9 **15.** 14 **17.** no solution **19.** 8 **21.** 6 **23.** 4 **25.** 7
27. 28 **29.** 6 **31.** 72 **33.** \$432, \$191, \$124, \$117

EXERCISE 8-2

1. $P = \dfrac{I}{RT}$ **3.** $V_0 = \dfrac{2S - at^2}{2t}$ **5.** $R = 2f$ **7.** $e = E - IR$ **9.** $V = \dfrac{W}{D}$ **11.** $V_2 = \dfrac{V_1 T_2}{T_1}$
13. $P_2 = \dfrac{V_2 P_1}{V_1}$ **15.** $R = \dfrac{rW}{F}$ **17.** $n = \dfrac{W}{F}$ **19.** $s = \dfrac{Pt}{F}$ **21.** $R_2 = \dfrac{RR_1}{R_1 - R}$

REVIEW EXERCISES FOR CHAPTER 8

1. 6 **3.** no solution **5.** 4 **7.** $\frac{3}{4}$ **9.** 936 mi **11.** $q = \dfrac{pf}{p - f}$ **13.** $t_1 = \dfrac{NT_1 T_2}{nt_2}$
15. $r = \dfrac{WR - 2PR}{W}$

EXERCISE 9-1

1. $9\,\ell$ **3.** 2 m **5.** 38 lb **7.** cannot be completed, dimensions are different **9.** 3 in.^{2}
11. cannot be completed, units are different **13.** 4 in. **15.** 2 hr 56 min 46 sec
17. 13 tons 435 lb **19.** 147 lb 8.0 oz **21.** 5.0 tons 534.0 lb **23.** 36 yd 8.0 in. **25.** 3 lb 5 oz
27. 4 tons 816 lb **29.** 6 hr 34 min 38 sec

EXERCISE 9-2

1. 133 in. **3.** 63.3 lb **5.** 4.41 mi **7.** 1.37 mi **9.** 6.35 kg **11.** 2.16 gal **13.** 46.36 cm
15. 99.48 m **17.** 650 in.^{3} **19.** 2.37 ft^2 **21.** 49.8 in.^{3} **23.** 20.5 m^3 **25.** 9.72 ft^2
27. 170,000 in.^{3} **29.** 1.6 gal **31.** 1930 ft/min **33.** 2.25 kg/cm^2 **35.** 59 ft/sec **37.** 1.16 ℓ/sec
39. 72.4 km/hr **41.** 11.3 g/cm^3 **43.** 88 kph **45.** 148 ℓ/sec

EXERCISE 9-3

1. 33 m^3 **3.** 3.0 m^3 **5.** 0.82 m **7.** 10 **9.** 180ft-lb **11.** 234 lb/ft^2 **13.** 170 ft/sec **15.** 380 m
17. $4\tilde{0}$ lb/in.^{2} **19.** 230 kg m^2/sec^2

REVIEW EXERCISES FOR CHAPTER 9

1. cannot be completed, the units are different **3.** 5 mi **5.** 14 ℓ **7.** 280 ft^3 **9.** 6 **11.** 7 lb
13. 19 yd 2 ft 2 in. **15.** 7 gal 2 qt 1 pt **17.** 5 hr 17 min 15 sec **19.** 90 lb 2 oz
21. 6 tons 737 lb **23.** 430 ft/min **25.** 2.95 kg/cm^2 **27.** 80.5 km/hr **29.** 22.6 gal
31. 2.71 g/cm^3 **33.** 144 lb/ft^3 **35.** 6.7 kg/cm^2 **37.** 24 km **39.** $7\tilde{0}$ lb/in.^{2}

EXERCISE 10-1

1. rate **3.** ratio **5.** rate **7.** $\frac{9}{2}$ **9.** $\frac{27}{352}$ **11.** $\frac{7}{144}$ **13.** $\frac{3}{1250}$ **15.** $\frac{1}{5}$ **17.** $1:4$ **19.** $\frac{2}{5}$

EXERCISE 10-2

1. true **3.** true **5.** true **7.** true **9.** not true **11.** true **13.** $x = 3$ **15.** $x = 2$ **17.** $y = 6$
19. $x = 5$ **21.** $x = 12$ **23.** $d = 6$ **25.** $P = \frac{1}{3}$

EXERCISE 10-3

1. \$1.75 **3.** $6\tilde{0}$ gallons **5.** 3.0 hr **7.** 12.0 lb **9.** 786 ft **11.** 6 ft **13.** 14 in. **15.** \$1.12
17. 2.25 g

EXERCISE 10-4

1. $d = km$ **3.** $T = \dfrac{k}{B}$ **5.** $F = \dfrac{kQU}{L}$ **7.** $H = \dfrac{k}{v}$ **9.** $C = \dfrac{kE}{R^2}$ **11.** $M = kN^2 R$ **13.** $d = kt^2$
15. $I = kE$ **17.** $k = 7$, $P = 63$ **19.** $k = 2$, $B = 10$ **21.** $k = 4$, $d = 12.5$
23. $k = 2$, $Q = 4\frac{2}{3}$ **25.** $k = 3$, $T = 12.5$ **27.** $W = \frac{8}{5}$ lb **29.** $20\tilde{0}$ lb **31.** 4 in. **33.** 12 ft
35. 142 lb **37.** \$5040

REVIEW EXERCISES FOR CHAPTER 10

1. ratio **3.** $\frac{1}{20}$ **5.** $\frac{1}{6}$ **7.** $x = 2$ **9.** $k = 3$ **11.** $d = 1$ **13.** 9.0 mℓ **15.** 18 lb **17.** 0.45 in.
19. 35 days **21.** 0.72 in.

EXERCISE 11-1

1. 8, 7 **3.** $H(-6, 8)$, $I(3, 4)$, $J(-5, -3)$, $K(7, 1)$, $L(0, -5)$, $M(7, -5)$, $N(-3, 4)$

5.

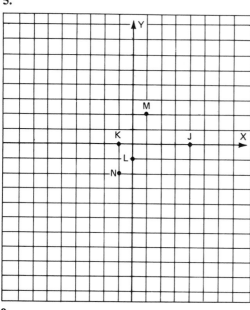

7.

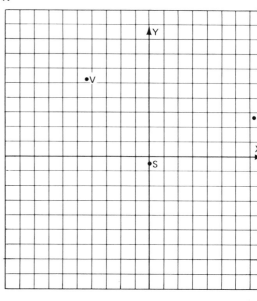

9.

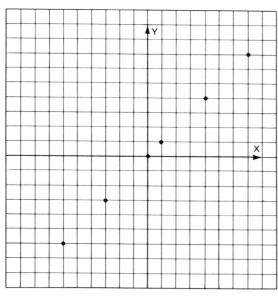

11.

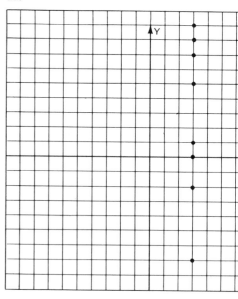

Answers to Odd–Numbered Exercises

13.

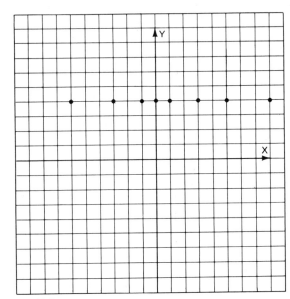

Exercise 11-2

1.

x	y
0	0
−2	−2
5	5

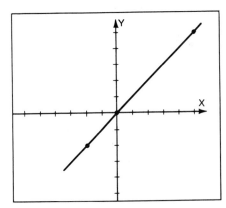

3.

x	y
0	−1
2	5
−2	−7

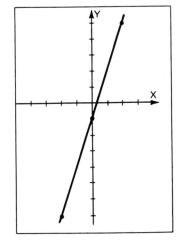

5.

x	y
0	4
3	10
−3	−2

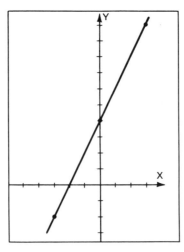

7.

x	y
0	−2
3	−1
−3	−3

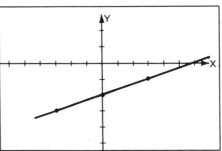

9.

x	y
0	3
2	4
−4	1

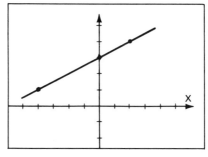

11.

x	y
0	2
4	−1
−4	5

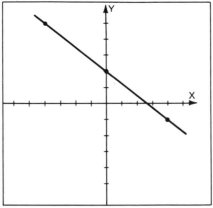

13.

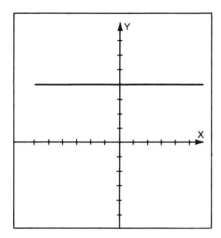

15.

x	y
-3	11
-2	2
-1	-3
0	-4
1	-1
2	6
3	17

17.

x	y
0	0
1	2
2	8
3	18
4	32
5	50

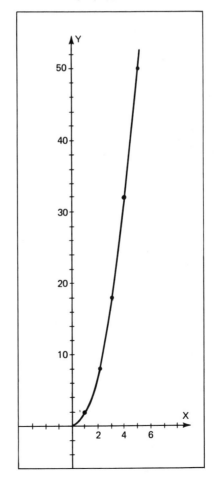

1.

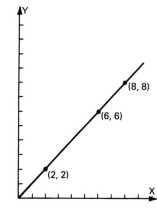

$m = 1$

3.

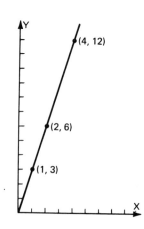

$m = 3$

5.

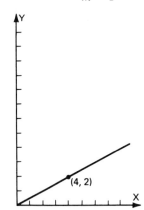

$m = \frac{1}{2}$

7.

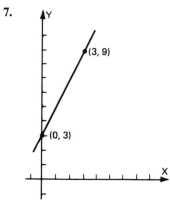

$m = 2$

9.

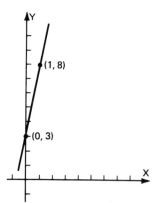

$m = 5$

11. $m = 3$ **13.** $m = 7$ **15.** $m = \frac{2}{3}$

EXERCISE 11-4

1. (a) yes, (b) $\frac{w}{v} = 15$, (c) 15

(d)

v	w
0	0
1	15
3	45

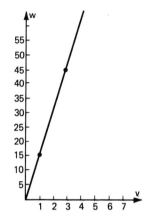

$m = 15$

3. (a)

w	e
0	0
2	3
6	9

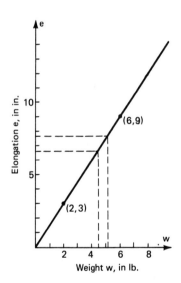

(b) 1.5 in./lb
(c) $e = 6.7$ in. (approx.)
(d) $w = 5.25$ lb (approx.)

5. (a)

$I(A)$	$V(v)$
0	0
2	40
5	100

(b)

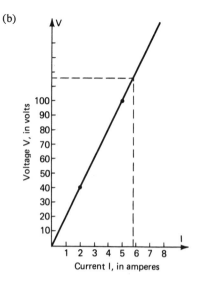

Voltage V, in volts

Current I, in amperes

(c) 5.75 A (approx.)

7. (a)

h	P
0	0
5	312
10	624

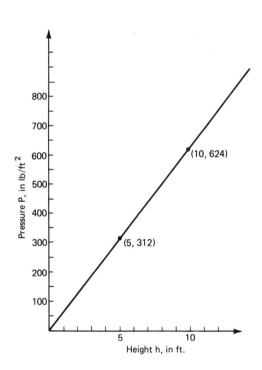

Pressure P, in lb/ft²

(10, 624)

(5, 312)

Height h, in ft.

(b) 62.4
(c) 62.4 lb/ft²

9. (a)

I	P
0	0
0.1	0.2
0.2	0.8
0.3	1.8
0.4	3.2
0.5	5.0
0.6	7.2
0.7	9.8
0.8	12.8
0.9	16.2
1	20.0

(b)

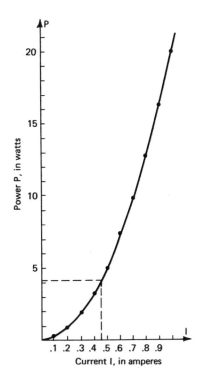

(c) 4.1 W (approx.)

11. (a)

t	V
0	0
1	3
2	10
3	21
4	36
5	55
6	78

(b) 17 V (approx.)
(c) 4.5 sec (approx.)

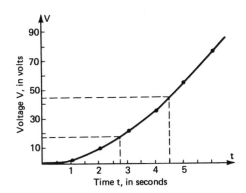

1. (a) 9 (b) 3

3.

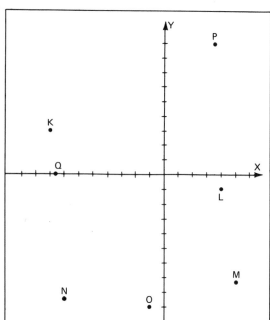

5.

x	y
0	−2
3	7
−2	−8

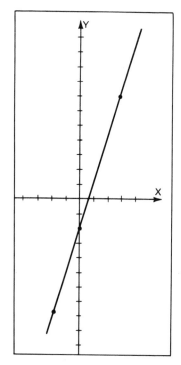

7.

x	y
0	5
3	1
9	−7

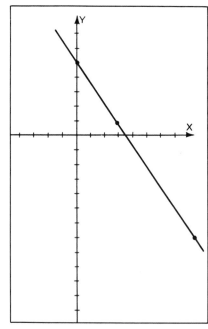

Answers to Odd–Numbered Exercises

9.

x	y
−3	20
−2	10
−1	4
0	2
1	4
2	10
3	20

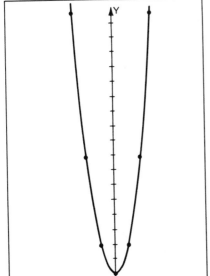

11.

m	k
0	0
1	8
2	32
3	72
4	128
5	200

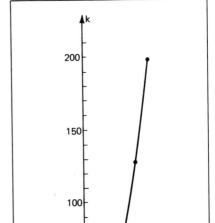

13.

x	y
0	2
1	9
−1	−5

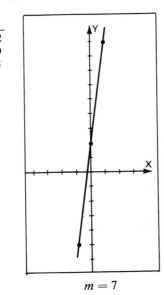

$m = 7$

15. $m = \frac{3}{4}$ **17.** $m = \frac{1}{4}$

19. (a)

I	E
0	0
2	6
5	15

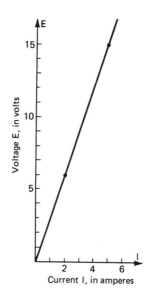

Voltage E, in volts

Current I, in amperes

(b) 3
(c) 3
(d) 3

21. (a)

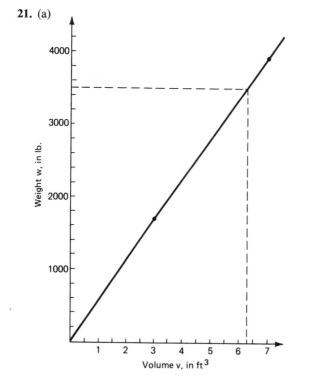

Weight w, in lb.

Volume v, in ft³

(b) density 555 lb/ft³
(c) 3480 lb (approx.)

EXERCISE 12-1

1. $\frac{1}{8}$ **3.** 27 **5.** $\frac{16}{9}$ **7.** $\frac{1}{5184}$ **9.** 26,244 **11.** $\frac{1}{6}$ **13.** 2 **15.** $\frac{1}{9}$ **17.** $\frac{1}{16}$ **19.** $\frac{a^2b^3}{12}$ **21.** $\frac{y}{5000}$ **23.** $\frac{3y}{xz}$

EXERCISE 12-2

1. 5.67×10 **3.** 7.06×10^2 **5.** 9.1×10^{-2} **7.** 4.9×10^8 **9.** 2.785×10^2 **11.** 1.7×10^{-4}
13. $5,100,000$ **15.** 0.00109 **17.** 0.0000067 **19.** 9.0×10^{-4} **21.** 9.70×10^{-4} **23.** 1.70×10^{-2}
25. 3 **27.** 2 **29.** 3

EXERCISE 12-3

1. 4.6 **3.** 2.1×10^{-6} **5.** 2.1×10^2 **7.** 1.7×10^7 **9.** 1.5×10^7 **11.** 2.6 ft

EXERCISE 12-4

1. 3 **3.** 12 **5.** 17 **7.** -2 **9.** $(-a)^2$ **11.** no real root **13.** $7^{1/2}$ **15.** $c^{3/5}$ **17.** $t^{1/5}$ **19.** $\sqrt{x}$
21. $\sqrt[3]{k^2}$ **23.** $\sqrt[7]{h^6}$ **25.** 32 **27.** 125

EXERCISE 12-5

1. $3\sqrt{2}$ **3.** $3\sqrt{5}$ **5.** $2\sqrt[3]{5}$ **7.** $3d\sqrt{2d}$ **9.** $15g^3\sqrt{3}$ **11.** $b\sqrt[3]{ab^2}$ **13.** $15M\sqrt[3]{10M^2N}$
15. $2g\sqrt{2eg}$ **17.** $14d^2\sqrt[3]{7c}$ **19.** $bz^2\sqrt{a}$ **21.** $6GRV\sqrt[3]{2R^2V^2}$ **23.** $42Y\sqrt{5VYZ}$
25. $2X^2\sqrt[3]{B^2C}$

EXERCISE 12-6

1. $7\sqrt{3}$ **3.** $-4\sqrt{13}$ **5.** $9\sqrt[3]{7}$ **7.** $3\sqrt{6} + 2\sqrt[3]{6}$ **9.** $3\sqrt[3]{12} - 2\sqrt{5}$ **11.** $6\sqrt{15} + 2\sqrt{11}$
13. $-3\sqrt{\frac{7}{8}}$ **15.** $-4.8\sqrt{2}$

EXERCISE 12-7

1. $2\sqrt{6}$ **3.** $2\sqrt{21}$ **5.** $2\sqrt[4]{2}$ **7.** $6B$ **9.** $35 - 7\sqrt{5}$ **11.** $30VW^2\sqrt[3]{2}$ **13.** $-8 - 2\sqrt{7}$
15. $18 - 2\sqrt{3}$ **17.** $10 - 4\sqrt{6}$ **19.** $2 + 3\sqrt{C} + C$ **21.** -23 **23.** -23 **25.** 38
27. $2\sqrt[3]{5}$ **29.** $\frac{3}{2}$ **31.** $20\sqrt{t}$ **33.** $\frac{\sqrt{15}}{3}$ **35.** $\frac{5\sqrt{a}}{a^2}$ **37.** $3\sqrt{2} + 3$ **39.** $\frac{14\sqrt{2} + 24}{23}$
41. $\frac{N + 15 + 8\sqrt{N}}{N - 9}$

REVIEW EXERCISES FOR CHAPTER 12

1. $\frac{1}{32}$ **3.** 3 **5.** $5e^2y$ **7.** s^2t^2 **9.** 3.20×10^4 **11.** 2.1×10^5 **13.** 2 **15.** 2.7×10^3
17. 2.6×10^{12} **19.** 9 **21.** $3^{3/4}$ **23.** $\sqrt[4]{B^3}$ **25.** $\sqrt[3]{T}$ **27.** $\sqrt{h}$ **29.** $11\sqrt{3}$ **31.** $RT^2\sqrt[3]{aR^2T}$
33. $\sqrt{7} + 11\sqrt[3]{7}$ **35.** $20a\sqrt{7}$ **37.** $5t^2u\sqrt{3v}$ **39.** 2 **41.** $\frac{\sqrt{14}}{4}$ **43.** $\frac{\sqrt{R}}{S}$ **45.** $\frac{11 - 4\sqrt{6}}{-5}$

EXERCISE 13-1

1.

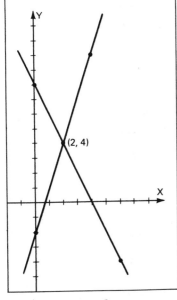

$x = 2$
$y = 4$

3.

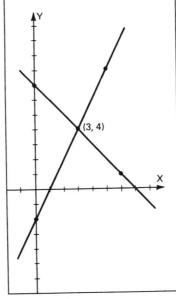

$x = 3$
$y = 4$

5.

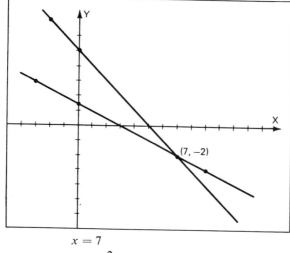

$$x = 7$$
$$y = -2$$

7. $x = 7$, $y = 5$ **9.** $s = 4$, $t = 8$ **11.** $r = 4$, $s = 1$ **13.** $v = 6$, $w = 7$ **15.** $x = 7$, $y = 3$
17. $x = 10$, $y = 4$ **19.** $x = 100$, $y = 300$ **21.** $x = 2$, $y = 6$
23. $x = 3$, $y = 9$ **25.** $x = 11$, $y = 4$ **27.** $x = 4$, $y = 3$ **29.** $x = 12$, $y = 8$

Exercise 13-2
1. 5, 7 **3.** $\frac{4}{15}$ **5.** 2 hr, 100 mi **7.** $250.00, $320.00
9. ball bearings $18.00/doz, roller bearings $16.00/doz
11. 180-lb unbleached, 120-lb whole wheat **13.** 6000 ℓ/hr, 7000 ℓ/hr

Exercise 13-3
1. $x = 2$, $y = 3$, $z = 4$ **3.** $x = 10$, $y = 2$, $z = 3$ **5.** $x = 3$, $y = 4$, $z = 6$
7. $x = 1$, $y = 3$, $z = 5$ **9.** A: $180, B: $60, C: $240

Review Exercises for Chapter 13
1. $x = 11$, $y = 12$ **3.** $x = 8$, $y = 7$ **5.** $x = 5.3$, $y = 7.1$ **7.** $x = 3$, $y = 1$, $z = 5$
9. 53°, 37° **11.** bushes $18.00 each, labor $12.00/hr
13. 300 ft³/hr, 350 ft³/hr **15.** ring stand $2.75, flask $3.15, bunsen burner $4.25

Exercise 14-1
11. $5x^2 - x - 2 = 0$ **13.** $8x^2 - 9x - 7 = 0$ **15.** $2x^2 - x - 15 = 0$
17. $x^2 + 3x - 2 = 0$ **19.** $5x^2 + x + 60 = 0$

Exercise 14-2
1. $x = -7$, $x = 1$ **3.** $x = -1$, $x = 5$ **5.** $x = -6$, $x = -5$ **7.** $x = \frac{1}{2}$, $x = -3$
9. $x = \frac{2}{3}$, $x = 2$ **11.** $x = -\frac{1}{2}$, $x = \frac{1}{3}$ **13.** $x = 0$, $x = 2$ **15.** $x = 0$, $x = -2$
17. $x = -\frac{5}{2}$, $x = \frac{5}{2}$ **19.** $x = -\frac{9}{5}$, $x = \frac{9}{5}$

Exercise 14-3
1. $x = 3$, $x = -3$ **3.** $x = 2$, $x = -2$ **5.** $x = 5$, $x = -5$ **7.** $x = 0.51$, $x = -0.51$
9. $x = 2.6$ (approx.), $x = -2.6$ (approx.) **11.** $x = 2.2$ (approx.), $x = -2.2$ (approx.)

Exercise 14-4
1. $x = 7$, $x = -1$ **3.** $x = \frac{3}{2}$, $x = \frac{2}{3}$ **5.** $x = 5$, $x = 3$ **7.** $x = 17$, $x = \frac{2}{3}$

9. $x = 0.281$ (approx.), $x = -1.781$ (approx.)
11. $x = 0.768$ (approx.), $x = -0.434$ (approx.) **13.** $x = \frac{3}{4}, x = -\frac{5}{2}$
15. $x = 2.885$ (approx.), $x = 0.116$ (approx.) **17.** $x = \frac{4}{3}, x = -6$
19. $x = 1.175$ (approx.), $x = -0.425$ (approx.)

EXERCISE 14-5
1. 4 **3.** 10.6 in. **5.** $L = 80$ cm, $W = 50$ cm **7.** 4 sec **9.** 3 in. **11.** 1.5 in.
13. length $= 24.6$ ft, width $= 8.20$ ft **15.** 0.45 cm

EXERCISE 14-6
1. $x = 3$ **3.** $x = 8$ **5.** $x = 6$ **7.** $x = 4$ **9.** $x = 19$, $x = 3$ **11.** $x = 4$

REVIEW EXERCISES FOR CHAPTER 14
1. $x = 0$, $x = -\frac{5}{3}$ **3.** $x = 0$, $x = 3$ **5.** $x = -\frac{4}{9}, x = \frac{1}{2}$ **7.** $x = 2$, $x = -2$
9. $x = 1.7$ (approx.), $x = -1.7$ (approx.) **11.** $x = 7$, $x = \frac{1}{2}$ **13.** $x = -\frac{4}{3}, x = \frac{4}{5}$
15. $x = 1.866$ (approx.), $x = 0.134$ (approx.) **17.** 11, 13 **19.** 4 **21.** 1 in. **23.** $x = 4$
25. $x = 2$

EXERCISE 15-1
5. ray EF **7.** angle 5 **9.** straight line c **11.** $\angle a$ **13.** $\overline{AB}, \overline{BC}, \overline{CD}, \overline{DE}, \overline{EA}$ **15.** vertex
17. $62° 17' 35''$ **19.** $76° 23' 33''$ **21.** $45° 44' 40''$

EXERCISE 15-2
3. right angle **5.** adjacent angles **7.** $55° 42' 55''$ **9.** $43°$

EXERCISE 15-3
3. $80°$, $100°$, $100°$ **5.** $18° 23' 48''$, $161° 36' 12''$, $161° 36' 12''$
7. $\angle a = \angle b = \angle e = \angle f = 77°$, $\angle c = \angle d = \angle g = \angle h = 103°$
9. $\angle 1 = \angle c' = \angle c$, $\angle 2 = \angle b$, and $\angle 3 = \angle a' = \angle a$. Since $\angle 1 + \angle 2 + \angle 3$
$= 180°$, then $\angle a + \angle b + \angle c = 180°$

REVIEW EXERCISES FOR CHAPTER 15
1. $65° 46' 27''$ **3.** $\angle a$ and $\angle d$ acute, $\angle b$ and $\angle c$ obtuse
5. $24° 28' 53''$, $155° 31' 7''$, $155° 31' 7''$

EXERCISE 16-1
1. $78°$ **3.** $54°$ **5.** $c = 13$ in. **7.** $c = 7.6$ cm **9.** $a = 12.0$ cm **11.** 6.96 m **13.** 8.9 ft

EXERCISE 16-2
1. triangles are similar, $b' = 9.00$, $c = 3.37$ **3.** triangles are similar, $a = 3.2$, $c' = 3.8$
5. $DE = 2.5$ ft **7.** $\angle b = \angle c = 39°$, $\angle e = 51°, f = 7.81$ cm **9.** 17 ft

EXERCISE 16-3
1. 5.0 cm, 51 cm² **3.** 61.0 ft, 222 ft² **5.** 68 mm, 290 mm² **7.** 110 ft² **9.** 330 in.² **11.** 21 in.²

EXERCISE 16-4
1. 13.2 mm **3.** 95.76 cm **5.** 17.6 in. **7.** 80.1 in. **9.** 1200 **11.** 10.2 cm

EXERCISE 16-5
1. 460 ft³ **3.** (a) 19.4 gal, (b) 1500 in.² **5.** 1.6 ft³ **7.** 5300 gal **9.** 3 050 cm³ **11.** 960 ft²
13. 620 in.³ **15.** $2,570.00

Review Exercises for Chapter 16
1. 7.5 mi **3.** 12 km² **5.** 270 in.³ **7.** 3000 gal

Exercise 17-1
1. sin A = 0.80, cos A = 0.60, tan A = 1.3, cot A = 0.75
3. sin A = 0.741, cos A = 0.672, tan A = 1.10, cot A = 0.907 ·
5. sin B = 0.823, cos B = 0.570, tan B = 1.44, cot B = 0.693
7. sin A = 0.514, cos A = 0.856, tan B = 1.67, cot B = 0.600
9. sin B = 0.415, cos B = 0.911, tan A = 2.19, cot A = 0.456

Exercise 17-2
21. 25° 10′ **23.** 38° 30′ **25.** 43° 20′

Exercise 17-3
1. 0.5445 **3.** 1.1184 **5.** 0.8307 **7.** 0.3870 **9.** 6° 17′ **11.** 14° 58′ **13.** 51° 43′ **15.** 44° 34′

Exercise 17-4
1. 17° 18′ **3.** 84° 12′ **5.** 2° 48′ **7.** 0.6847 **9.** 0.7455 **11.** 0.5621 **13.** 71.30° **15.** 73.55°
17. 66.45° **19.** 31.68°

Exercise 17-5
1. B = 72°, a = 7.5, c = 24 **3.** A = 68°, a = 0.85, b = 0.34
5. B = 46° 30′, a = 357, c = 518 **7.** B = 42° 10′, b = 38.3, c = 57.1
9. A = 38° 50′, a = 7.02, b = 8.72 **11.** B = 61° 42′, b = 78.62, c = 89.28
13. A = 42° 8′, b = 78.77, c = 106.2 **15.** A = 28° 17′, a = 3254, c = 6867
17. A = 36° 20′, B = 53° 40′, a = 3.03 **19.** A = 31° 49′, B = 58° 11′, c = 2.453
21. A = 34° 55′, B = 55° 5′, a = 9.193 **23.** B = 66.8°, a = 19.19, b = 44.78
25. A = 15.4°, a = 185.9, b = 674.9 **27.** B = 55.1°, b = 8783, c = 10,710
29. B = 63.7°, b = 34.01, c = 37.94 **31.** B = 61.6°, a = 0.7808, c = 1.642

Exercise 17-6
1. 56 m **3.** 37° 57′, 37° 57′, 104° 6′ **5.** (a) 23 ft, (b) 25 ft **7.** $\overline{AC}$ = 130 m, $\overline{BC}$ = 92 m
9. 320 ft **11.** r = 61.18 cm, perpendicular distance = 52.06 cm **13.** 39.9 ft

Review Exercises for Chapter 17
17. 0.7262 **19.** 2.6371 **21.** 29° 7′ **23.** 74° 23′ **25.** 21° 24′ **27.** 77° 48′ **29.** 0.8320
31. 0.8941 **33.** 72.28° **35.** 56.38° **37.** B = 61°, b = 58, c = 66
39. B = 55° 42′, a = 21.31, c = 37.80 **41.** A = 37.4°, a = 1.387, c = 2.283
43. A = 47° 11′, B = 42° 49′, b = 2.051 **45.** $\overline{AC}$ = 840 ft, $\overline{AD}$ = 970 ft, $\overline{CD}$ = 210 ft
47. l = 2.183 in.

Exercise 18-1
1. $\angle b$ **3.** $\angle AOB$

Exercise 18-2
1. 45° **3.** 34° **5.** 13° 20′ **7.** 16° 20′ **9.** 41° 43′ **11.** 21° 7′ **13.** 5° **15.** 0.3420 **17.** 0.4226
19. −0.7951 **21.** −22.904 **23.** 0.5407 **25.** 0.1736

Exercise 18-3
1. 114.6° **3.** 160.4° **5.** 212.0° **7.** 1.5129 **9.** 0.5026 **11.** 16.1081 **13.** 45° **15.** 120° **17.** 540°
19. 600° **21.** $\frac{\pi}{3}$ **23.** $\frac{3\pi}{2}$ **25.** $\frac{2\pi}{3}$ **27.** $\frac{17\pi}{36}$ **29.** $\frac{5\pi}{12}$

1. 21 in. **3.** 2.3 m **5.** 9.4 cm **7.** 1.3 revolutions **9.** 9600π in./min
11. (a) 9 rad/sec, (b) 86 RPM **13.** (a) 27.4 rad/sec, (b) 262 RPM

EXERCISE 18-5
5. amplitude = 3, period = 3π **7.** amplitude = 1, period = π
9.

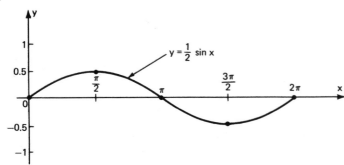

11.

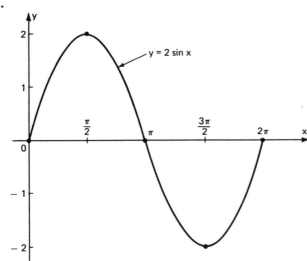

13.

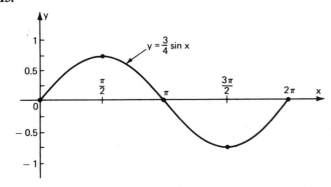

15.

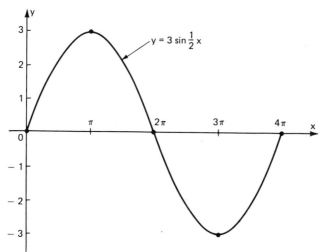

$y = 3 \sin \frac{1}{2} x$

17.

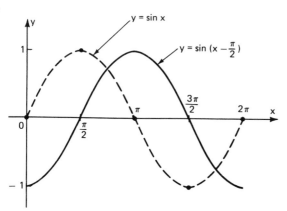

$y = \sin x$

$y = \sin \left(x - \frac{\pi}{2} \right)$

19.

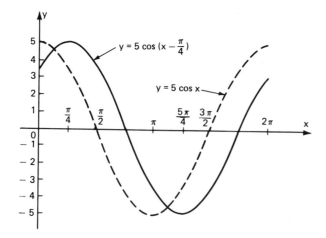

$y = 5 \cos \left(x - \frac{\pi}{4} \right)$

$y = 5 \cos x$

21.

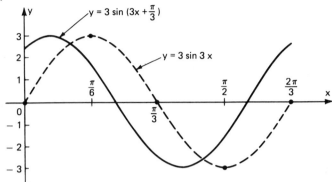

Review Exercises for Chapter 18

1. 62° **3.** 16° 10′ **5.** −1 **7.** −1 **9.** 292.2° **11.** 212° **13.** 0.2565 **15.** 2.0556 **17.** 390°

19. 140° **21.** $\frac{\pi}{10}$ **23.** $\frac{13\pi}{30}$ **25.** 8π rad/sec **27.** (a) 69 rad/sec, (b) 660 RPM

29. amplitude $= 4$, period $= \frac{\pi}{2}$

31.

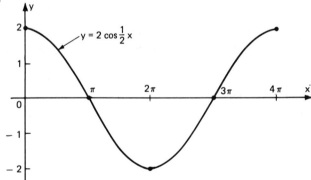

33.

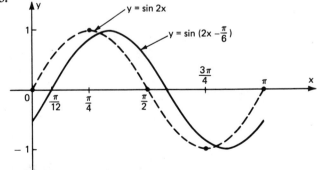

35.

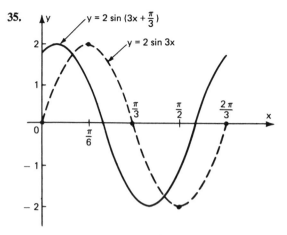

$y = 2 \sin (3x + \frac{\pi}{3})$

$y = 2 \sin 3x$

EXERCISE 19-1

1. $C = 73° \, 0'$, $b = 276$, $c = 278$ **3.** $A = 97° \, 40'$, $a = 132$, $c = 113$
5. $B = 42° \, 50'$, $b = 4.05$, $c = 5.74$ **7.** $C = 79° \, 40'$, $b = 39.7$, $c = 40.3$
9. $B = 54° \, 30'$, $a = 7.97$, $b = 21.2$ **11.** $C = 95° \, 32'$, $b = 833$, $c = 929$
13. $B = 43° \, 49'$, $b = 121$, $c = 139$ **15.** $C = 63° \, 50'$, $a = 13.5$, $c = 13.6$
17. 4.67 km

EXERCISE 19-2

1. $B = 50° \, 10'$, $C = 88° \, 20'$, $c = 199$; $B' = 129° \, 50'$, $C' = 8° \, 40'$, $c' = 30.0$
3. $B = 22° \, 50'$, $C = 32° \, 20'$, $c = 3.17$ **5.** $B = 19° \, 50'$, $C = 131° \, 50'$, $c = 52\overline{0}$
7. no triangle **9.** $B = 71° \, 30'$, $C = 57° \, 10'$, $c = 4.72$; $B' = 108° \, 30'$, $C' = 20° \, 10'$, $c' = 1.94$

EXERCISE 19-3

1. $A = 107° \, 00'$ **3.** $a = 3.8$ **5.** $A = 81° \, 10'$ **7.** $b = 93.8$ **9.** $c = 8.35$ **11.** $B = 87° \, 50'$
13. 91.0 cm **15.** 1.76 mi

EXERCISE 19-4

1. $H = 66 \, \text{lb}$, $V = 35 \, \text{lb}$ **3.** $H = 349 \, \text{lb}$, $V = 239 \, \text{lb}$ **5.** 711 lb **7.** 366 lb
9. $R = 521$, $\angle \, R_a = 14° \, 10'$

EXERCISE 19-5

1. $R = 8.8$, $\angle \, R_a = 11°$ **3.** $R = 373$, $\angle \, R_a = 74° \, 20'$ **5.** $R = 47.7$, $\angle \, R_a = 222° \, 00'$

REVIEW EXERCISES FOR CHAPTER 19

1. $C = 75° \, 25'$, $b = 42.06$, $c = 41.49$ **3.** $A = 112°$, $B = 39°$, $C = 29°$
5. no triangle possible **7.** $r = 12.5 \, \text{m}$ **9.** 473 ft **11.** $R = 16\overline{0} \, \text{lb}$, $\angle \, R_a = 30° \, 00'$

EXERCISE 20-1

1. $5j$ **3.** $13j$ **5.** $3\sqrt{5} \, j$ **7.** $\frac{2}{5}j$ **9.** $4\sqrt{5} \, j$ **11.** -1 **13.** -1 **15.** j **17.** $-j$ **19.** $5 - 3j$
21. $-7 - 4j$ **23.** $-6 - j$ **25.** $5 + 3j$ **27.** 7 **29.** $x = -2$, $y = 6$ **31.** $x = 7$, $y = -3$
33. $-4 + 5j$ **35.** -3 **37.** $8 + 3j$ **39.** -11

EXERCISE 20-2

1. $3 + 5j$ **3.** 25 **5.** $-8 - 5j$ **7.** $7 - 10j$ **9.** $12 - j$ **11.** $7 - 6\sqrt{2} \, j$ **13.** $-2 - 2j$ **15.** 6

17. $-6 - 3j$ **19.** $-2j$ **21.** $27\sqrt{2}\,j$ **23.** $-\frac{7}{25} - \frac{26}{25}j$ **25.** $\frac{1}{5} + \frac{2}{5}j$ **27.** $7 - 24j$ **29.** $\frac{6}{5} + \frac{8}{5}j$
31. $14 - 4j$ **33.** $6 - 5j$ **35.** $8 - j$ **37.** $2 - 3j$ **39.** $5 - 4j$ **41.** $\frac{2}{3}j$ **43.** $11 - 10j$ **45.** $\frac{1}{2} - \frac{1}{2}j$

EXERCISE 20-3

1.

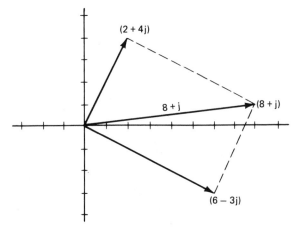

3.

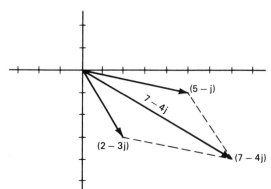

5.

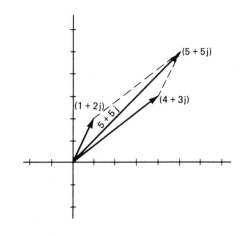

7.

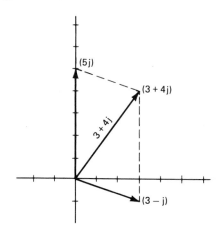

9.

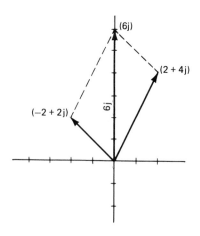

EXERCISE 20-4

1. $5\sqrt{2}$ (cos 135° $+ j$ sin 135°) **3.** 2 (cos 120° $+ j$ sin 120°) **5.** 9 (cos 0° $+ j$ sin 0°)
7. 4 (cos 330° $+ j$ sin 330°) **9.** $2\sqrt{13}$ (cos 43° 54′ $+ j$ sin 43° 54′) **11.** $\frac{3}{2}\sqrt{2} + \frac{3}{2}\sqrt{2}\,j$
13. $2\sqrt{2} - 2\sqrt{2}\,j$ **15.** $-2j$ **17.** $-2\sqrt{3} + 2j$ **19.** $-1.6008 + 1.1990j$

EXERCISE 20-5

1. $1.3917 + 9.9027j$ **3.** $-6.7614 - 1.8116j$ **5.** $-3.6320 + 7.1280j$ **7.** $-2.2944 - 3.2768j$
9. $-1.8471 + 2.364j$

EXERCISE 20-6

1.

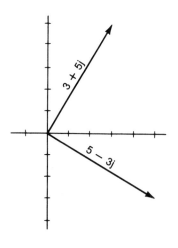

3.

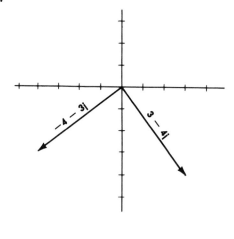

5.

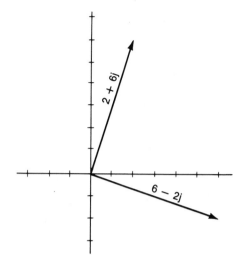

7.

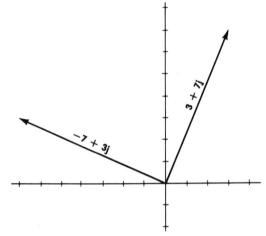

REVIEW EXERCISES FOR CHAPTER 20

1. $6j$ **3.** $3\sqrt{5}\,j$ **5.** j **7.** 1 **9.** $8 + 4j$ **11.** $-5 + 3j$ **13.** $12 + 9j$ **15.** -19 **17.** 13
19. $7 - 6j$ **21.** $-\frac{1}{10} + \frac{17}{10}j$ **23.** $-5 + 13j$

25.

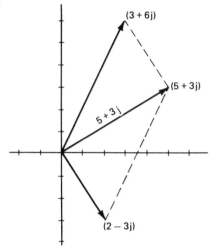

27.

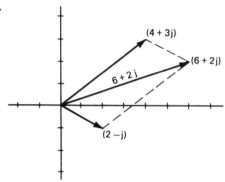

29. $\sqrt{26}\,(\cos 258°\ 41' + j\sin 258°\ 41')$ **31.** $-2\sqrt{2} - 2\sqrt{2}\,j$ **33.** $-2.6990 - 4.2090j$

35.

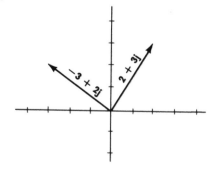

Index

Base:
exponents, 40, 77, 81
trapezoid, 272
triangle, 260
Base angles of an isosceles triangle, 260
Binomial(s):
definition, 77
multiplication, 96
Bisector, 261
Braces, 80
Brackets, 80

Cancelling, 20, 114, 120-21
Central angle, 279
Chord, 279
Circle, 278
Circumference, 279
Coefficient, definition of, 77
Cofunction, 300
Combined variation, 166
Common denominator, 12, 17
Common factor, 104
Common multiple, 13
Commutative law:
addition, 61, 79
multiplication, 61, 79
Complementary angles, 251
Complex algebraic fractions, 122
Complex fractions, 24
Complex number(s):
absolute value, 380
addition, 373
algebraic form, 369
amplitude, 380
argument, 380
complex plane, 376
definition, 369
division, 373, 384
graphic representation, 376
interpreted as vectors, 378
modulus, 380
multiplication, 373, 384
polar form, 381
rectangular form, 376
subtraction, 373
trigonometric form, 380
Component method, 363
Component of a vector, 358
Concentric circles, 279
Cone, 285
Cone, altitude of, 285
Congruent figures, 261
Congruent triangles, 261
Conjugate, 210, 370
Constant, 75
Constant of proportionality, 165
Conversion factors, computations with, 47
Conversion of units, 149
Coordinates, 174
Coordinate system, rectangular, 174
Corresponding angles, 266

Corresponding sides, 266
Cosine, 296
Cosine, law of, 356
Cotangent, 296
Coterminal angles, 321, 330
Cross-multiplication, 140, 160
Cube, 283
Curve:
amplitude, 339
cycle, 339
$y = A \sin Bx$, 341
$y = A \sin (Bx + C)$, 345
$y = \cos x$, 339
$y = \sin x$, 339
Curved line, 247
Cycle of curve, 339
Cylinder, 283
Cylinder, hollow, 284

Decimal degrees, 306
changing from decimal degrees to degrees, 307
changing from degrees to decimal degrees, 307
Decimal point, 26-27
Decimals, 26
addition, 27
changing to fractions, 39
division, 30
multiplication, 29
subtraction, 27
Degree, angular, 248
Degree, decimal, 306-307
Degree of equations, 173
Denominate numbers, 147
Denominator, 3
least common denominator (LCD), 14, 117
Density, 55
Dependent variable, 174
Depression, angle of, 314
Diagonal, 272
Diameter, 278
Difference of two squares, 105
Dimension(s):
of a denominate number, 147
in geometry, 146
in physics, 146
Dimensional analysis, 146
Directed numbers, 62
Displacement:
angular, 334
linear, 359
Distance, linear, 334
Distributive law, 61, 79
Division:
algebraic expressions, 102
algebraic fractions, 120
complex numbers, 373, 384
decimals, 30
fractions, 22
numbers in scientific notation, 204

Point(s):
    coordinates, 174
    decimal, 26
    geometric, 247
    intersection, 254
    plotting, 175, 339
Polar form of a complex number, 381
Polygon, 260
Polynomial(s):
    addition, 92
    definition, 77
    degree, 78
    division, 102
    factoring, 103-10
    multiplication, 95
    subtraction, 93
Positive numbers, 62
Precision of numbers, 33, 35
Prefixes, metric system, 45
Prime expression, 103
Prime factorization, 9
Prime numbers, 8
Principal square root, 205, 241
Proper fraction, 4
Proportion, 160
    properties, 160
    terms, 160
Proportionality constant, 165
Pure imaginary number, 369
Pythagorean theorem, 262, 299, 380

Quadrant, 174
Quadrantal angles, 324, 330
Quadratic equations, graphing of, 182
Quadratic equations in one variable, 230
    principle of zero, 231
    solution by factoring, 231
    solution by the quadratic formula, 235
    solution by the square root method, 234
Quadratic equations in two variables, 173
Quadrilaterals, 272
Quotient, 30

Radian, 331
Radical(s), 205
    addition, 208
    division, 209
    multiplication, 209
    simplification, 207
    subtraction, 208
Radical equations, 240
Radical form, 206
Radical sign, 40
Radicand, 205
Radius, 278
Rates, 158, 162-63
Ratio(s), 157
    proper, 158
    terms, 157
    trigonometric, 295
    with no units, 158

Rationalizing factor, 211
Rationalizing the denominator, 211
Rational numbers, 61, 63
Ray, 247
Real numbers, 63, 79
Reciprocals, 24, 126
Rectangle, 272
    area, 275
Rectangular coordinate system, 174
Rectangular solid, 282
Reducing to lowest terms, 8, 113
Reference angles, 324
Relation, 174
Resolution of vectors, 359, 364
Resultant of vectors, 359
Rhombus, 273
Right angle, 251
Right circular cone, 285
Right circular cylinder, 283-84
Right triangle(s):
    definition, 260
    labeling, 262, 295
    solution, 308
Rise, 159
Root(s):
    of equations, 125
    of fractions, 41
    of negative numbers, 205, 235, 368
    of numbers, 40
Rounding off numbers, 36
Run, 159

Scientific notation, 202
    computations in, 203
    and significant digits, 202
Second degree equations in two variables, 173
Semicircle, 279
Sense of a vector, 358
Side(s):
    adjacent, 295
    of an angle, 248
    corresponding, 266
    of an equation, 125
    initial, 321
    opposite, 295
    terminal, 321
Sign(s):
    algebraic, 74
    equality, 74
    functions of angles in the four quadrants,
      table of, 328
    inequality, 62, 74
    "not equal to," 74
    trigonometric functions, 327-28
Signed numbers, 62
    addition, 64, 67
    division, 70
    irrational, 63
    multiplication, 68
    rational, 63
    subtraction, 66-67
Significant digits, 33